LASER

ELECTRO-OPTICS TECHNOLOGY

SERIES

COMPONENTS

CORD

CENTER FOR OCCUPATIONAL RESEARCH AND DEVELOPMENT

ISBN 1-55502-024-0

CORD COMMUNICATIONS
324 Kelly Drive
P.O. Box 21206
Waco, Texas 76702-1206

CONTENTS

LASER/
ELECTRO-OPTICS

MODULE 6-1
OPTICAL TABLES AND BENCHES

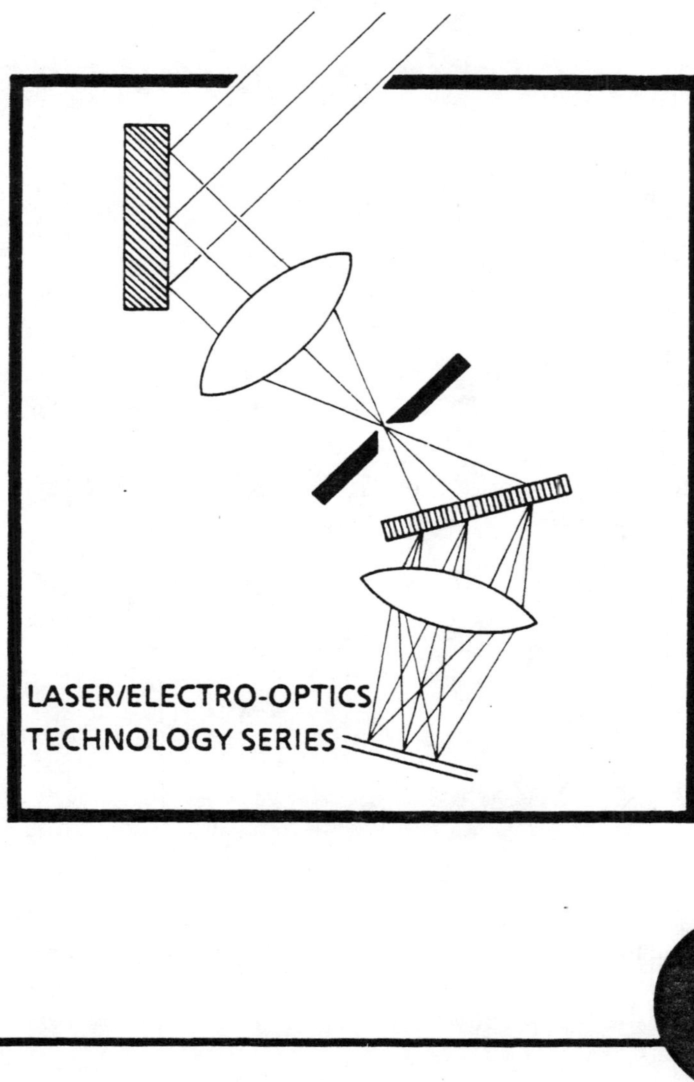

CORD

The Center for Occupational Research and Development
601 C Lake Air Drive
Waco, Texas 76710

Printed in the U.S.A.

ISBN 1-55502-024-0

OPTICAL TABLES AND BENCHES

MODULE 6-1

In electro-optical laboratory environments, few considerations are more critical than the way optical components and associated equipment are mounted.

In Module 6-2, "Component Support," we'll discuss the actual devices that are used to mount components such as mirrors, lenses, prisms and diffraction gratings. In this module we will focus our attention on the surface on which the optical mounts are affixed. We will direct most of our discussion toward the requirement of nonpermanent optical assemblies. Most of the optical assemblies that you'll construct in research laboratories are of the "experimental" nature. They are meant to be dismantled after serving a particular purpose. Permanent optical assemblies, such as the optics in a supermarket bar-code scanner, are custom-designed. They are not part of our discussion—although most of the same principles that we'll discuss still apply.

The main purpose of this module will be to help you to make a reasoned judgment of what type of device to select as the "foundation" of a particular optical system. Depending on the particular geometry of an optical setup, various options will be more favorable than others.

For example, if you need to assemble an optical system that lies along a single straight axis, an *optical rail* or *optical bench* might be an appropriate choice on which to attach the various optical mounts. On the other hand, if the optical setup will by necessity occur in a plane, or even several planes, then an *optical table* or *optical breadboard* will usually be a more convenient choice.

We'll also discuss the major requirements of critical optical assemblies such as mechanical and thermal stability and vibration isolation. Most of our discussion will be about modern optical benches, tables and breadboards that have become de facto industry standards in laboratories that use lasers and related optical and electro-optical apparatus.

Before you start this module, you should understand algebra, general laboratory and electrical safety, waveform analysis and the use of transducers, oscilloscopes and low-power helium-neon lasers.

OBJECTIVES

When you complete this module, you should be able to do the following.

1. Name four classes of optical benches. List one advantage and one disadvantage of each as given in the text.
2. Explain the following expressions in your own words:
    - Vertical vibrations in a table
    - Horizontal vibrations in a table
    - Vibration isolation from table support
    - Internal table damping from acoustically induced vibration
    - Thermally induced movement
    - Table rigidity
    - Table flatness
3. List the advantages and disadvantages as given in the text (at least two each) of these optical tables:
    - Lab workbench
    - Lab workbench covered with thick metal plate with matrix of tapped mount holes
    - Concrete slab on inner-tube support
    - Granite slab with insert matrix, mounted on steel frame with rubber shock mounts
    - Granite slab with insert matrix, mounted on pneumatic vibration isolators
    - Steel table with magnetic mounts
    - Metal honeycomb table, with stainless steel skin and matrix of tapped holes, vibration isolation
4. Align an optical bench parallel to an optical axis established by a laser beam.
5. With an accelerometer and oscilloscope, investigate the vertical and horizontal vibrations in an optical table, as follows:
    a. Using a spherical object dropped to the table surface from a constant height, measure approximate frequencies of the

vibrations thus excited. Compare such responses at various impact points.

b. Determine peak-to-peak amplitudes of the major resonances for the table, in terms of the patterns displayed on the oscilloscope.

c. Determine the time required for each major resonance to decay to half of its peak amplitude.

d. Compare the vibrational responses of the table in two horizontal directions, 90° apart, with those in the vertical direction.

e. By comparing peak amplitudes of vibrations on the table with those on the floor at the table base—both excited by identical impulses applied to the floor—appraise relative effectiveness of the table in damping floor-induced vibrations.

DISCUSSION

Optical Benches

An *optical bench*—sometimes called an optical rail—gives a stable, in-line surface where optical system components and supports can be installed and adjusted to various relative positions and then held rigidly in place. The optical bench is most useful when the optical axis of the experimental arrangement or system lies along a straight line or in a vertical plane.

Technicians use most optical benches for one or more of these applications.

a. Alignment of an optical system (such as a laser).
b. Evaluation of an optical system (such as measurement of the modulation-transfer function of a lens).
c. Support of a breadboard for an optical system.

In each case the bench or rail is the stable base on which you mount the component supports. The mounting devices are called "carriages." The component support is mounted either directly onto the carriage or at an adjustable height above the carriage on a rod. These elements are all shown in Figure 1.

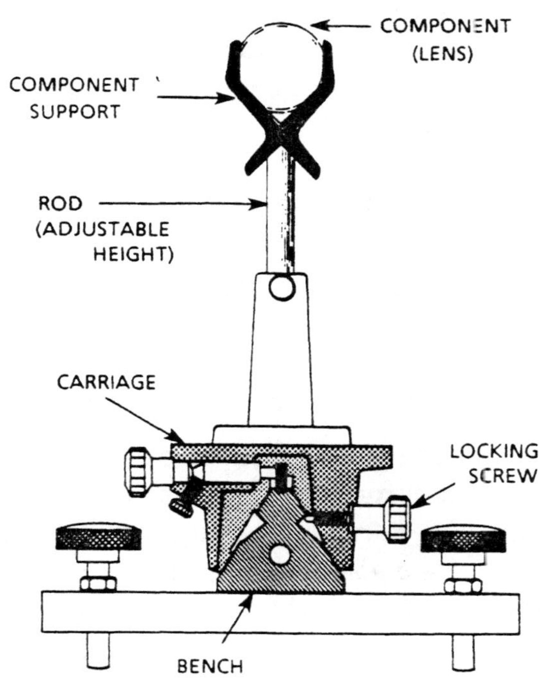

Fig 1. Elements of an optical-bench setup.

This description of optical rails or benches will be divided into four broad types:

Lightweight aluminum rails
H-beams
Triangular rails
Flat-bed or "lathe-bed" benches

Generally speaking, this list is shown in increasing order of quality and expense. You can get a good-quality one-meter-long aluminum rail for less than $200. But, certain complete flat-bed systems will cost more than $60,000.

The quality of an optical bench is measured in terms of how well it satisfies these two questions:

1. How rigidly can it hold component supports in place?

   Rigidity is controlled by the material used in the bench and the size and shape of the bench cross section. For example, a granite bench is typically more rigid than a metal one. The larger and more massive benches usually are more rigid, and various-shaped cross sections (H-beam, triangular and dovetail) are used to increase rigidity.

2. How accurately can optical components be translated along an optical axis when guided by a bench?

   The answer to this question lies in the "straightness" of the bench. Metal and granite benches are machined flat and straight. They can be accurate (on large lathe-bed benches) to a thousandth of an inch.

Lightweight Aluminum Rails

A typical example of a *lightweight aluminum rail* is shown in Figure 2. Stock lengths of rail are available from about 15 centimeters up to two meters, with either English or metric mounting holes. Attached to the rail are two typical carriages or carriers that accept rod-mounted optical components.

Aluminum rails like these usually are designed with slotted mounting holes to allow secure attachment to optical tables, as shown in the figure. Many manufacturers also provide coupling plates so that two or more optical rails can be firmly attached at various angles to accommodate optical setups that require more than a single axis.

A variety of carriages are also available. The carriages let you mount a wide variety of components and equipment such as lens holders, translators and rotators. A more complete discussion of these mounts is in Module 6-2.

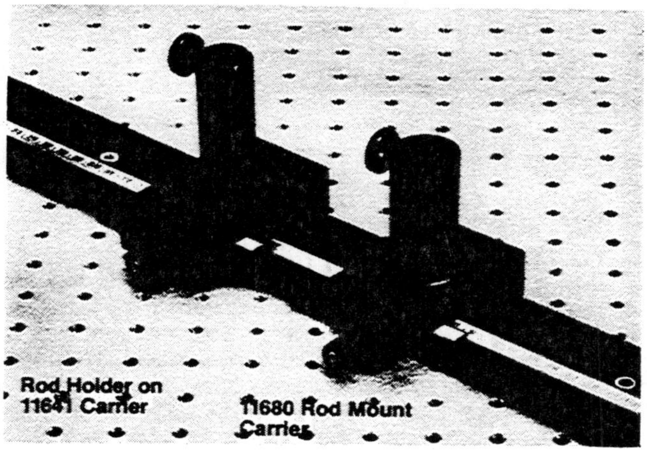

Fig. 2 A typical aluminum rail with carriages shown attached to an optical table. Photograph courtesy of Oriel Corporation.

For added convenience, many optical rails have linear scales running along their length. The linear scale helps you to position equipment accurately. Vernier scales are also often attached to the carriages. These give an even more accurate determination of the position of components mounted on the rail.

H-Beam

An *H-beam optical bench* is used for rugged breadboards of heavy electro-optical or laser systems. One is shown in Figure 3a.

The bench is made from a structural H-beam of either steel or aluminum. Channel often is used also. The top surface is milled flat. Components are either bolted down to the H-beam or held rigidly with C-clamps.

The H-beam bench is relatively inexpensive and can be made in the shops. Its main disadvantage is that elements can't be translated along the beam and remain in optical alignment. However, where you need this feature, you can modify an H-beam bench as shown in Figure 3b by milling edges and clamping the component to the side with a setscrew.

In the early days of the laser industry, such "homemade" optical rails were quite common. In many cases, the optical components were held in mounts, affixed to the H-beam with double-sided tape! With the improved

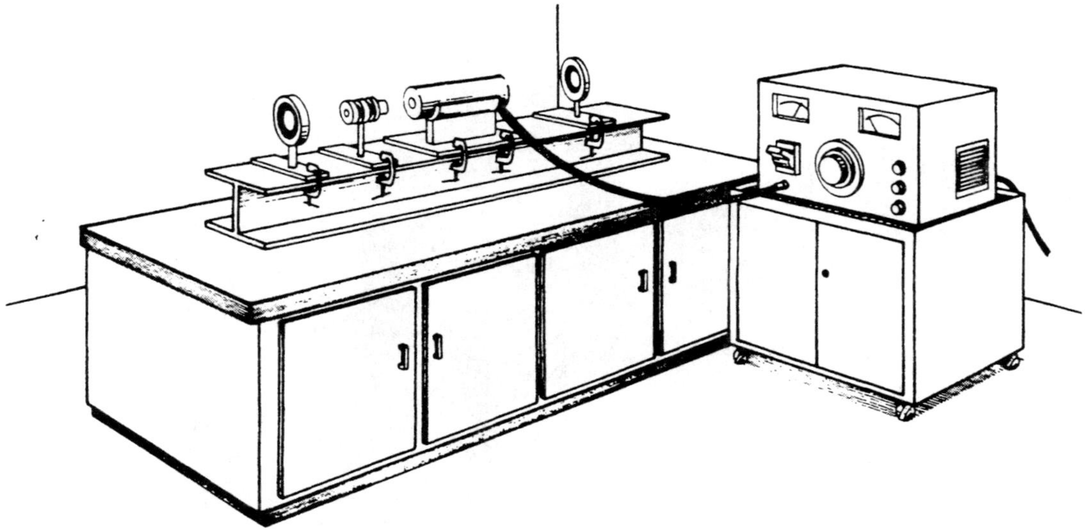

a. Laser breadboard on H-beam bench with C-clamps to hold component supports.

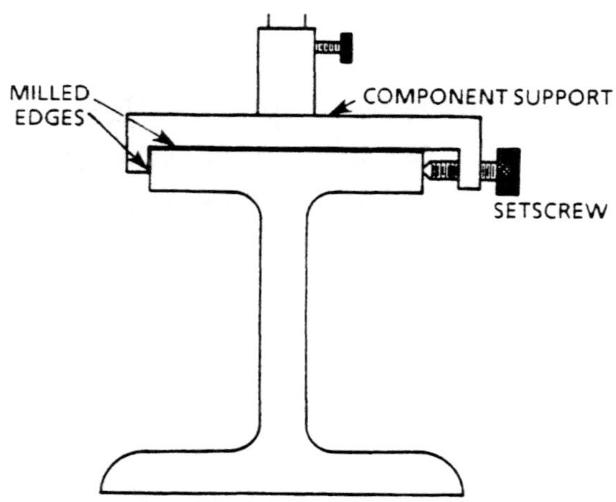

b. Modified H-beam bench to maintain component alignment during translation.

Fig. 3 H-beam benches.

convenience and availability of commercial aluminum rails and ferromagnetic optical tables and breadboards, such improvised mounting systems are an increasingly rare sight in industrial and university laboratories.

Triangular Rail

A *triangular optical rail* (Figure 4) is the most widely used optical bench. The rail usually is made of aluminum, cast iron or steel. It has an equilateral triangular cross section (about 2 inches per side) with grooves on the two top sides. The rails can be made straight to 0.005 inch. The grooves running down the length of both sides are used to lock the carriages in place with a screw. (See Figure 1.) Triangular benches are fixed directly to optical tables. Or, they're mounted on cross-feet for precise leveling.

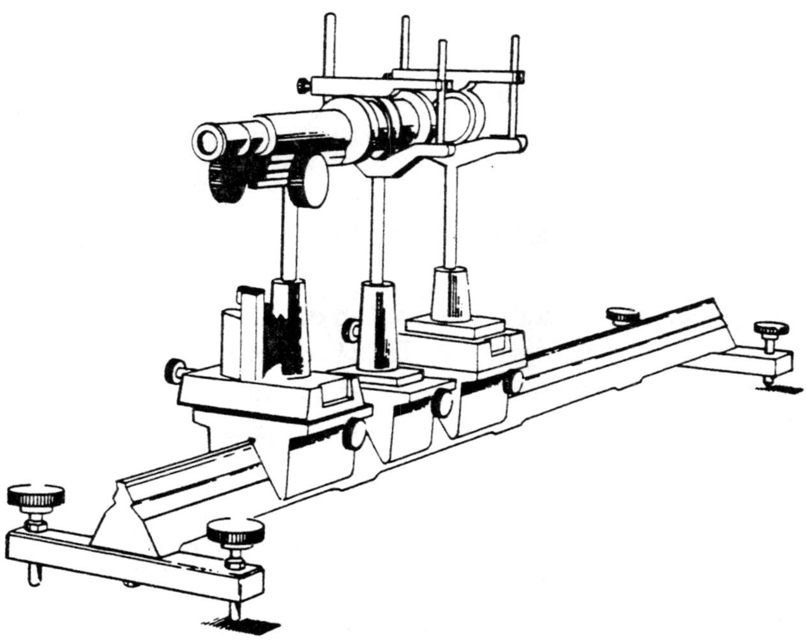

Fig. 4 Triangular optical rail.

Flat-bed Benches

Flat-bed optical benches (Figure 5) offer the ultimate in rigidity and straightness in an optical bench. They can be made from high-grade stainless steel or granite. The bed usually has a dovetail cross section for accurate, stable guiding and clamping of optical carriages. The benches are extremely heavy and expensive—they're used for very special applications where high precision is required.

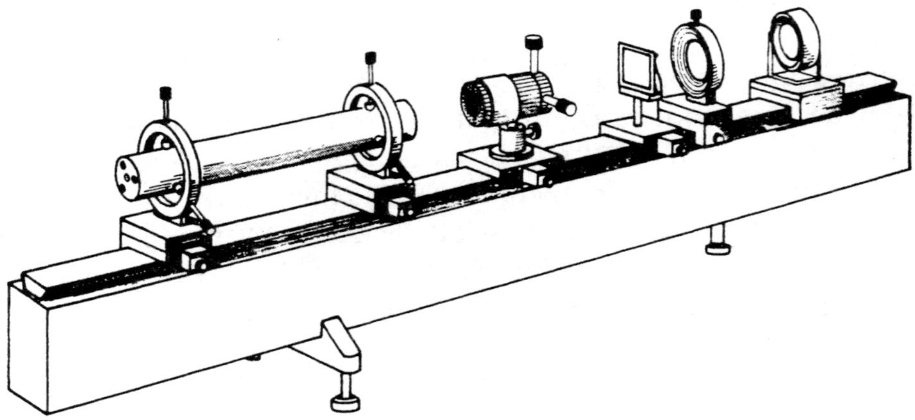

Fig. 5 Flat-bed optical bench.

Thermal Effects on Bench Length

Spacing between elements mounted on an optical bench varies if the temperature of the bench changes. This is due to thermally induced expansion and contraction of the bench material. Thermal expansion is discussed in greater detail in the section on Optical Tables.

Care of Optical Benches

Optical benches must be cared for to prevent warping, rusting and nicking. A long bench should be supported from beneath at points approximately every two feet along its length. Such support will prevent the bench from bending under its own weight and the weight of the elements it supports. Use common sense to prevent loading a bench with elements that are too heavy.

Since most benches are made of steel, they can rust. Use a coat of light oil—applied sparingly and then wiped off—on the bench and the carriages every several months. **Do not use oil on the component supports.**

Be careful not to hit the bench surface with a carriage or other object. This can cause nicks and grooves that will prevent the carriages from moving smoothly along the surface and will frustrate attempts to maintain alignment.

Aligning a Bench to an Optical Axis

Frequently a test will require that an optical bench be aligned parallel to an existing optical axis or light beam. If a laser or light beam defines the optical axis, you accomplish the alignment procedure by using an iris diaphragm mounted on the bench with a carriage. You adjust the bench position until the laser beam passes through the center of the iris without perpendicular carriage adjustment at any position along the optical bench.

If a laser beam does not define the optical axis, you should use one of the following two procedures.

1. Set up a laser beam to establish the optical axis and proceed as described above.
2. Set up an autocollimator/telescope aligned with the optical axis. Mount a cross target on the optical bench with a carriage. Adjust the bench position until the cross target is in alignment with the autocollimator/telescope at any position along the optical bench.

When Does an Optical System Require a Table or Breadboard?

A table (Figure 6) is desirable for breadboarding an optical system, test or experiment if any one or more of the following conditions exists or is required.

1. More than one optical path in the setup
2. Optical path runs in more than one horizontal direction
3. Vibration isolation is desired
4. Components are numerous or are relatively massive

Typical examples of applications where you should use optical tables are listed below.

1. Multiple-path interferometers—Mach-Zehnder, Twyman Green, Michelson. Stability of the mirror separation is critical to a fraction of a micron. Vibration isolation is also desirable.
2. Holography. Long-term vibration isolation is required. Rigidity and maintenance of specific separation between components also are required. See Figure 7. Tables for this use usually contain small inner tubes that provide both support and vibration isolation.
3. Research laser systems. The laser mirrors require rigid mounts to ensure stable output. Associated devices such as Q-switches, frequency-

Fig. 6 Example of a high-quality optical table. This table has pneumatic isolation legs, a lower accessory shelf (for mounting a laser) and beam-steering mirrors. Photograph courtesy of Newport Corporation.

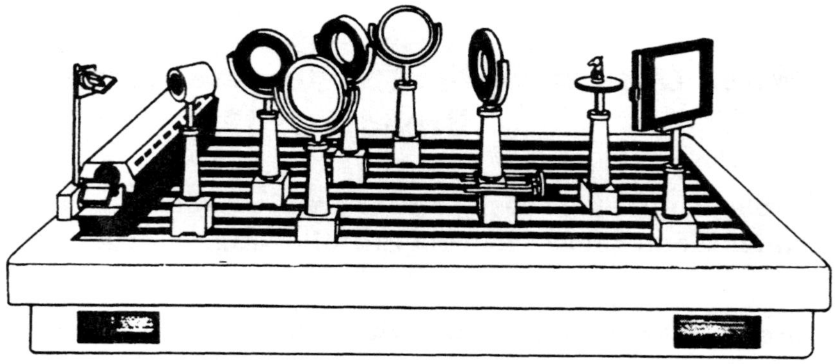

Fig. 7 A holographic setup mounted on a small multi-rail optical table.

doubling crystals, amplifiers and optical power meters require a stable mounting surface. For convenience, components often are mounted at angles to the original laser beam path.

4. Electro-optical systems development (examples: IR scanner, scanning imaging spectroradiometer, star trackers, laser radars). Requires support of massive components, stability, location of electronic and thermal equipment immediately adjacent to the optical setup. See Figure 8.

Fig. 8 Electro-optical system test on an optical table.

Desirable Characteristics of Optical Tables

Generally, an optical table should provide a stable plane surface where optical components and supports can be rigidly placed or moved to any position or angle. The ability of a table to satisfy these requirements depends on the degree to which it has the following characteristics.

1. **Flatness.** Table flatness is desirable so that if a mirror support is moved from one position on the table to another, the mirror isn't "tilted" in angle. Flatness also is needed to prevent "rocking" of the component support because its bottom plane and the tabletop are not matched.

   A surface is defined to be flat to within ±0.00X inch if no point on the surface is more than "X" thousandths of an inch from an imaginary reference plane through the surface. Modern honeycomb optical tables are flat to about ±0.005 inch over lengths of 8 or more feet.

2. **Rigidity.** The static measure of rigidity is the "stiffness" of a table surface. Rigidity describes the degree to which a table bends under its own weight and the weight on the components it supports. Rigidity is determined by the type of support under the table, the composition and structure of the table material, and the table thickness.

   It's obvious that a 0.5-inch-thick aluminum table surface five feet square, supported by four legs (located near the corners) will flex in the center due to its own weight and the weight of anything it supports. However, it may not be as obvious that four-foot by eight-foot granite tables will have noticeable differences in rigidity if they're only eight inches thick instead of fourteen inches thick. Rigidity in bending improves as the square of the table thickness.

Optical tables with metal honeycomb cores and ferromagnetic stainless steel top surfaces have become essentially a de facto industry standard for most applications in electro-optical research laboratories. Rigidity of such tables isn't as good as a granite table of similar dimensions. But, it's usually more than adequate for the vast majority of typical laboratory applications. The graph in Figure 9 shows the static deflection of the center of a 4-foot-wide honeycomb table as a function of both table thickness and table length. For five-foot-wide tables multiply all deflections by 0.80. This deflection is caused solely by the weight of the tabletop. It doesn't take into account additional load applied by equipment mounted on the table surface.

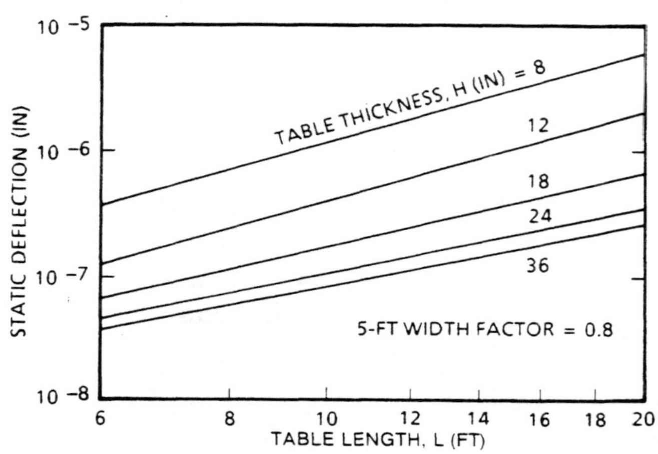

Fig. 9 Plot of static deflection of a 4-ft wide steel honeycomb-core optical table. Courtesy of Newport Corporation.

Figure 10 shows how a table surface will deflect when it experiences a point load at its center. Such a load occurs when heavy equipment—such as lasers and spectrometers—is mounted on the table.

Although equipment doesn't apply load to the table at a single point, the point-load model is useful in estimating the actual deflection that can be expected. You can figure the additional deflection of the center of a honeycomb table due to a point load at the center from Equation 1.

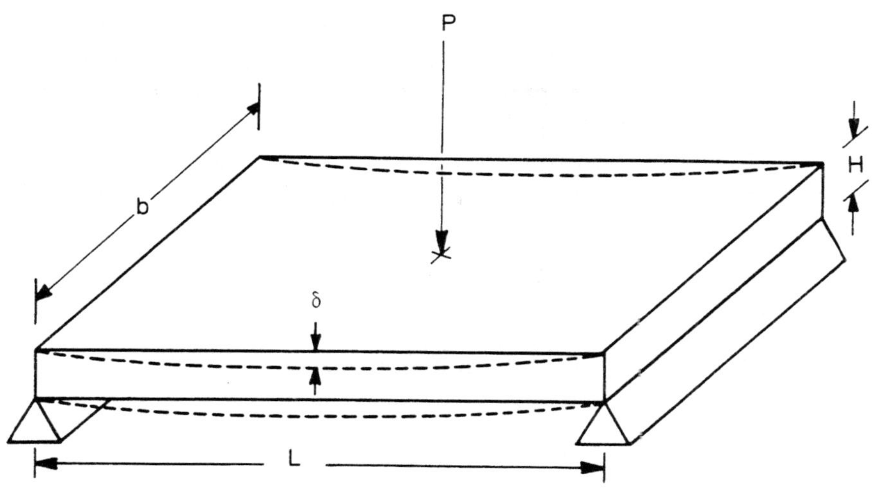

Fig. 10 Deflection of the center of a honeycomb optical table under a central point load. Figure courtesy of Newport Corporation.

$$\delta \simeq \frac{PL^3}{24\,EbTH^2} + \frac{PL}{4GHb} \qquad \text{Equation 1}$$

where:

- P = Force exerted by the point load
- L = Length of the panel
- b = Width of the panel
- H = Thickness of the panel
- T = Thickness of the skins
- E = Young's modulus for the skin material
- G = Shear modulus for the core

Example A: Deflection of an Optical Table

Given: A piece of equipment with a weight of 100 pounds (mass of 45.4 kg) is placed in the middle of a 4-ft x 10-ft steel honeycomb table. The stainless steel "skins" of the table are 3/16 inch thick. Total panel thickness is 8.0 inches. Assume that the equipment produces a point load at the center of the table.

Find: Maximum deflection of the center of the table.

Solution: For a steel honeycomb

$$E \simeq 2.90 \times 10^7 \frac{lb}{in^2} = 4.18 \times 10^9 \frac{lb}{ft^2}$$

$$G \simeq 2.25 \times 10^5 \frac{lb}{in^2} = 3.24 \times 10^7 \frac{lb}{ft^2}$$

$$H = 8.0\ in = 0.667\ ft$$

$$P = 100\ lb$$

$$T = \frac{3}{16}\ in = 1.56 \times 10^{-2} ft$$

$$\delta \simeq \frac{(100)(10)^3}{(24)(4.18 \times 10^9)(4)(1.56 \times 10^{-2})(0.667)^2} + \frac{(100)(10)}{(4)(3.24 \times 10^7)(0.667)(4)}$$

$$\simeq 3.88 \times 10^{-5} ft = 1.18 \times 10^{-2}\ mm$$

Deflection in this realistic example is quite small. But, when aligning certain optical systems, this small movement could be critical. For example, in holography, motion on the order of a fraction of a wavelength of light can cause significant problems. Compare the table deflection to the wavelength of a HeNe laser to get a better feeling for the potential significance of the motion.

$$\frac{\delta}{\lambda_{HeNe}} = \frac{1.18 \times 10^{-2} mm}{6.33 \times 10^{-4} mm} = 18.7$$

Therefore our table deflection represents a motion of almost 19 wavelengths of light. This degree of motion could ruin the results of a sensitive experiment such as holographic image recognition.

3. **Thermal expansion.** When a material is heated, it will expand according to the relationship,

$$\Delta L = \alpha L_0 \Delta T \qquad \text{Equation 2}$$

where:
L_0 = Original length of the material
ΔL = Change in length of the material
ΔT = Temperature change, °C
α = Coefficient of linear expansion for particular materials, $°C^{-1}$

The coefficient of linear expansion, α, varies with different materials. Typical values of α are given in Table 1.

Table 1. Typical Values of α

Material	α (in $°C^{-1}$)
Aluminum (type 6061-T6)	23.4×10^{-6}
Stainless steel (type 400)	10.4×10^{-6}
Super Invar (31% Ni, 5% CO, 64% Fe)	-0.19×10^{-6} (a 20°C)
Cast iron	10.8×10^{-6}
Quartz	0.55×10^{-6}
Pyrex	3.2×10^{-6}
Granite	7.2×10^{-6}

A comparison of the values of α shows that metal tables are more susceptible to thermal expansion than granite. The significance of this is that temperature fluctuations of a centigrade degree in a laboratory could change a one-foot optical-path length in an interferometer on an aluminum table by about 11 wavelengths.

Example B: Thermal Expansion of Laser Cavity

Given: An "open-cavity" HeNe laser is comprised of a Brewster-angle plasma tube and the HR and OC mirrors whose mounts are attached 1.00 meter apart to the top of a stainless steel optical table. Room temperature increases by 1.00°C.

Find: How many longitudinal modes of the laser cavity the laser will "hop" through.

Solution:

$$\Delta L = \alpha L_0 \Delta T = (1.04 \, x \, 10^{-5})(1.00)(1.00)$$

$$\Delta L = 1.04 \, x \, 10^{-5} \text{ meters}$$

$$\textit{New longitudinal laser mode will form every time } \Delta L = \frac{\lambda}{2}$$

$$\frac{\lambda}{2} = \frac{6.33 \, x \, 10^{-7}}{2} \text{ meters} = 3.17 \, x \, 10^{-7} \text{ meters}$$

$$\textit{Number of "mode hops"} = \frac{\Delta L}{\frac{1}{2}\lambda} = \frac{1.04 \, x \, 10^{-5} \text{ meters}}{3.17 \, x \, 10^{-7} \text{ meters}}$$

$$\simeq 33$$

where: λ = *wavelength of laser output for HeNe*
λ = *6.33 x* 10^{-7} *µm*

Therefore, the longitudinal modes in the laser cavity will "hop"—in other words longitudinal modes will form, collapse and reform—about 33 times as the optical table heats up. If we monitored the output of the laser with an optical power meter, we'd notice a periodic fall and then rise in the laser output.

In laboratories where thermal stability is particularly important the optical tables often are made of Invar, a low-expansion alloy whose expansion coefficient is only – 0.20 x 10^{-6} °C^{-1}. That's about 55 times better than typical stainless steel!

4. **Vibration isolation.** Vibrations induced into the surfaces can cause slight but significant movement in the table, and consequently the optical components. Vibrations can be induced in the table from the floor through the support structure by mechanical vibrations, or through the air by acoustic disturbances.

Vibrations are *vertical vibrations* (where table ends flex up and down) and *horizontal vibrations* (where the table moves laterally). Generally, the vibrations of concern are relatively low-frequency (less than 100 hertz). In addition, the tabletop itself can vibrate in a number of torsional and bending modes as shown in Figure 11.

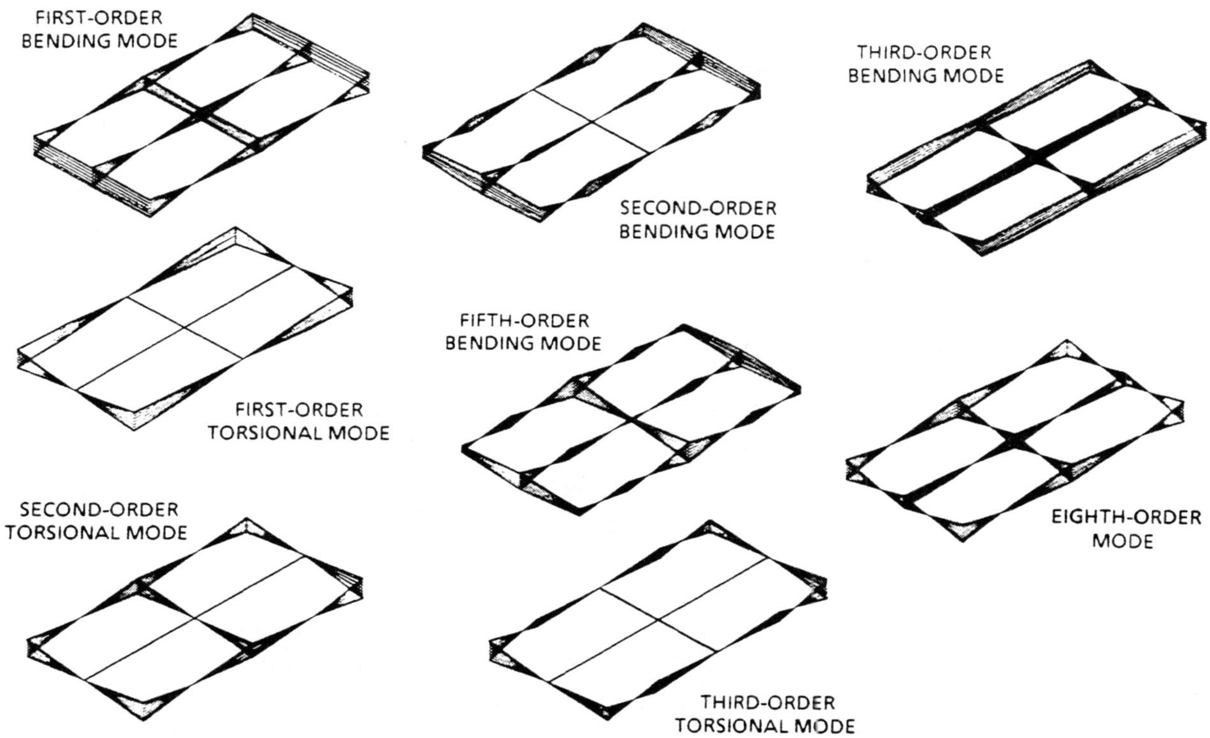

Fig. 11 Low-order bending and torsional modes of a uniform rectangular plate. Figure courtesy of Newport Corporation

Bending and torsional modes of the table surface can be reduced greatly in amplitude by constructing the core of the table from an appropriate damping material. Most modern optical tables use various metal honeycombs for this purpose.

Honeycomb construction was developed in great part for use in aircraft construction. Honeycombs provide excellent internal damping. And, they provide good mechanical strength with relatively little weight. Figure 12a shows the response of a 4-foot x 10-foot x 8-inch solid granite slab to an impulse. Note that after 0.30 second there's still an appreciable amplitude to the "ringing" of the tabletop. By comparison, a similar impulse is applied to a same-size steel honeycomb tabletop. The resulting "ringing" is damped out in approximately 30 milliseconds (Figure 12b). That's a dramatic difference in damping rates.

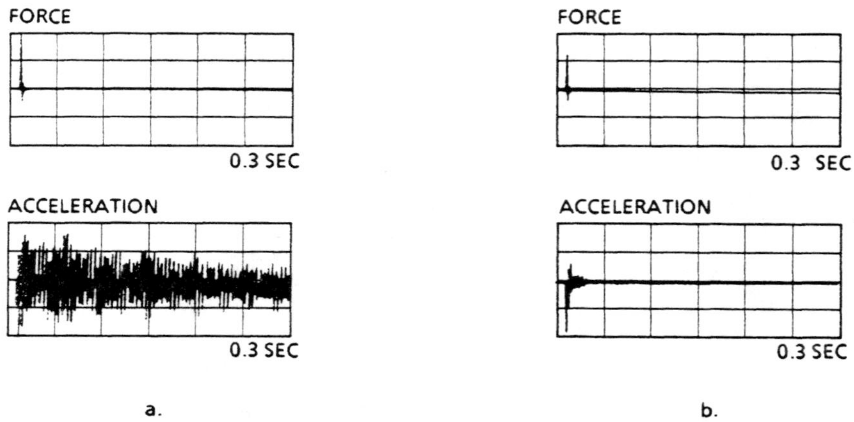

Fig. 12 Responses of solid granite plate (a) and metal honeycomb optical tabletop (b) are shown in lower plots when they're subjected to impulse forces shown in upper traces. Figures courtesy of Newport Corporation.

Vibrations that result from motion of the laboratory floor usually occur in approximately the 5-Hz to 30-Hz range. Modern isolation optical table systems usually are suspended on pneumatic legs that are designed to resonate mechanically at about 1 Hz when loaded with a tabletop. The mechanical resonance is far lower in frequency than typical floor vibrations. So, little energy is transferred from the floor to the table surface.

The pneumatic isolation support system combined with the metal honeycomb damping materials make modern optical tables an excellent surface on which to conduct vibration-sensitive experiments. For these reasons—in addition to convenience and cost—steel honeycomb table systems with pneumatic supports have replaced solid metal or granite tabletops in most electro-optical laboratories.

5. **Ability to attach a component holder firmly in place.** Optical components normally are held by mounts that have precision-flat bases. This allows the experimentalist to slide the base around on the table until the component is positioned correctly. When the mount is in the proper position, it's desirable to firmly attach it to the table.

   As we'll see in Module 6-2, several methods of attaching mounts to surfaces commonly are used in electro-optical laboratories. The most common types of attachment are: bolts into threaded holes drilled in the table or bench surface, inertial mounts (heavy bases) and magnetically held mounts. Modern honeycomb-construction optical tables and breadboards usually are supplied with a standard pattern of threaded bolt holes and a ferromagnetic upper surface, to facilitate the mounting of components.

Types of Tables—Methods of Fixture Attachment

Several types of optical tables are used. Some are very crude (and inexpensive) and some are very elaborate (and expensive). It's important to know how much quality (flatness, rigidity, vibration isolation and low expansion) you need for a specific application.

If an optical system doesn't require high-precision alignment and stability, it's economically wasteful to specify a high quality. Conversely, if high-precision alignment and/or stability are required, a crude setup will waste the experimentalist's time and possibly cause erroneous results.

The following list describes most optical tables that are commonly used, along with the method of attachment of fixtures to the table.

1. **Workbench or office table.** These crude surfaces often are used for simple, noncritical optical alignment or tests in most laboratories. Typical tests that are conducted on these surfaces are: a rough check on the paper output of gas laser (see Figure 13a); frequency response of LED communication system; and measurement of the time description of the output of a pulsed laser.

   Optical fixtures and components usually are mounted in separate stands, or on lab jacks, books or boxes. Components can be held somewhat rigidly by double-sided tape. Considerable vibration can be introduced into this setup because fixtures don't rest flat on the table, table legs don't rest uniformly on the floor and table support is usually unsteady. And, many elements may be cantilevered.

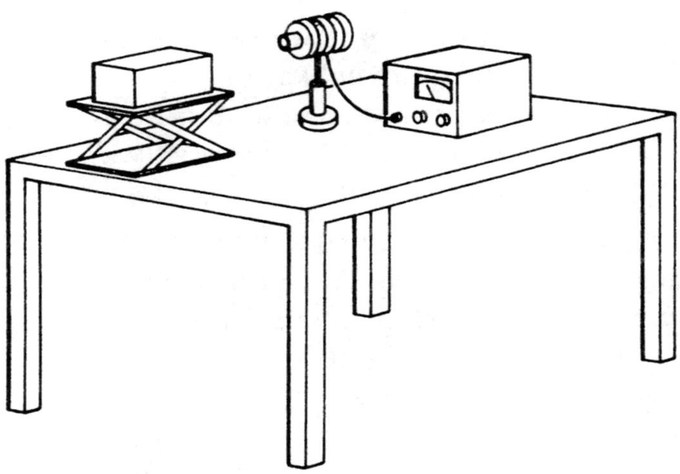

a. Laser power measurement on a workbench.

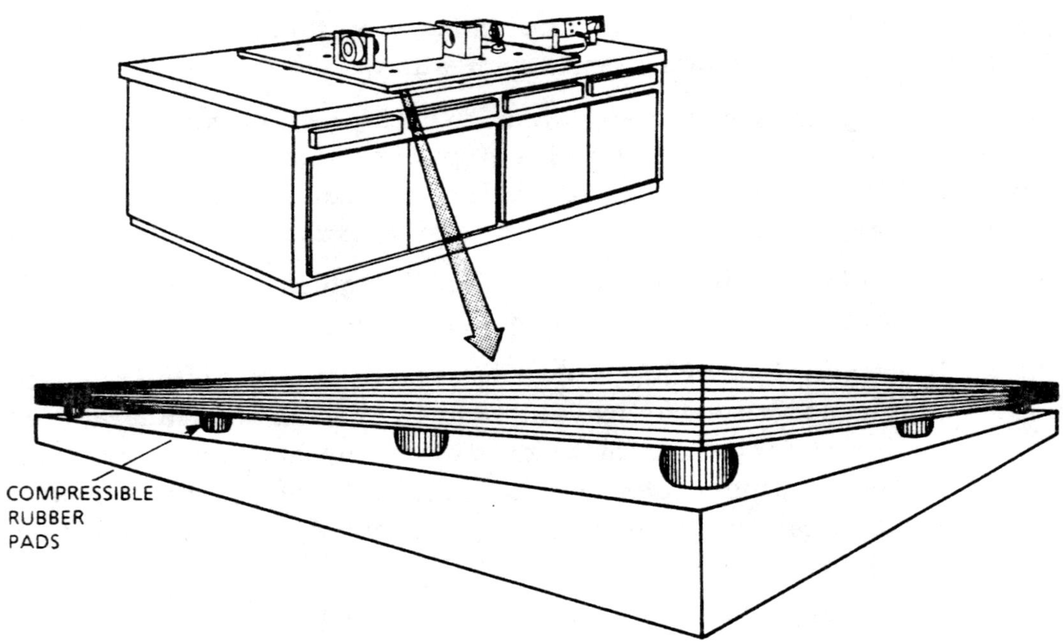

b. Laser breadboard on workbench with metal surface plate.

Fig.13 Laser workbench setups.

2. **Workbench with metal surface plate.** An inexpensive way to improve the optical table quality of a workbench is to mount a precision-flat, heavy metal plate on the top of the bench. (See Figure 13b.) This technique will give a rigid, flat surface and allow the component fixtures to be firmly attached by either mechanical or magnetic mountings. (We'll describe these mounting techniques under "Metal Optical Tables.") Significant vibration isolation at frequencies greater than 100 hertz can be achieved by mounting the plate on compressible rubber pads.

3. **Smooth granite table.** The classical form of optical table is a smooth granite slab (Figure 14a). These tables can be obtained in surface areas greater than 6 feet by 12 feet, thicknesses of 14 inches, and flatness of ± 0.000025 inch. Component holders are attached to the smooth granite surface by suction holders. (See Figure 14b.)

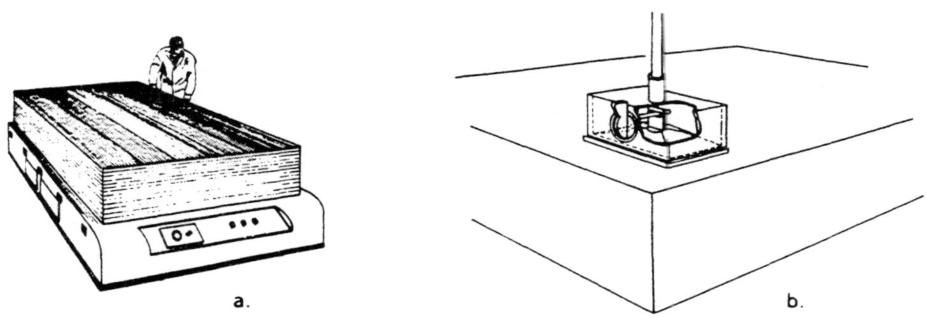

Fig. 14 Smooth granite table and suction-type component holder.

So-called "inertial mounts" sometimes are used on smooth surfaces such as granite or steel. An inertial mount is a heavy base—it usually weighs at least 5 pounds—that contacts the table surface at three points. Such mounts are inexpensive and can be repositioned easily. Inertial mounts (or "gravity bases") are useful in optical setups where the components are unlikely to be bumped or otherwise inadvertently moved.

Desirable properties of granite are:

- Naturally noncorrosive
- Resistance to warping or distortion
- Dense structure with low porosity
- Low thermal distortion

- Uniform hardness, exceeding that of hardened steel
- No crater raised by the impact of a sharp point or edge

4. **Granite table with threaded insert holes.** A more rigid tie-down of component supports to a granite table is achieved with a matrix of holes in the top surface. The holes are filled with steel inserts, tapped for machine screw threads. (See Figure 15a.) The screw size for the inserts is usually 1/4 - 20. You can attach component supports to the table by screwing them directly into the table inserts, or by clamping them with bars screwed into the inserts. (See Figure 15b.) The matrix of inserts may be located as close together as 2 inches or as far apart as 6 inches, depending on the application.

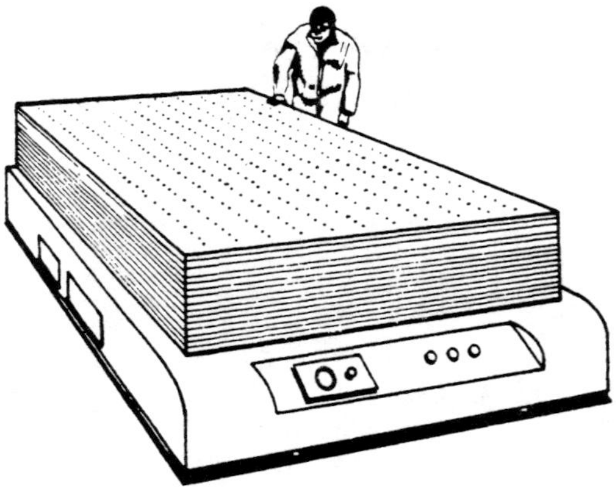

a. Granite table with threaded insert holes.

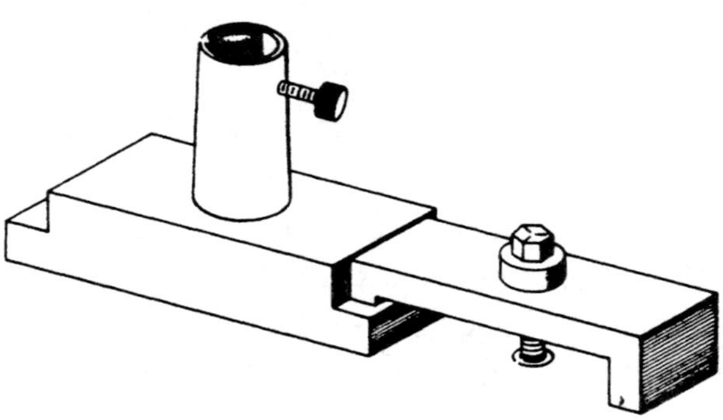

b. Attachment for component support.

Fig. 15 Granite table.

5. **Metal tables with tapped mounting holes.** These tables are used similarly to the granite tables with inserts. Solid metal tables are

usually not as thick or expensive as comparable granite tables. In certain electro-optic applications, it may also be desirable to use a metal table so that it can act as an electrical grounding plane for the system electronics.

As we said earlier, most metal (steel or aluminum) tables aren't solid. They have a metal "honeycomb" core structure covered with a thin metal skin (1/8 inch to 3/16 inch thick). (See Figure 16.) This composition provides increased rigidity and good damping of lateral vibrations.

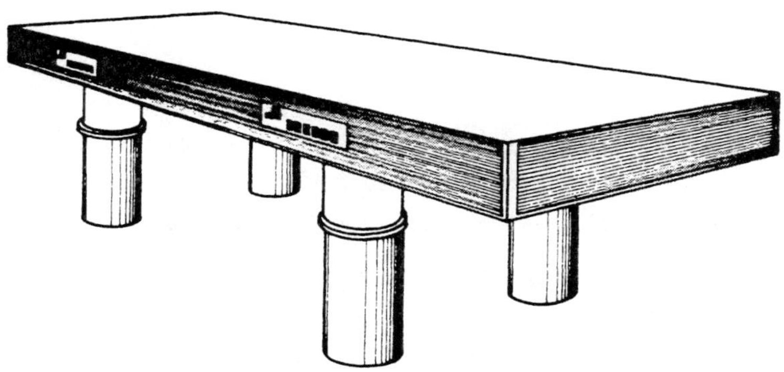

Fig. 16 Metal honeycomb table with thin metal surface skin.

Similar tables are also available that are composed of a phenolic-impregnated paper honeycomb core and solid surface with a matrix of tapped holes.

6. **Smooth metal tables with magnetic mounting.** Either solid or honeycomb-core tables are available with smooth working surfaces of ferromagnetic metal. Component supports are attached to these tables by magnetic mounts. (See Figure 17.)

   These magnetic bases are switched "on" and "off" by rotating a strong permanent magnet within the base. In the "off" position the magnetic flux travels almost entirely through the metal of the base. In the "on" position, the permanent magnet is rotated so that virtually all of the magnetic flux travels through the table surface.

   Depending on their size and cost, such magnetic bases can have holding forces of from 24 to 300 pounds. These bases work best on unpainted, ferromagnetic surfaces (such as type-410 or -430 stainless steel) with a skin thickness of at least 5 mm (3/16 in).

Fig. 17 Magnetic mounts to attach component supports to metal optical tables.

This technique should not be implemented where the presence of strong magnetic fields might temporarily distort or destroy an electro-optic component (such as image-converter tube assemblies).

7. In many applications, excellent vibration isolation isn't needed, but the convenience of a ferromagnetic mounting surface and predrilled holes is desired. In such cases, optical "breadboards" often are used as a lower-cost alternative to honeycomb isolation tables. Figure 18 shows one example of an optical breadboard.

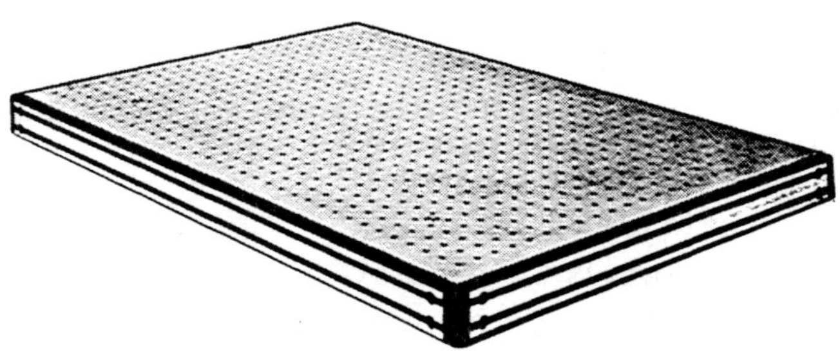

Fig. 18 A modern ferromagnetic optical breadboard.

Like honeycomb isolation tables, these breadboards come in a variety of stock and custom sizes and hole patterns. Typical hole patterns are 1/4-20 holes on 1-in centers. But, metric holes (usually M6 on 25-mm centers) are also available. The 2-ft x 3-ft breadboard shown in Figure 18 also has a slot along each side to permit equipment to be "hung off" the side using bolts and captive nuts.

Some manufacturers also will finish the breadboard surface with epoxy-based semigloss black paint to reduce stray reflections. Such finishes, while attractive, can be fairly easily marred. And they somewhat reduce the holding power of magnetic bases.

If you need some degree of vibration isolation, you can suspend breadboards on firm tabletops with small-diameter inner tubes or sheets of foam rubber. In some cases, these "budget" isolation surfaces make sense as a significantly lower-cost alternative to pneumatic isolation tables.

Vibration Isolation

As we said earlier in this module, optical tables are susceptible to *induced vertical and horizontal vibrations.* These vibrations are undesirable in applications that involve very precise optical alignment systems because they result in *relative movement between the optical components mounted to the table.*

You can understand *vertical vibrations* within the table surface by visualizing a *very small, periodic bending motion*, or flexing, in the table—similar to a diving board. (See Figure 11.) These vibrations originate *as floor or building resonant vibrations. They're induced through the table supports.* The *frequencies of these vibrations typically range from 5 to 30 Hz.*

You can isolate some vibration in a rigid-frame table support by placing *rubber pads* between the table legs and the floor and between the support and the table surface. You'll get much more effective isolation by using *air pistons* in the table legs. (See Figure 19.) These pistons actually "float" the tabletop and provide isolation from vertical floor vibrations. Vibration-isolation systems are similar to a high-frequency cutoff filter, insofar as they pass low-frequency vibrations and isolate all frequencies above some design value. *Air pistons provide vertical isolation for frequencies above 5-10 Hz*, and generally require a source of compressed air in the laboratory.

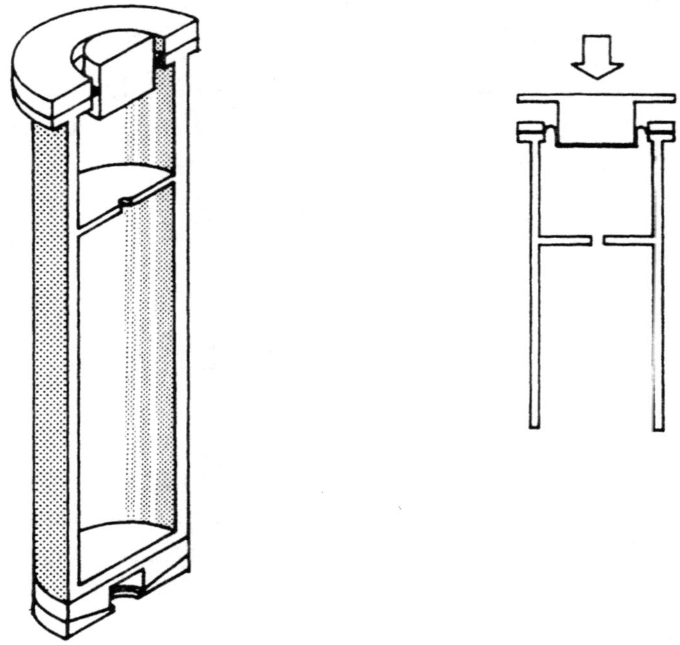

Fig. 19 Air piston for optical table vibration isolation.

Selecting the Correct Table

The design and construction of a vibration-isolated optical table takes considerably more theoretical insight and working experience than we can give you in this module. The important points to learn are:

- A general understanding of the kind of performance you can expect of crude and sophisticated optical tables
- An estimate of the optical table performance required by the experiment or alignment to be performed
- Knowledge of how and what to specify when you need an optical table

The lab tasks in this module should provide some experience for the first point. An understanding of the optical experiment or alignment should provide data for the second point. Address the third point by answering these questions:

- How much working area do you need?
- How accurately must you maintain component spacings?
- What maximum temperature variations do you expect in the laboratory?
- What degree of table surface flatness is required?
- What vibration isolation (if any) is required? How much vibration can be tolerated, and at what frequencies?

- Do you need to fasten components rigidly to the table surface?
- Would the fields produced by magnetic support attachments distort the operation of the electro-optical system?
- Would a conductive work surface (ground plane) be beneficial or detrimental to the electro-optical system operation?
- What's the "long-term" use of the table? How much versatility do you need?
- How much money is in the budget for this equipment?

Quite often, an early consideration of the last item forces compromises in system requirements for the optical table.

EXERCISES

1. Name four classes of optical benches. List one advantage and one disadvantage of each.
2. Define the following terms:
   - Vertical vibrations in a table
   - Horizontal vibrations in a table
   - Vibration isolation from table support
   - Internal table damping from acoustically induced vibration
   - Thermally induced movement
   - Table rigidity
   - Table flatness
3. List the advantages and disadvantages (at least two of each) of the following optical tables:
   - Lab workbench
   - Lab workbench covered with thick metal plate with matrix of tapped mount holes
   - Concrete slab on inner-tube support
   - Granite slab with insert matrix, mounted on steel frame with rubber shock mounts
   - Granite slab with insert matrix, mounted on pneumatic vibration isolators
   - Steel table with magnetic mounts
   - Metal honeycomb table, with stainless steel skin and matrix of tapped holes, vibration isolation.
4. In your own words, outline the procedures used in the laboratory to align an optical bench parallel to an optical axis established by a laser beam.
5. Outline the specifications of two optical tables of comparable size. Use the criteria given in the section, Selecting the Correct Table, and

catalogs from suppliers of optical tables. From your outline, evaluate the effectiveness of the table in isolating low-frequency vibrations.

6. How much will the center of a 4-ft x 8-ft x 12-in steel honeycomb table deflect when loaded with a 150-lb spectrometer at its center. Skin thickness is 1/8 inch. The spectrometer applies a point load to the center of the table. Get any more data that you need from the example worked earlier in this module. Express your answer in mm.

7. In an experiment that's particularly sensitive, you can allow the center of your optical table surface to deflect downward by an additional distance equal to only one wavelength of light (633 nm). Assume that you use the table from Exercise 6 and that the only significant load on the table is a point load in the center. What's the maximum point load allowed? Express your answer in pounds.

8. During a day of use, the air temperature in an optics laboratory increases by 2.50 C°. Compute the thermal change in length of three 8.00-ft-long optical tables used in the lab. The tables are made respectively of aluminum, type-400 stainless and super invar. Express your answers in multiples of the red HeNe laser wavelength.

MATERIALS

Workbench
2-ft x 3-ft x 1-in shock-mounted metal plate to fit on workbench
4 rubber shock mounts
Pneumatic vibration-isolated optical table, 4-ft x 8-ft working surface
Baseball or tennis ball
Sandbag, about 15-25 pounds of mixed sand and various-sized rocks
Aluminum cube, 3-in sides, tapped to accept accelerometer mounting stud
HeNe laser
Autocollimator/telescope
Optical bench
Cross target
Carriage
Accelerometer
Oscilloscope

PROCEDURES

1. Align an optical bench parallel to a given optical axis defined by a laser beam. Use the technique described in the discussion.
2. Determine the response of a table in each of the three axis directions to an induced vibration.
   a. Arrange accelerometer, cable and scope as shown in Figure 20. The accelerometer should be in the vertical position (circular end facing up), and directly over the approximate center of the tabletop. This will now indicate vibration in the Z-axis direction. The accelerometer connecting cable shouldn't be kinked (damage can result) or looped (inductive pickup can occur). Some scopes are also sensitive to vibration. They should be supported on vibration-damping pads. Foam rubber is effective.

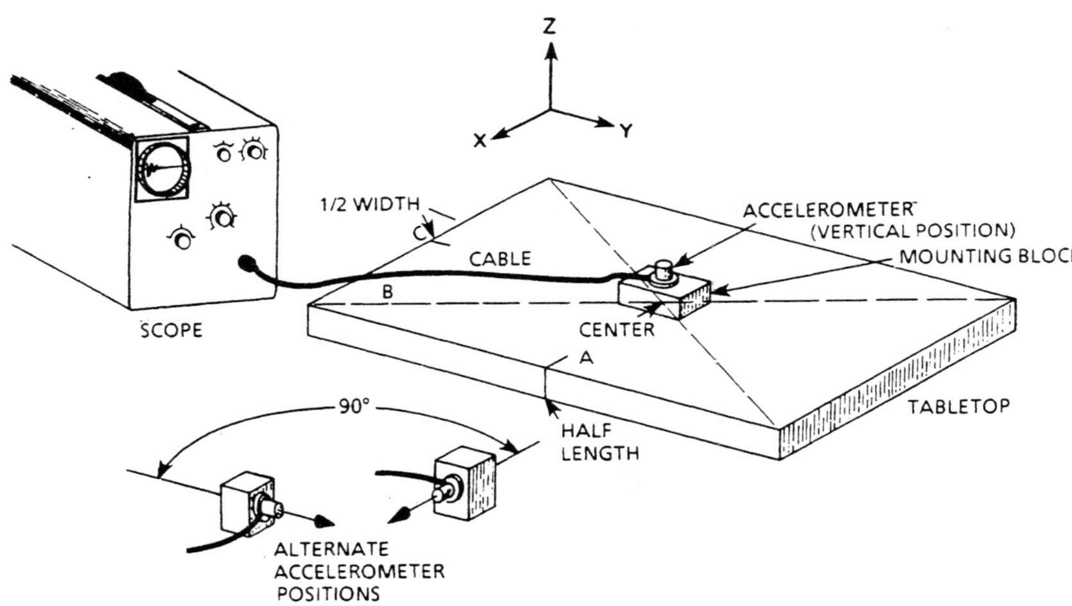

Fig. 20 Experimental arrangement for measuring table vibrations.

   b. Adjust the vertical amplifier of the scope for maximum sensitivity (5-mV/cm desired). Set the sweep speed at 0.1 msec/cm. Adjust trigger controls to initiate a sweep when a fingertip is brushed *lightly* across the top of the accelerometer. Triggering level will be near *zero*, stability control is *Preset*, slope control on *Internal +*, and trigger mode on *AC slow*. If you can't adjust the trigger level to activate on the weak signal generated by touching the accelerometer, you can set the trigger mode on *automatic.*

c. Hold the ball accurately 30 cm above the middle of the long edge of the table (point A). Release it (be prepared to catch it on the first bounce) while you watch the scope trace pattern. If possible, reduce room light or shade the scope screen so that you can see a persisting image.

d. If the pattern height is less than one or two cm, increase the height from which you drop the ball. Make subsequent drops from the same height. Record this height as (h).

e. Examine the scope pattern to estimate the general range of frequencies present. The number of full waves (distance between successive peaks) per centimeter, divided by the time per cm, will give an approximate frequency value. For example: the pattern shown in Figure 21 has about 2.5 full waves in a length of 1 cm. With the sweep rate at 0.1 sec per cm, the vibration frequency would be 2.5/0.1 - 25 hertz. If higher frequencies are indicated by the presence of complex wave forms and fine structure, increase the sweep rate while you excite the table (ball drop) until a measurable and reasonably repeatable pattern is produced. Compute its approximate frequency. Record all data in Data Table 1.

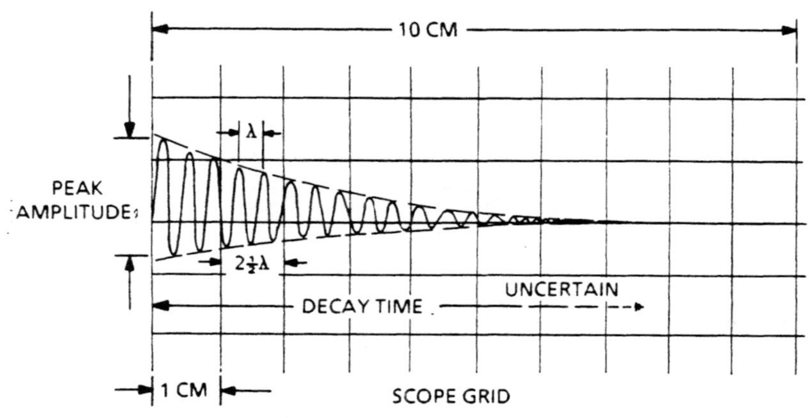

Fig. 21 Analysis of vibration data.

f. Repeat Steps c, d and e, releasing the ball from the same height (h) over points B and C on the table. Determine and record frequencies, as well as any significant difference in the amplitudes or number of frequencies identified.

g. Set the sweep rate at a lower value, so that the signal amplitude decays to about half its peak height in one sweep. Measure the maximum peak-to-peak height of the pattern. Convert to mV, and record as peak amplitude for that frequency.

h. Using a slower sweep, determine and record the time required for the signal to decay to half its peak amplitude.

i. Turn the mounting block on its side, so that the accelerometer top is parallel to the long side of the table. The accelerometer now will measure vibration in the Y-axis direction. Record data in Data Table 2.

j. Repeat the above procedure after moving the accelerometer to measure vibration in the X-axis direction. Record data in Data Table 3.

3. Determining effectiveness of table in damping floor-induced vibrations.

a. Position the accelerometer as in Part 2, Step 1.

b. Locate a point on the floor, a few (2-3) meters from the table legs or base, preferably on a line bisecting the floor contact points or area, and perpendicular to either width or length of table support.

c. Set the vertical sensitivity of the scope to maximum. While you observe the pattern, drop the sandbag from a height that will cause a scope pattern of at least 2-cm peak-to-peak height. Note location of the drop point (X) and value of the drop height. Use these for subsequent drops.

d. Determine major resonances in pattern, their peak amplitudes and decay times (half-amplitude decay). Record for each frequency.

e. Relocate the accelerometer and block (accelerometer vertical) on the floor near the base at point B or C in Figure 22 or near the legs (point A) of the table support.

f. Measure and record accelerometer location. Reduce vertical sensitivity by a factor of 100 (0.5 V/cm). Drop the sandbag on the selected point (X) from the selected height while you watch the scope pattern. Adjust sensitivity to produce a pattern with peak-to-peak amplitude of 4 to 10 cm.

g. Identify and measure major resonances of floor, as you did for the tabletop. Convert peak amplitudes to millivolts. Record all data in a table that you prepare.

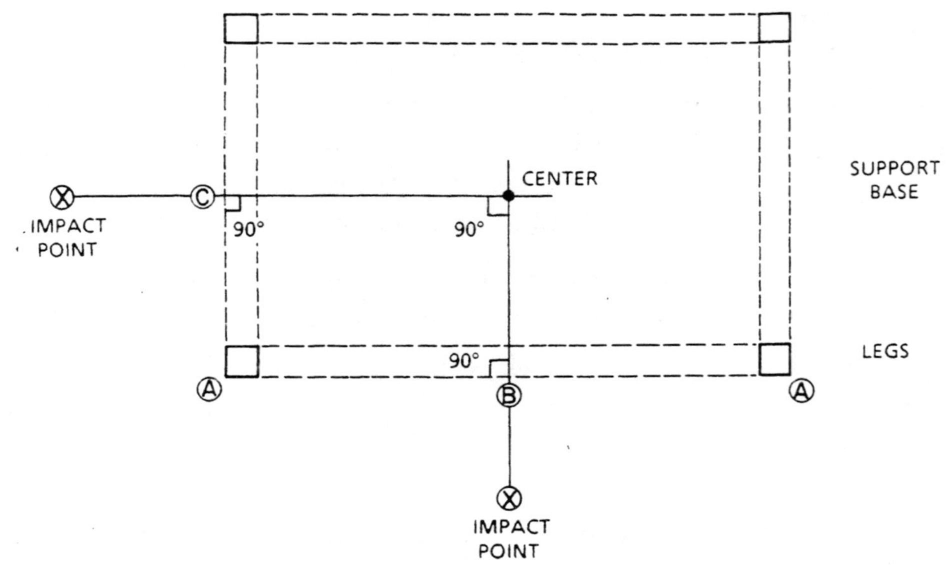

Fig. 22 Experimental arrangement.

h. Compare floor and tabletop vibrations by computing the ratio of peak-to-peak amplitudes at both floor and table accelerometer locations. This ratio will be a measure of the effectiveness of the table in reducing vibrations of the frequencies observed.

4. If you have time, compare the effectiveness of three different tables in damping floor-induced vibrations. Repeat procedures outlined in Section 3. What's your conclusion?

Data Tables

Data Table 1. Accelerometer Adjusted for Z Axis

Height of drop______________
Type of drop ______________

	Frequency or Frequencies Observed	Peak-to-Peak Height	Time for Half Peak Amplitude	Comments (Differences)
Drop point A				
Drop point B				
Drop point C				

Data Table 2. Accelerometer Adjusted for Y Axis

Height of drop______________
Type of drop ______________

	Frequency or Frequencies Observed	Peak-to-Peak Height	Time for Half Peak Amplitude	Comments (Differences)
Drop point A				
Drop point B				
Drop point C				

REFERENCES

Ealing Electro-Optics Product Guide. Ealing Electro-Optics, Inc., 22 Pleasant Street, South Natick, MA 01760, 617/655-6029.

Melles Griot Optics Guide 3. Melles Griot, 1770 Kettering Street, Irvine, CA 92714, 714/261-5600

The Newport Corporation Catalog of Precision Laser/Optics Products, No. 100. Newport Corporation (formerly NRC), 18235 Mount Baldy Circle, Fountain Valley, CA 92728-8020, 714/963-9811.

Oriel Corporation Catalog Vol. I of Tables, Benches, Micropositioners, Optical Mounts. Oriel Corporation, 250 Long Beach Boulevard, Stratford, CT 06497-0872, 203/377-8282.

Specification sheet and application data on accelerometers.

LASER

ELECTRO-OPTICS

MODULE 6-2
COMPONENT SUPPORTS

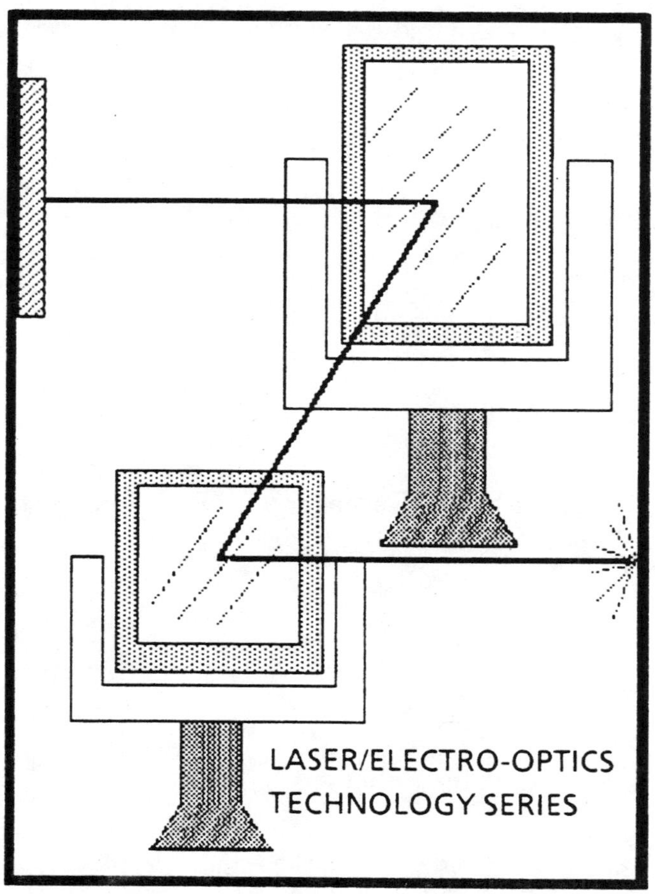

CORD

The Center for Occupational Research and Development
601 C Lake Air Drive
Waco, Texas 76710

Printed in the U.S.A.

ISBN No. 1-55502-024-0

COMPONENT SUPPORTS

MODULE 6-2

A component support is any device that holds optical components in the right position in an optical setup. The simplest example of a component support is your hand. When you take a magnifying glass in your hand and use it to examine a small object, your hand is the component support. The hand isn't steady enough or strong enough for many applications, so mechanical supports have been developed. The purpose of a component support is to provide a stable and rigid mounting position and to allow controlled movement of a piece of optical equipment.

This module will help you to select, recognize and properly use the many types of component mounts available.

You should have an understanding of algebra and of general laboratory safety. You also must be familiar with the use of low-power helium-neon lasers.

OBJECTIVES

When you complete this module, you should be able to:

1. From an examination of drawings of component supports, identify and describe the function of the following:
   - Translator table
   - Rotary tables
   - Tilting trough
   - Scissors jack
   - Focusing holder
   - Two-axis translator

Two-axis mirror mount
Laser cavity mirror mount
Laser beam director
Six types of lens holders
Filter holders
Prism holders
Laser mounts
Translator plus rotator
Hold-down clamp
Vertical translator
Sine table

2. Select the proper component support to perform these functions:
   - alter height and horizontal direction of a light beam
   - move a component along the optic axis
   - adjust a prism to its minimum deviation angle
   - adjust an aperture in two directions, both perpendicular to the optic axis
   - support unmounted lenses of several different diameters
   - align a reflector to return a beam exactly back to its source
   - support two unmounted sheet polaroids to form a polarizer and analyzer
3. Calculate the linear resolution of a piezoelectric translator, given its rate of movement and a minimum change signal.
4. Calculate the linear resolution of a differential-screw micrometer translator, given the screw constants and the minimum possible change in screw rotation.
5. Estimate the minimum distance that an optical beam can be deflected using a variety of rotational stages.
6. Set up an optical system to expand, collimate, reflect and disperse a light beam; then to reflect and focus the first-order diffracted beam of a specified wavelength onto a target to an accuracy of ± one-half of the smallest diameter of the resultant spot.
7. Compare quality and price of five optical component supports by referring to catalogs from suppliers of optical equipment.

DISCUSSION

Component supports are used to hold and position optical devices. To attempt an optical experiment or alignment if the component supports aren't stable enough, or if they don't provide for smooth enough adjustment of a component's position, can be very frustrating.

Linear Movement

The fine adjustment of position on a component support (whether linear or angular) generally is accomplished by the slow linear movement of a threaded rod as it's screwed through or against a plate. A typical example is in Figure 1.

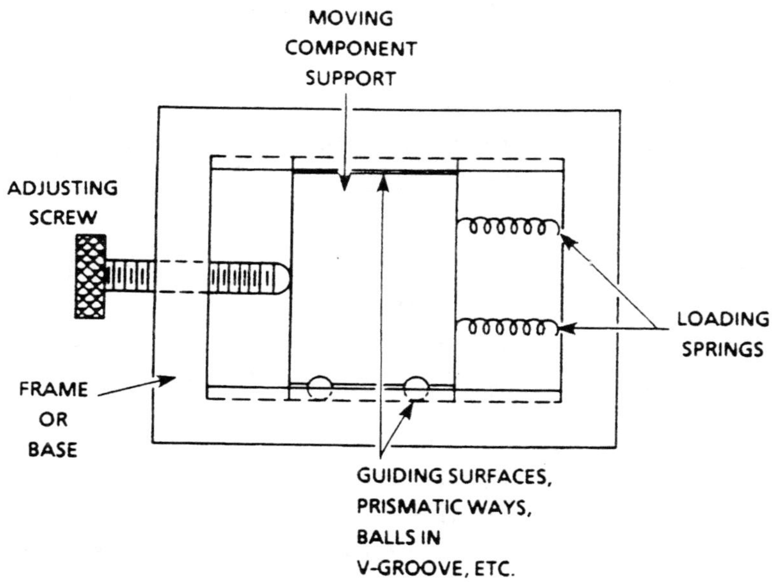

Fig. 1 One-dimensional linear translator.

The precision, smooth adjustment is possible because of the principle on which the screws work. As the threaded rod is turned, it pushes the component table a proportionate distance. The springs keep the adjustment screw and component support in close contact with one another. They cause the support to return when the screw direction is reversed. The increment or resolution of linear movement is limited by the least amount you can turn the screw and the pitch of the threads on the screw. For example, if the screw had 40 threads per inch, each turn will move the platform 1/40 = 0.025 inch/revolution. If the knob on the screw is about one inch in diameter, the average person can twist the screw a minimum of about 2 degrees (2 deg/360 deg/rev = 0.0055 revolution) or 0.0055 x 0.025 = 0.000138 inch. The increment of movement ΔS would be the product of these two. Metal rods seldom are threaded finer than 40 threads per inch.

You can get finer resolution by a device called the differential screw. This device is shown in Figure 2. The long adjustment screw passes

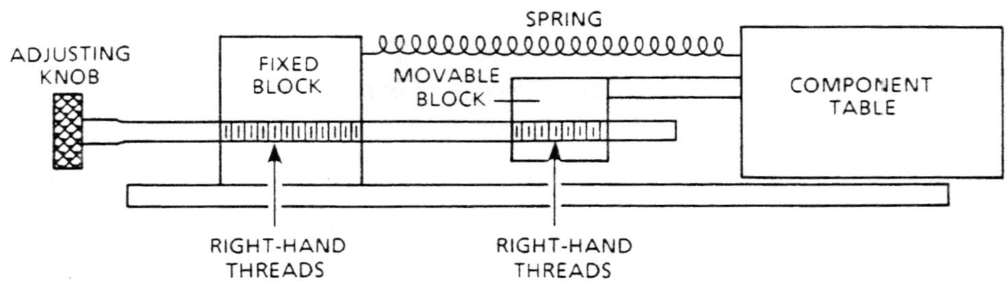

Fig. 2 The differential screw.

first through a fixed block at the left. In this block, the screw has right-hand threads. The screw also penetrates a "movable" block attached to the component table. The screw in the movable block **also** has **right-hand threads.**

When you rotate the adjusting knob in a clockwise direction, the screw moves forward, pushing the movable block and component table to the right. At the same time, the movable block moves to the *left* on the rotating screw, offsetting some of the rightward motion of the screw.

Suppose that the pitch is 30 threads per inch in the fixed block and 40 threads per inch in the movable block. Then, when you turn the knob clockwise one complete revolution, the screw advancement through the fixed block pushes the component table 1/30 inch = 0.0333 inch **to the right**, while the opposite motion of the movable blocks along the advancing screw pulls the table 1/40 inch = 0.0250 inch **to the left.** The **net** movement for one revolution is 0.0333 inch – 0.0250 inch = 0.0083 inch to the right.

If the knob on the adjusting screw is turned two degrees at a time, the increment of movement will be:

$$\frac{2^\circ}{360^\circ} \, x \, 0.0083 \, \frac{inch}{revolution} = 4.6 \, x \, 10^{-5} \, inches = 1.17 \mu m$$

or about 2λ for green light.

You can get very small linear movements on components (such as mirrors) by using piezoelectric translators. A piezoelectric device is a special crystal (such as barium titanate) that changes its physical dimensions when an electric field is applied across it. Normally, a piezoelectric crystal has 3 sets of parallel faces. One is coated with metal electrodes. When a voltage is applied to the electrodes, one of the dimensions expands. An increase in voltage causes a corresponding increase in the lateral dimension of the crystal. A typical piezoelectric sensitivity is 2×10^{-9} meters/volt. This means that a change in electrode voltage of 100 volts will cause a change in crystal size of (100 volts) x (2×10^{-9} meters/volt) = 2×10^{-7} meters or 200 nanometers, about $\lambda/2$ for violet light.

Piezoelectric translators are used to move a mirror that adjusts the length of a laser resonant cavity when "tuning" a longitudinal mode.

Angular Movement

You can make angular adjustments with a threaded rod in a configuration similar to the one in Figure 3. Movement of the screw causes the

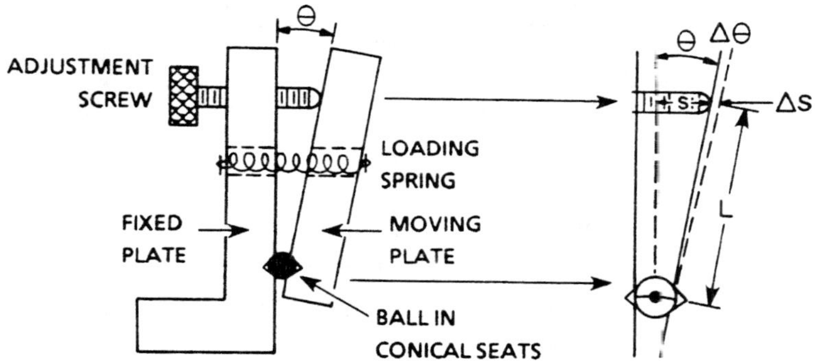

Fig. 3 Angular translator.

movable plate to rotate about the pivot ball. The angular change $\Delta\theta$ is, for small values of θ, approximately equal to the screw movement ΔS divided by the pivot screw point distance L. ΔS is, of course, the resolution of linear movement of the screw. The angular resolution is simply that value divided by L. Use the 40-pitch screw ($\Delta S = 1.38 \times 10^{-4}$ inches) and a 2-inch distance for L. Angular resolution ($\Delta\theta$) is:

$$\Delta\theta = (radians) = \frac{\Delta S}{L} = \frac{1.38 \times 10^{-4}\, inches}{2\, inches} = 6.9 \times 10^{-5}\, radians\ or\ 14.2\ arc\ seconds$$

where

$$1\ radian = \frac{360}{2\pi} \simeq 57.3^{\circ}$$

Possible Movements or Degrees of Freedom

The position or motion of a piece of optics can be described using six motions or six "degrees of freedom." Figure 4 shows the six possible motions or adjustments in position for a lens.

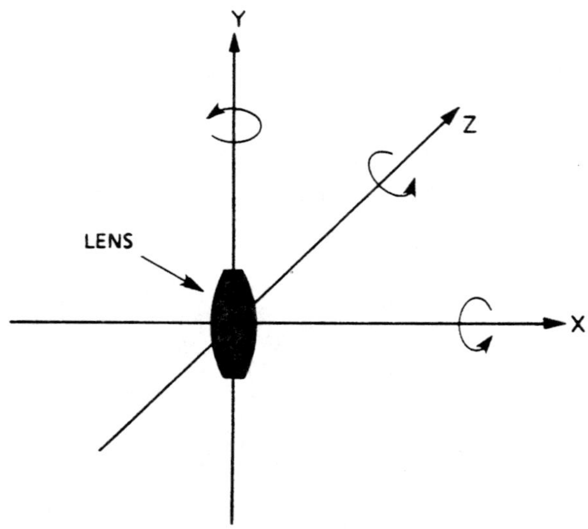

Fig. 4 Possible movements.

The six possible movements for the lens are:

a. Movement along X
b. Movement along Y
c. Movement along Z

all of which are classed as linear, and:

d. Rotation about X
e. Rotation about Y
f. Rotation about Z

which are all rotational.

By combinations of these six basic motions, you can put the lens—or other component—in any position you want. One or more of these six motions usually are provided by component supports.

Supports With One-Degree Freedom

The simplest type of adjustable component support is the one that provides one degree of freedom. This lets you move the component back and forth along, or rotate it about some line.

Figure 5 shows a translator table (translational stage) that supports components and provides one degree of freedom (linear). The translator table

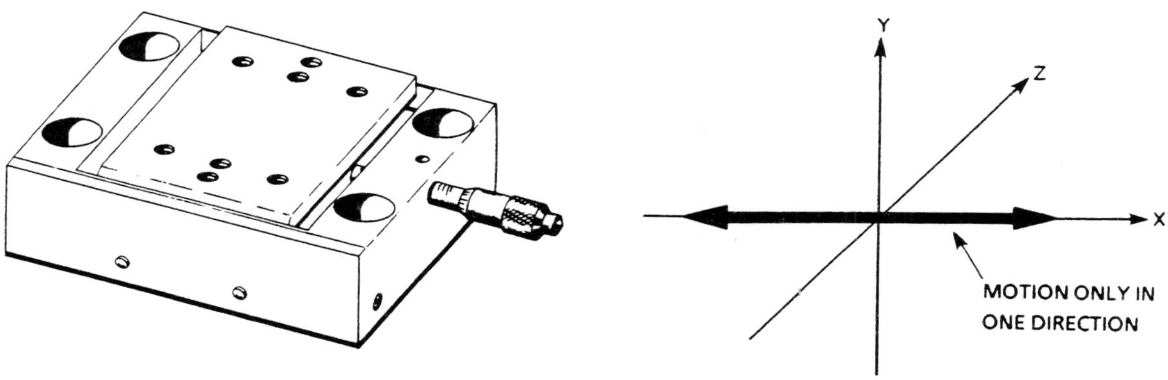

Fig. 5 Translator table.

gives controlled linear movements. With the micrometer, it permits position readout. You can use it along or transverse to the optical axis to position prisms, crystals, lenses and other components. The translator table is controlled by a micrometer pushing against a ball-bearing-supported, spring-loaded table.

Figure 6 shows examples of other supports that allow only one degree of freedom (rotational). The rotary table and rotational mount provide controlled rotational movements with high resolution and readout capabilities. They can be used to rotate prisms, polarizers and crystals.

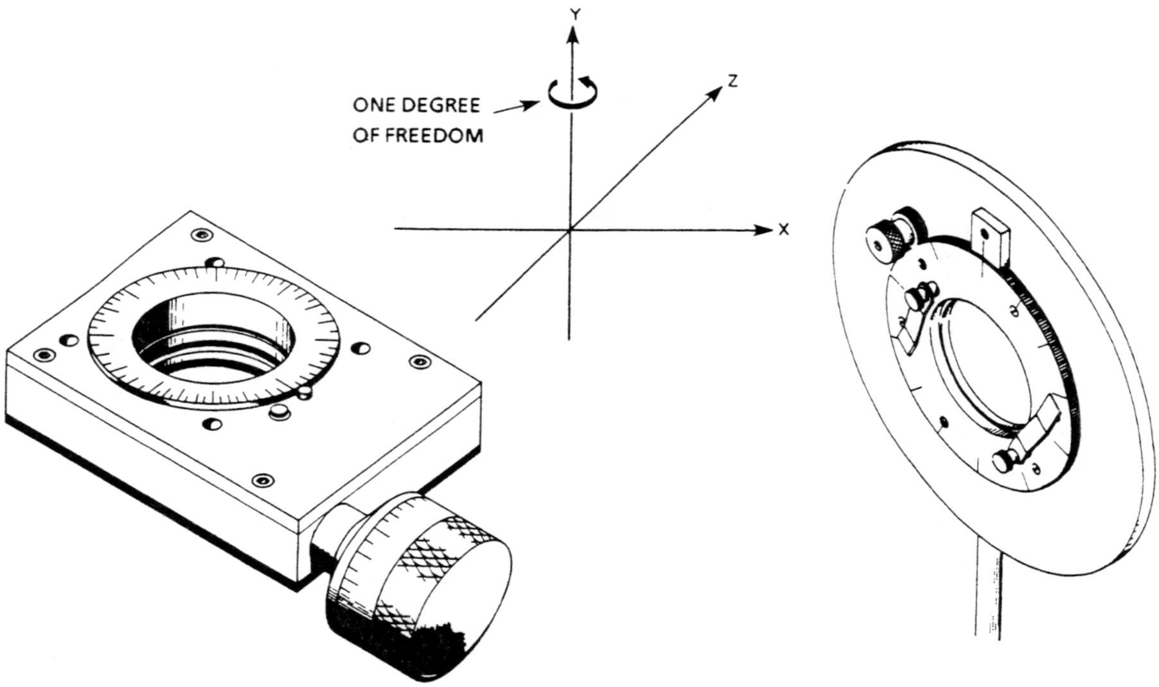

a. Rotary table (rotational stage)

b. Rotational mount for polarizer

Fig. 6 Rotational devices.

The rotary table (or "rotational stage") is controlled by a worm gear on the knob shaft that drives a circular ring gear through a full 360°. A simpler, and less expensive, rotational stage is shown in Figure 7. In this device,

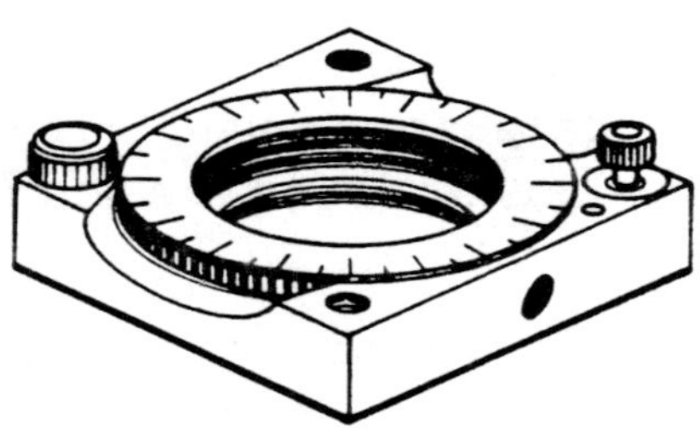

Fig. 7 An example of a rotational stage driven directly by a rotating shaft.

rotation is accomplished by a round shaft (attached to the knob) that rolls against the circumference of a central circular platform. The circular degree scale is provided for noncritical angular measurements. The knob on the right is used to lock the position of the stage. This particular stage is 3.5" square.

The central platform usually is supported by ball bearings to allow smooth rotation. You can purchase the platform of the stage with small tapped mounting holes or a larger-diameter access hole, with or without internal mounting threads.

You can determine the resolution of such a device by noting that the arc length along the circumference through which both the central platform and the shaft move is the same (if no slippage occurs). Thus you can determine relative angular motion from a ratio of shaft to platform circumference.

Example A: Rotational Stage

Given: A rotational stage similar to the type shown in Figure 7. The important dimensions are shown in the sketch on the right. The average user can turn the knob through a minimum repeatable angle of 5.0 degrees.

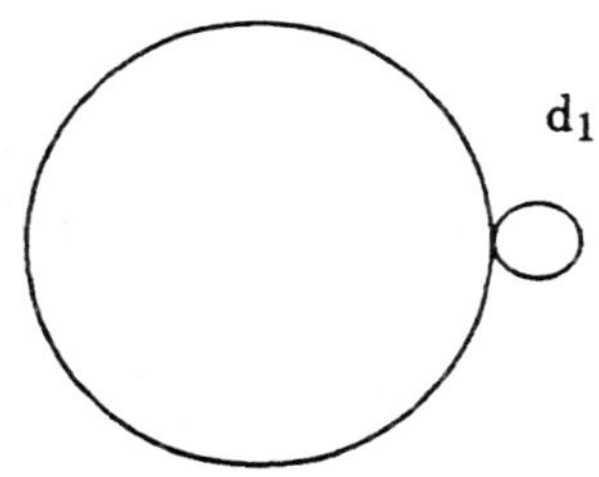

Find: Minimum angle through which you can be expected to rotate the stage platform.

Solution:

$$\text{Arc length } S_1 = r_1\theta_1 = \frac{1}{2}d_1\theta_1 \quad (\text{arc length formula})$$

$$\text{if no slipping } S_1 = S_2 = \frac{1}{2}d_2\theta_2$$

$$\therefore \theta_2 = \theta_1\left(\frac{d_1}{d_2}\right) = 5.0^\circ\left(\frac{1.02\ cm}{6.15\ cm}\right) = 0.83^\circ$$

Thus, the expected resolution of this stage is 0.83°.

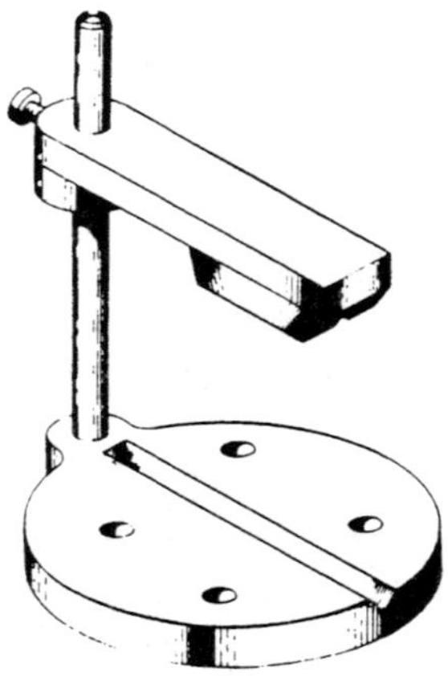

Fig. 8 Hold-down clamp.

A hold-down clamp like the one in Figure 8 often is used with a rotary table. But it is not limited to such usage. You can use it on the translator tables or any of the other mounts where you want to hold an irregular-shaped optical component in place.

The Scissors Jack (Lab Jack)

Another component support with one degree of freedom is the scissors jack. It's shown in Figure 9. This support looks like one type of auto jack. It's useful in changing the height of heavy components above a work surface. Its basic design, including a number of pivots and relatively thin load-bearing members, makes it relatively unstable. So, it's not very satisfactory when you need rigidity.

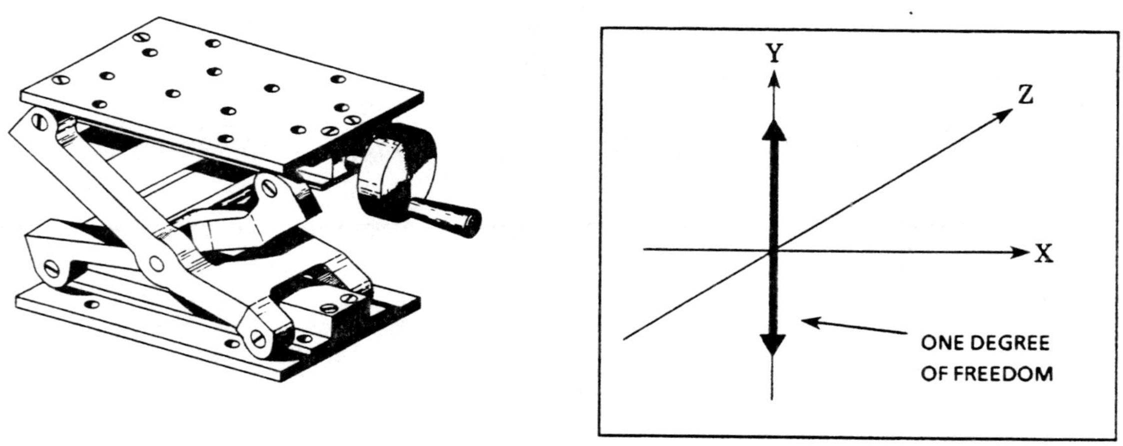

Fig. 9 The scissors jack.

If you need very fine control of the vertical motion, you can use a vertical translator equipped with a micrometer. One is shown in Figure 10.

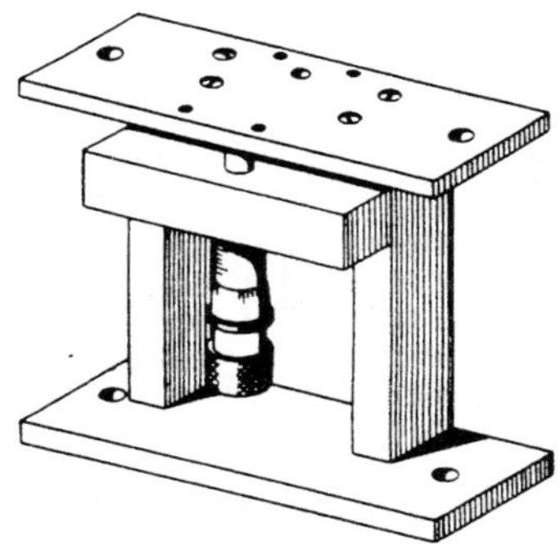

Fig. 10 High-resolution vertical translator.

The Sine Table

A controlled angular displacement about one axis is provided by the sine table shown in Figure 11.

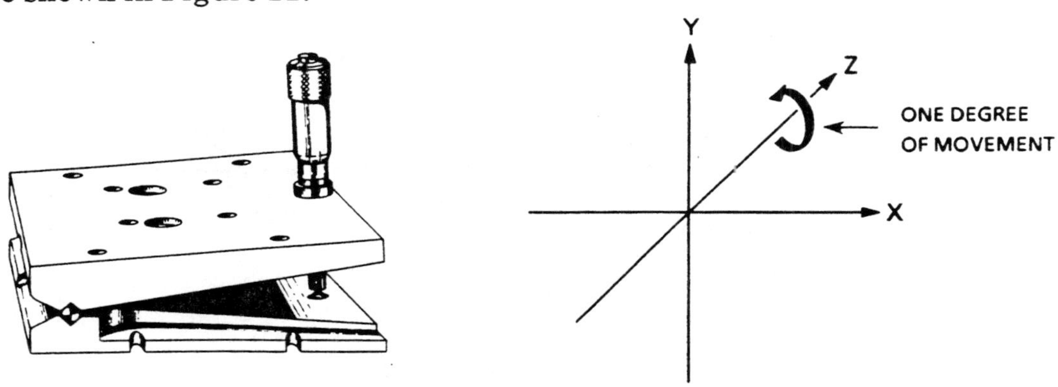

Fig. 11 The sine table.

The Tilting Trough

Movement of one degree of freedom is provided by telescopes, collimators, and other cylindrically shaped objects by the tilting trough support. The tilting trough shown in Figure 12 has about 10 degrees of fine pitch adjustment by turning a spring-loaded screw.

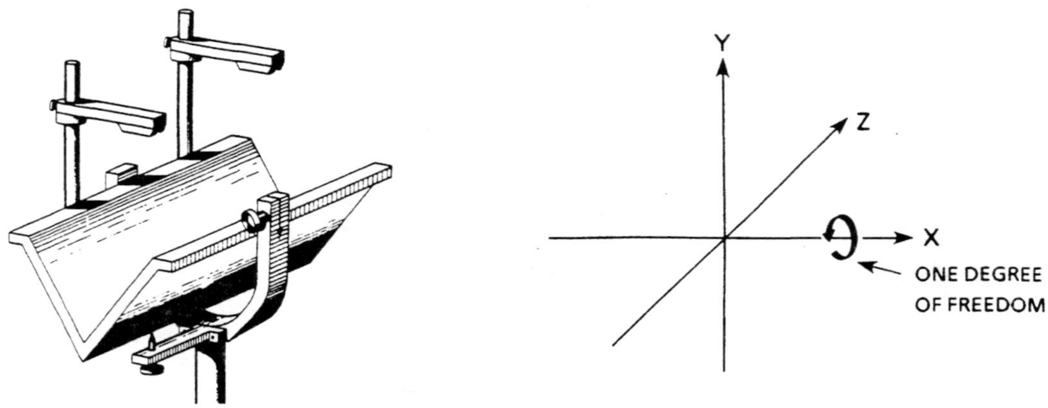

Fig. 12 The tilting trough.

Tube Holder with Focusing Motion

The last example of an optical support with one degree of freedom of motion is the focusing tube holder. This device (shown in Figure 13) is used to

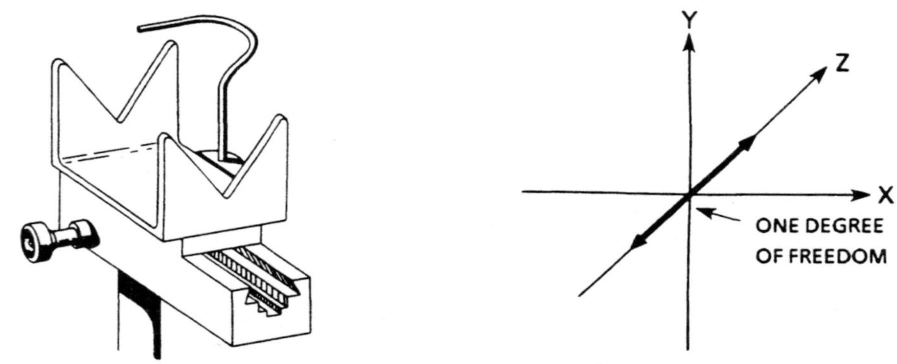

Fig. 13 Focusing holder.

hold microscopes, telescopes, spectroscopes, and other cylindrical units. The one controllable adjustment is used to bring these instruments to their focus position.

Multiple Degrees of Freedom

The component supports providing one degree of freedom can be combined to provide adjustments in many directions. Some of the combinations used more often will be discussed in this section.

Two-Axis Translation (X-Y Translator)

You can stack two separate translator tables in a way that will allow controlled adjustment along two axes. Figure 14 shows a two-translator

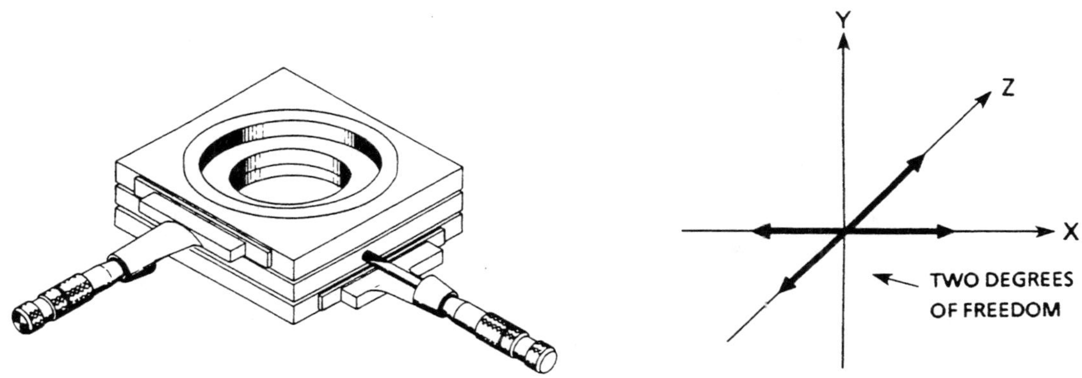

Fig. 14 The two-axis translator (X-Y translator).

arrangement and the movements that it makes possible. An additional translator can be bolted at right angles to the upper table to produce motion along three orthogonal axes. Such an assembly often is called an "X-Y-Z translator."

If you combine a translation unit and a rotational unit, you'll get a device similar to Figure 15, permitting the movements indicated. Such a device might be used to first center an object along the optical axis with the

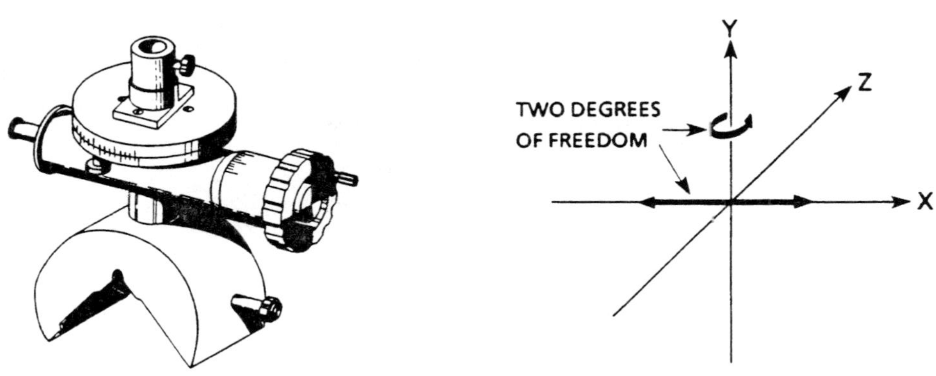

Fig. 15 Translator plus rotator.

translation stage and then view it from all rotational angles with the rotary table.

Two-axis Mirror Mount (Gimbal Mounts)

You'll need mounts with two rotational degrees of freedom to position a mirror for many uses. Figure 16 shows a two-axis mirror mount where the

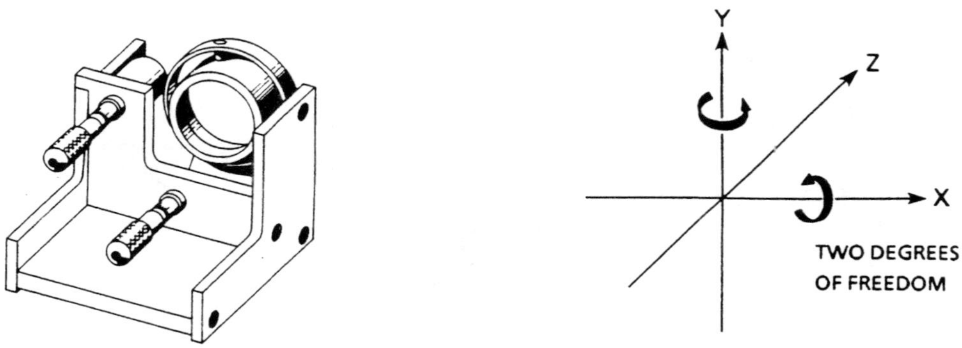

Fig. 16 Two-axis mirror mount (gimbal mount).

mirror is supported by two adjustable rings. A mount like this might be used to point a laser beam in a certain direction without having to move the entire laser.

A variation of the two-axis mirror mount is shown in Figure 17. This is the laser cavity mirror mount. It functions similarly to the two-axis mount

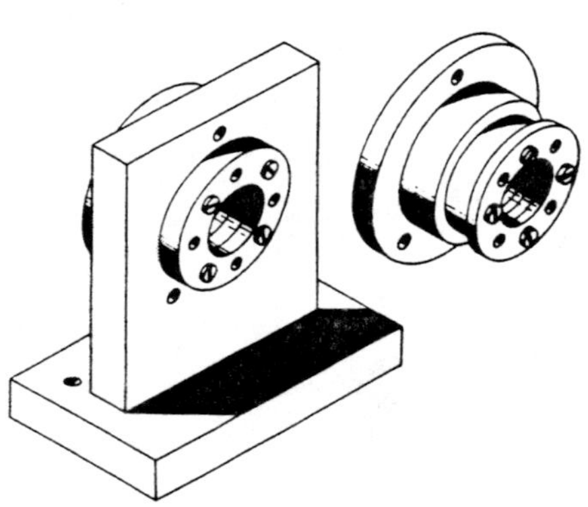

Fig. 17 Laser cavity mirror mount.

described above, but the adjustments are made with screws instead of micrometers. It's available in several forms, differing in the type of pivots and loading that are used to prevent "backlash."

Laser Beam Steering Device

In the laboratory you'll often need to redirect the laser beam without moving the laser. Laser-beam directors are made to fill this need. As shown in Figure 18, the laser-beam steering device lets you move a mirror in three

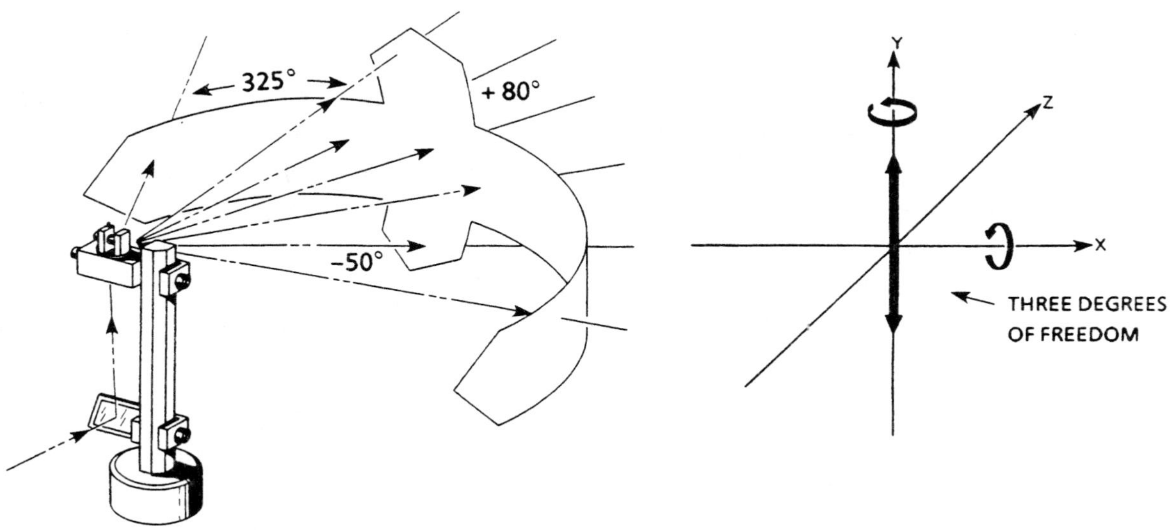

Fig. 18 Laser-beam steering device.

degrees of freedom so that a beam can be put anywhere in the field of view of that mirror. Both mirror carriages can be moved independently of the vertical support column so that you can position the laser beam to the desired height and direction easily. This use is shown in Figure 19.

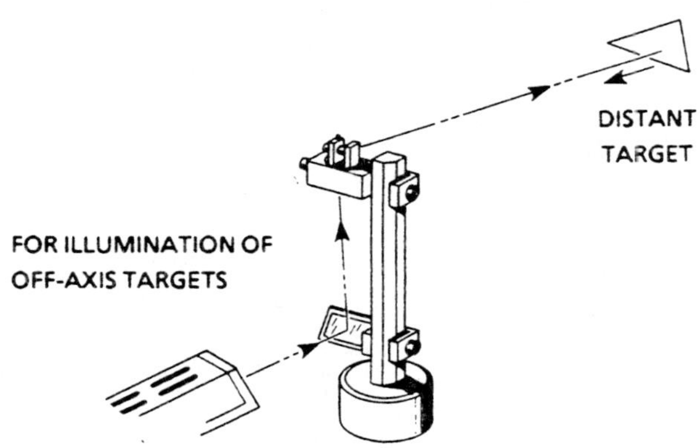

Fig. 19 One use of laser-beam steering device.

Individual Component Supports

Some optical components are used so often in optics work that special mounts are made for them. This section will discuss mounts for lenses, mirrors, prisms, and other commonly used components.

Lens Holders

Lenses are used in so many optical setups that there are many types of lens supports. The simplest type is the sliding-grip lens holder shown in Figure 20. The bottom and top supports have deep grooves for seating the lens. The top support slides on the two vertical rods and has a locking screw.

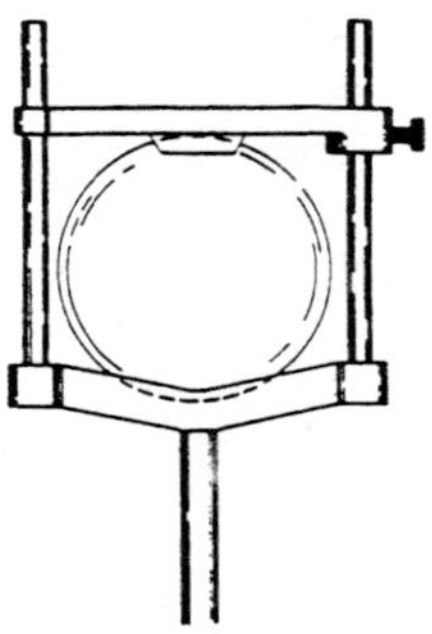

Fig. 20 Sliding-grip lens holder.

An inexpensive lens holder is the spring-grip lens holder shown in Figure 21. The lens is inserted between two jaws. The jaws are operated by pressing down two spring-loaded triggers.

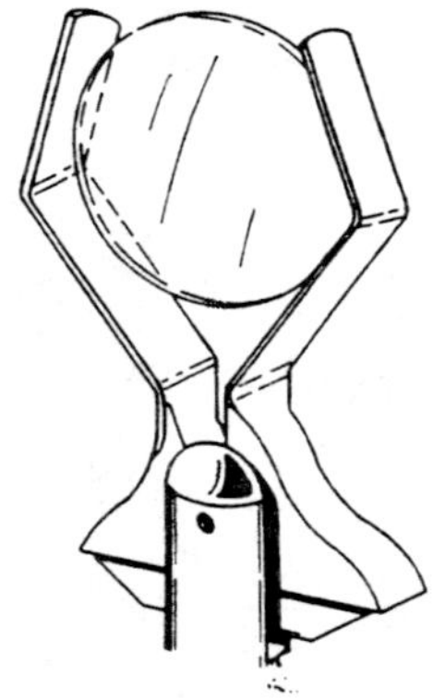

Fig. 21 Spring-grip lens holder.

The swing-arm lens holder is another variation of lens support. It's shown by Figure 22. This type of lens holder has a swing-up, spring-loaded arm to hold the lens. Both the bottom support and the arm have grooves for seating the lens.

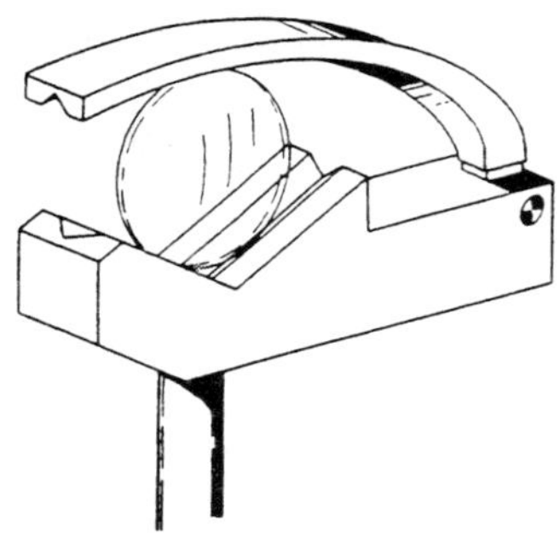

Fig. 22 Swing-arm lens holder.

For permanent or semipermanent lens-mounting applications, you should use the threaded lens holder. This mount must be used with a fixed size of lens, which is held in place by threaded rings. A threaded lens holder is shown in Figure 23. Adapter rings are usually available to let you mount more than one size of lens in a given size of holder.

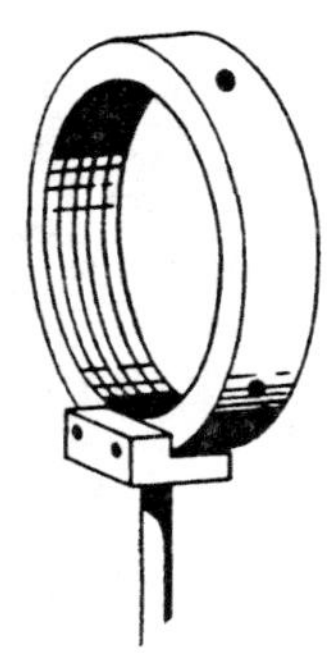

Fig. 23 Threaded lens holder.

When an optical setup calls for using lenses of various diameters interchangeably, you can use the self-centering lens holder. Figure 24 illustrates such a lens holder and shows why it's self-centering.

The three arms of this lens holder pivot on screws in a fixed ring. The outer ends of the arms have pins that ride in grooves in a spring-loaded, rotatable outer ring. When the top levers are pinched, the rotatable ring is

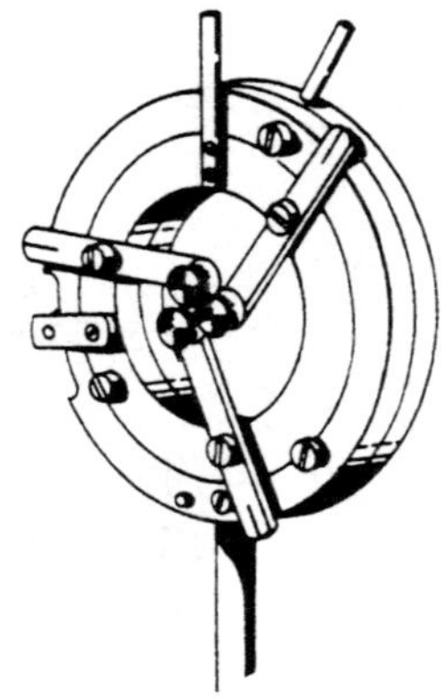

Fig. 24 Self-centering-lens holder.

moved and the jaws open. When the levers are released, the lens is gripped at three points by jaws with deep grooves, and the lens is automatically brought to the center of the holder. With this holder, the lenses can be changed quickly and automatically located in the optical axis without making further adjustments.

Another self-centering lens holder is the iris-diaphragm lens holder. It's shown in Figure 25. The leaves of the iris diaphragm are strong. They will

Fig. 25 Iris-diaphragm lens holder.

hold any lens that has flat edges. You merely place the lens in the center of the aperture and close the leaves on it. You operate the iris diaphragm by a large control knob that engages a gear drive. An added advantage of this type of lens mount is that no light is transmitted except through the lens.

Filter Holder

The simplest filter holder is shown in Figure 26. This holder is a spring clip mounted on the stand. Besides filters, this type of mount can hold ground or mirrored glass, beam-splitters, targets, beam-blocks, or paper.

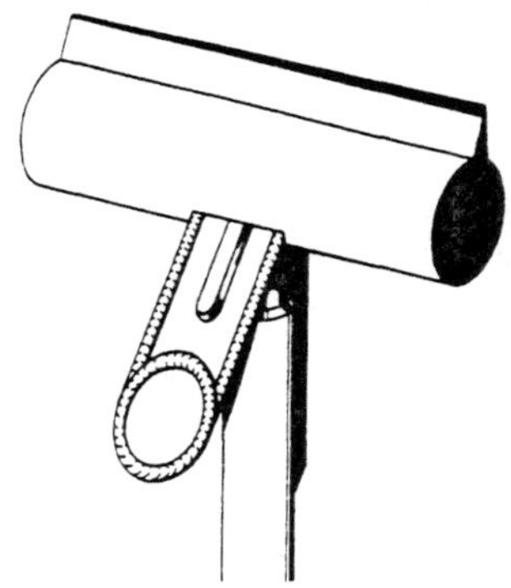

Fig. 26 Spring-clip filter holder.

A slightly more complicated filter holder is shown in Figure 27. This one has two spring clips that push against a square frame. It's used to hold filters

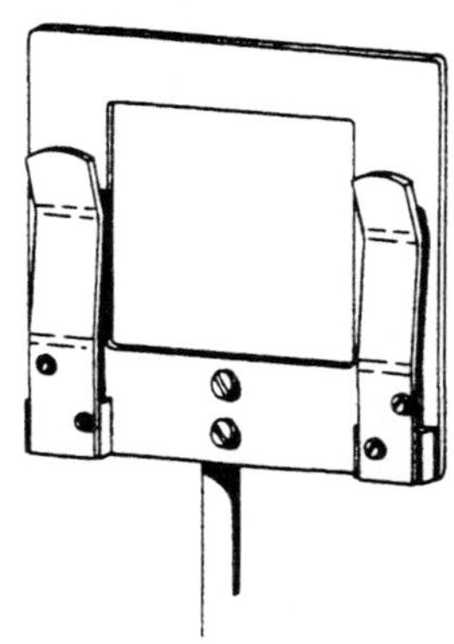

Fig. 27 Two-clip filter holder.

or targets that are too flexible to be held in the single-clip holder. The two-clip holder can hold filters, frosted glass screens, opaque screens, mirrors, mounted graph paper, targets, and so on.

If you're going to add filters one at a time until a desired density is reached, you'll use a multiple filter holder. Figure 28 shows a rack-type holder for holding up to four filters. You can build holders like this inexpensively by cutting off sections of linear 35-mm slide trays and attaching mounting rods. One side of this holder is mounted with thumb screws and

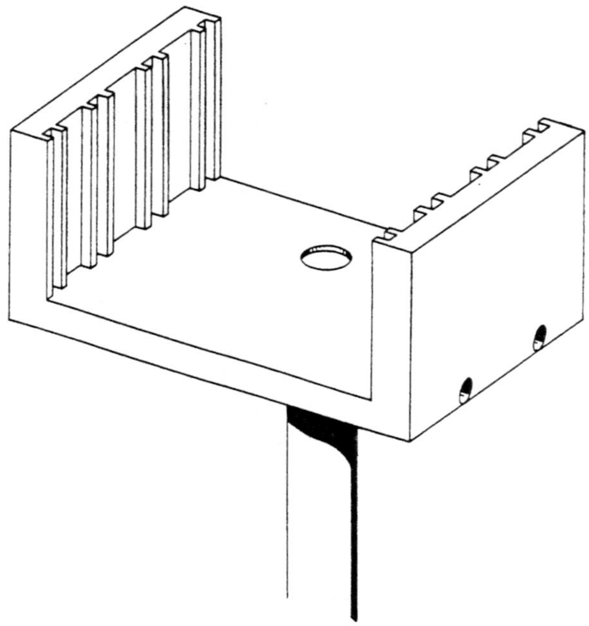

Fig. 28 Multiple-filter holder.

slotted holes so that you can move it to tilt the filters. With this feature, most internal reflections can be directed out of the optical path. In addition, after transmission it can displace the beam or change the angle or incidence.

Prism Holder

A special holder for mounting prisms or beam-splitter cubes is shown in Figure 29. The rubber coated, spring-loaded arm on this mount holds the prism firmly against a flat table.

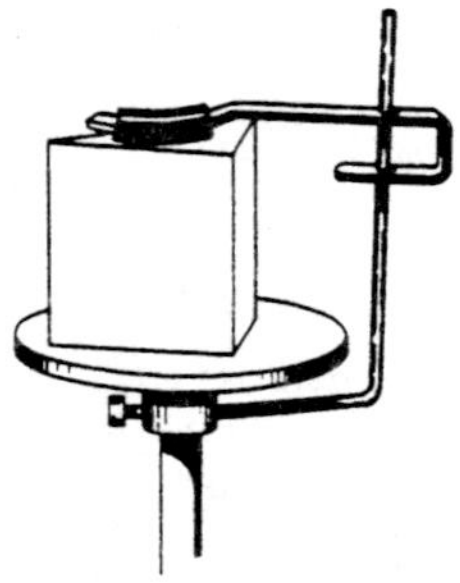

Fig. 29 Prism holder.

Laser and Collimator Mounts

You can mount lasers, telescopes and collimators a lot of ways. Some of the ways were discussed in the sections on the tilting trough, the hold-down clamp, and the scissors jack. Many lasers, however, are too large and heavy for these types of supports. A custom-made support for the particular laser must be designed. One popular design for circular laser tubes and collimators is the V-block mount. It's shown in Figure 30.

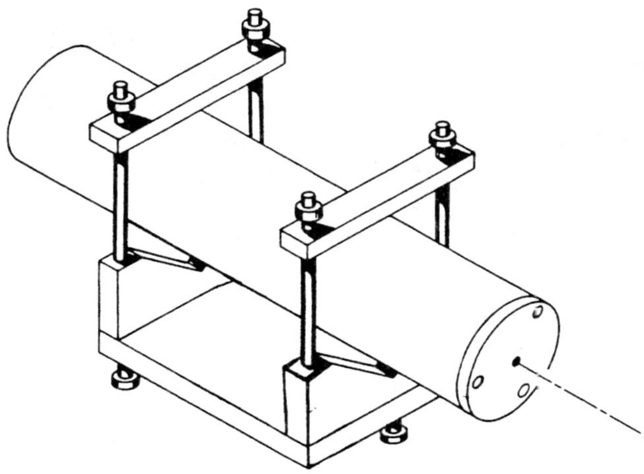

Fig. 30 V-block laser mount.

The V-block mount has three adjustable screws for feet. They let you level the laser and point the beam in the desired direction. Note here that three points are sufficient to level an instrument. You should avoid mounts with four points.

Another way to mount lasers is shown in Figure 31. In this case, the rubber feet normally supplied on the laser have been removed, and "T"-shaped

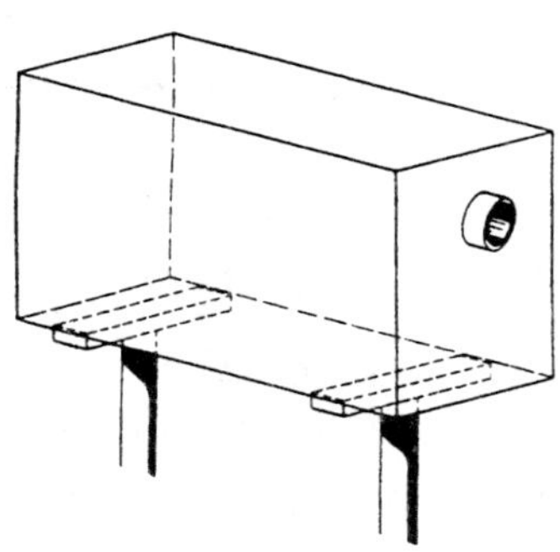

Fig. 31 "T"-bar laser mount.

bars have been attached. This lets you mount the laser to standard optical bench carriers.

Perhaps the most convenient method of mounting small cylindrical HeNe laser heads for general alignment use in the laboratory is shown in Figure 32. Figure 32a shows a typical cylindrical laser mount that's available

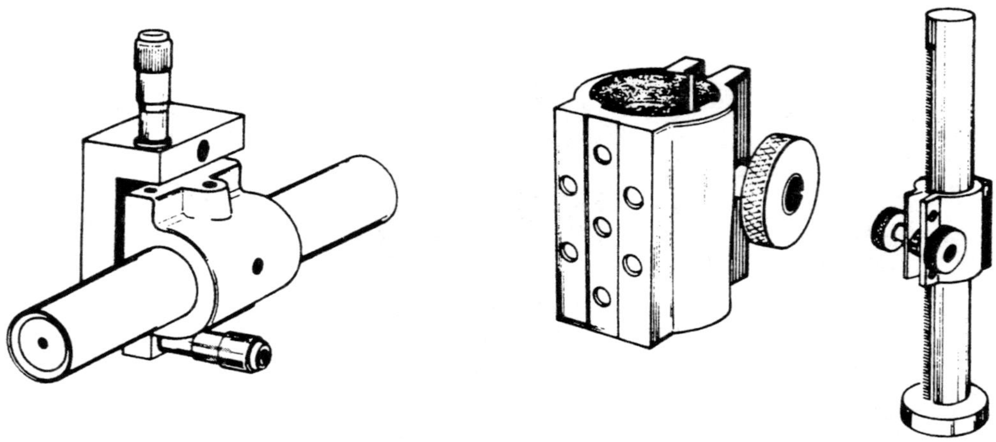

a. Typical cylindrical laser mount   b. Typical rod/clamp assembly

Fig. 32 Various laser mounts available for holding alignment lasers and other components.

with a variety of inner diameters. In Figure 32b a typical rod/clamp assembly is shown. A variety of component mounts can be bolted to the clamp assembly for vertical height adjustment.

Mounts like these can be equipped with micrometers that can tilt the laser in two directions with considerable accuracy and repeatability. For additional convenience, the laser mount can be attached to a vertical rod/clamp assembly as shown in Figure 32b. The clamp assembly has mounting holes and moves vertically on a rack-and-pinion gear. You can attach the mounting rod directly to the table or bench with mounting bolts, or attach it to a magnetic base for easy repositioning. For even more flexibility, translational stages can be bolted between the clamp assembly and the laser mount to give micrometer-driven resolution in one or several directions.

When possible, it's always a safe policy to buy your optical mounts from a single vendor. Experience shows that the mounting holes from various manufacturers seldom align. That means you would need to fabricate adapter plates to align the holes.

Design Your Own

You can see from all the types of components shown so far that there's no single way to mount optical components. Many times you'll have to custom-make part of the setup or rearrange it so that it will work right. In mounting optical hardware, you have to consider all aspects of the problem. One aspect is which supports are available and which components are to be mounted. You should consider the range over which you want the mounts to be adjustable, and the degree of rigidity required for the component weight. Advanced planning of the intended use is important. You'll need to incorporate for positioning, translating, focusing, splitting, or reflecting the light through the required ranges.

An Example of Designing Your Own

A need arose during testing of the Apollo laser altimeter that called for a custom-designed mount. The design approach for this mount shows some of the things you should consider when you build or assemble component supports.

The laser altimeter was mounted on a table along with a pointing telescope. The arrangement is shown in Figure 33. Two major problems in mounting the two components had to be overcome.

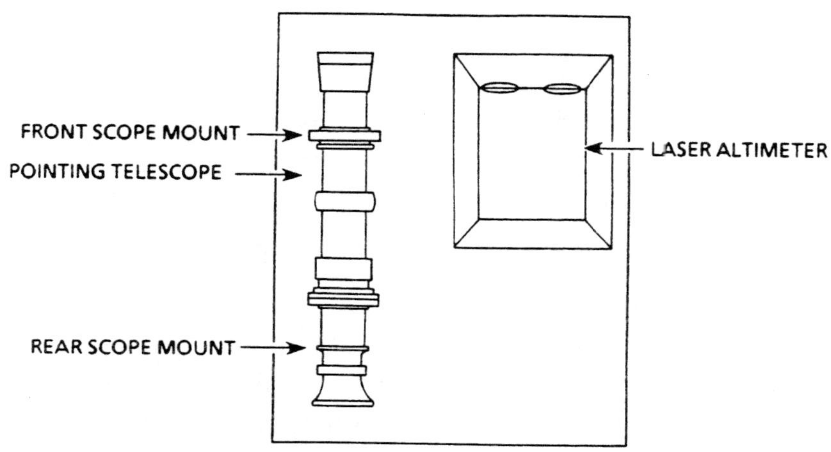

Fig. 33 Laser altimeter test mount.

The first problem was to mount the telescope on the table so that it could be adjusted to point in the same direction as the laser altimeter. The altimeter was fastened firmly to the table. It wasn't adjustable.

The second problem was to mount the entire table so that it could point the telescope (and thus the laser) precisely at a small target 30 miles away.

The elements considered in the solution of the first problem were:

1. The telescope mount had to have sufficient adjustment in two degrees of freedom to allow the optical axis of the telescope to be made parallel to the laser altimeter.
2. The adjustments had to be made smoothly so that alignment could be made easily.
3. Once aligned, the mount had to maintain that alignment.

The pointing telescope mounting problem was solved by a custom-made mount shown in Figure 34. This design called for two custom-made stands

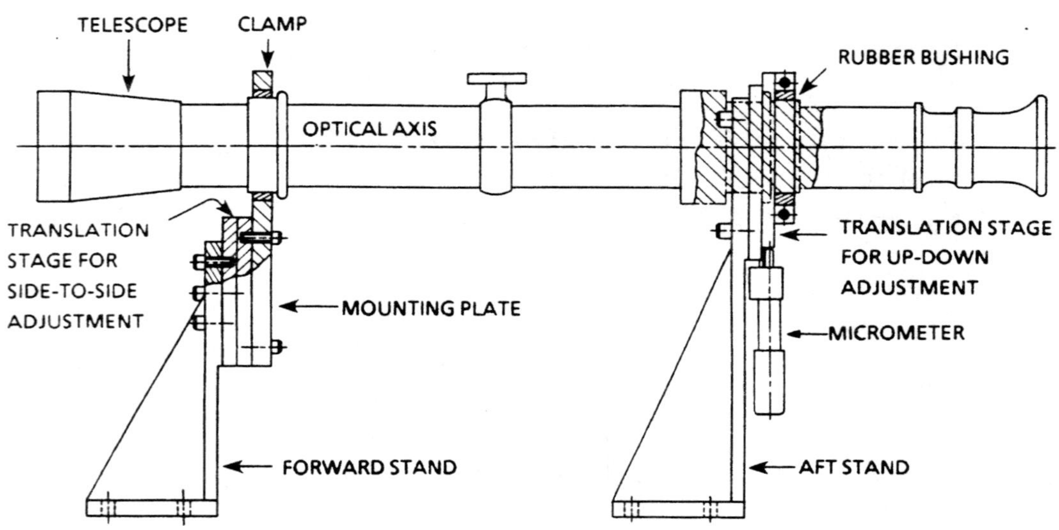

Fig. 34 Pointing-scope mount.

with a commercially available translation stage mounted on each one. The micrometers on the translation stages could be locked in position when the desired adjustment was reached. The second mounting problem was solved by supporting the entire table by a second table, as shown in Figure 35.

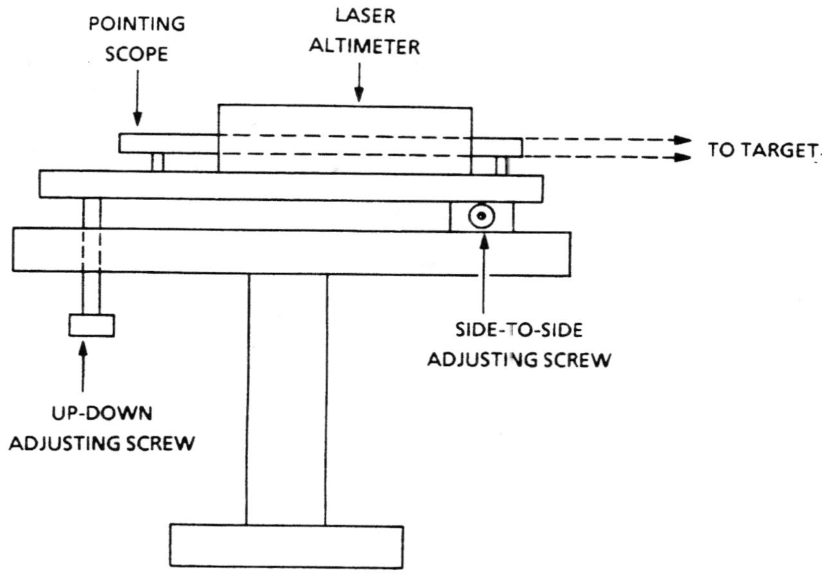

Fig. 35 Laser-altimeter pointing mount.

This way of supporting the laser and telescope was used to aim the laser (by looking through the telescope) at the small target 30 miles away. Therefore, care had to be taken to make the threads of the adjusting screws fine enough. They were made so that a full turn of one of the screws moved the laser one-fourth of the full beam width. The table and adjusting screws also were made strong enough that when alignment could be maintained, accurate aiming was achieved.

One final note. In this module we usually have limited our discussion to manually driven translators and rotators. Virtually any micrometer now can be replaced with a motorized version. Drive resolutions for single-step motions of 0.02 micrometer are available. The additional cost of motorized drives may be warranted if you need high repeatability or remote operation. Figure 36 shows a joy-stick controller and a variety of motorized micrometer heads. Controllers are also available that connect to microcomputers for data-acquisition and process-control applications. If even more resolution is required, piezoelectric drivers—which were discussed at the beginning of this module—can be attached to optical components to produce adjustments on the order of a small fraction of a wavelength of light.

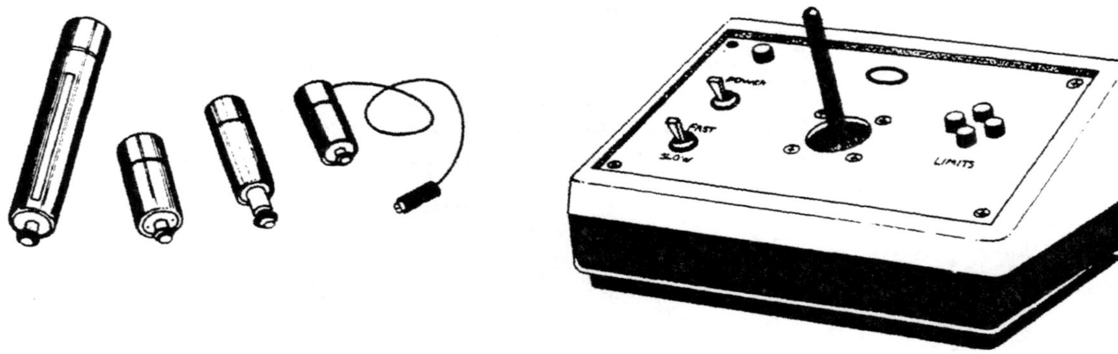

Fig. 36 A variety of motorized micrometers with a joy-stick controller.

EXERCISES

We show problems 1 through 26. For each one, identify the component supports by writing the name of the component support in the appropriate blank. Describe its intended use in the space provided below its name.

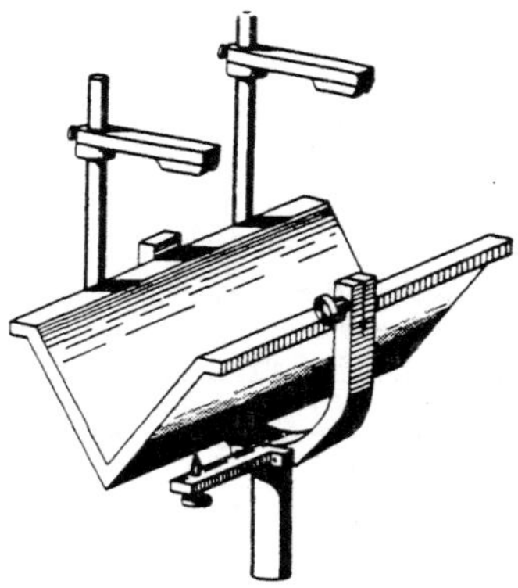

1. Tilting trough
1° of freedom moves about x-axis

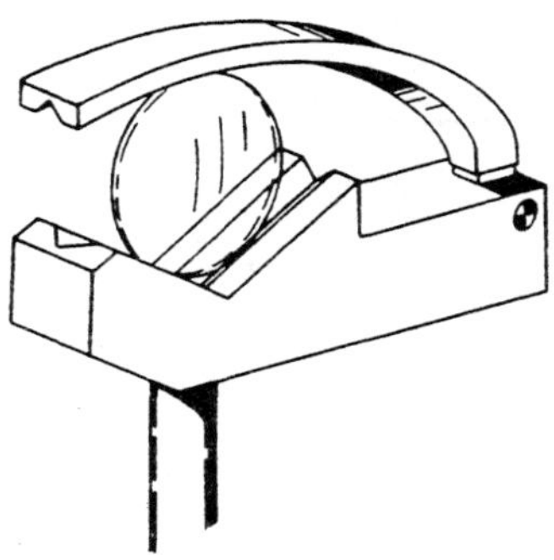

2. Swing arm lens holder
0° of freedom

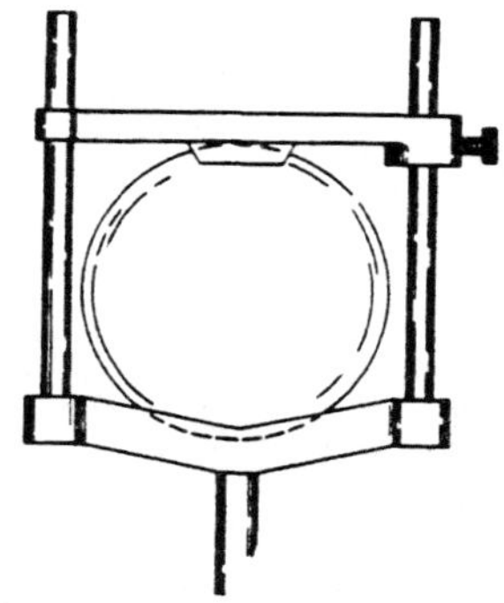

3. Sliding Grip lens Holder
0° of freedom

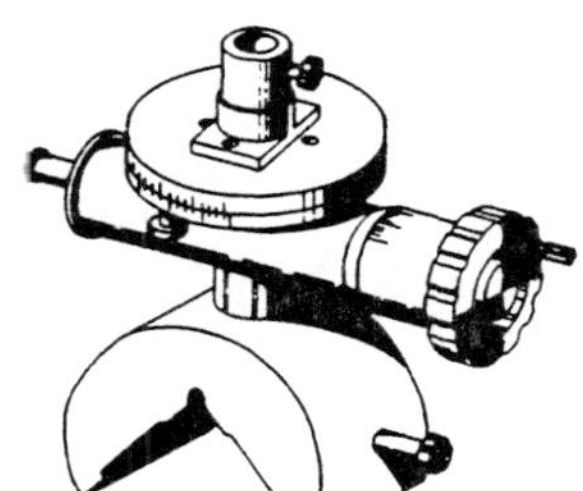

4. trans later rotater
2° of freedom

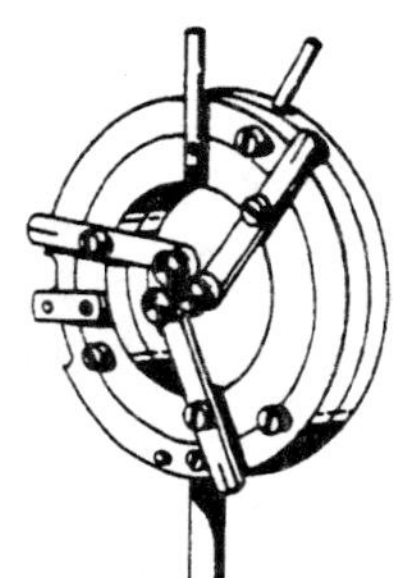

5. Self centering lensholder
0° of freedom

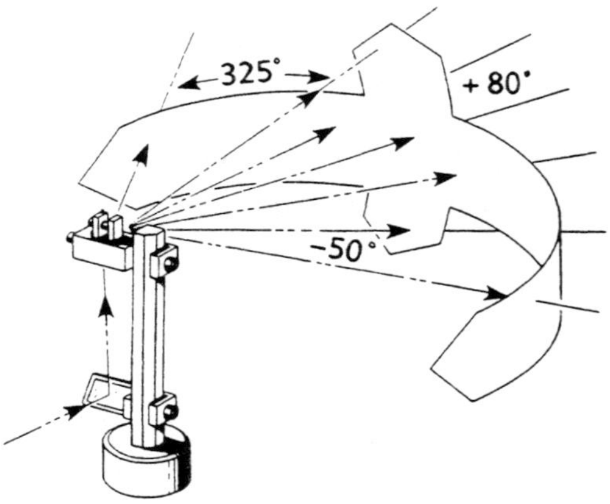

6. Laser Beam steering device
3° of freedom

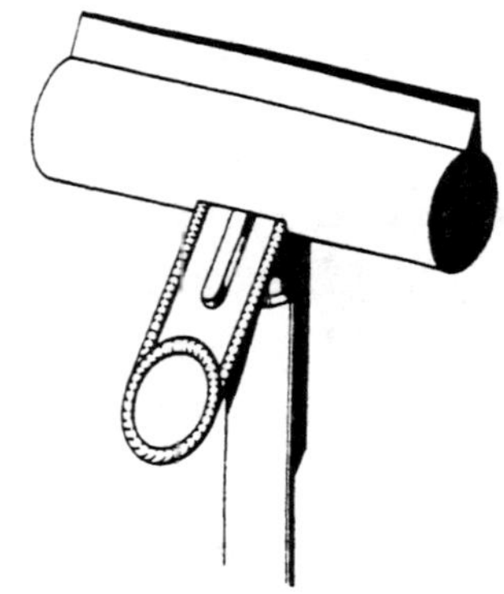

8. Spring clip filter holder
0° freedom

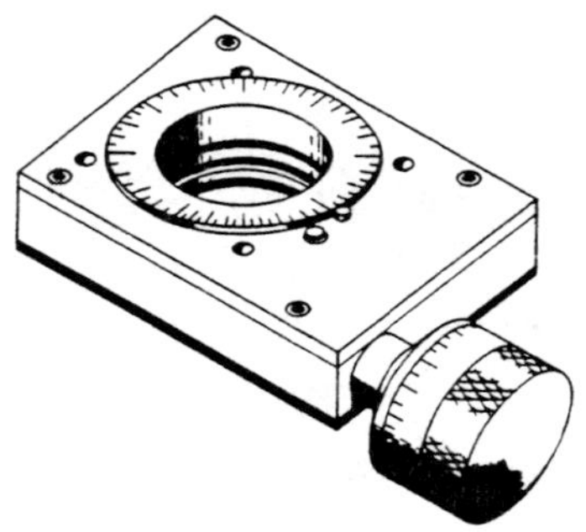

9. Rotary table
1° freedom

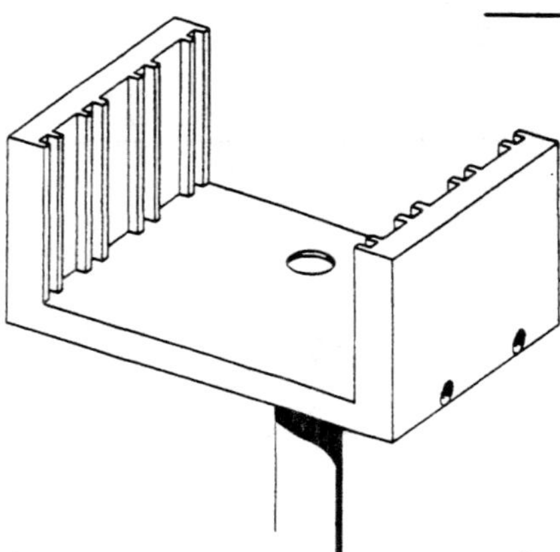

7. multiple filter holder
0° freedom

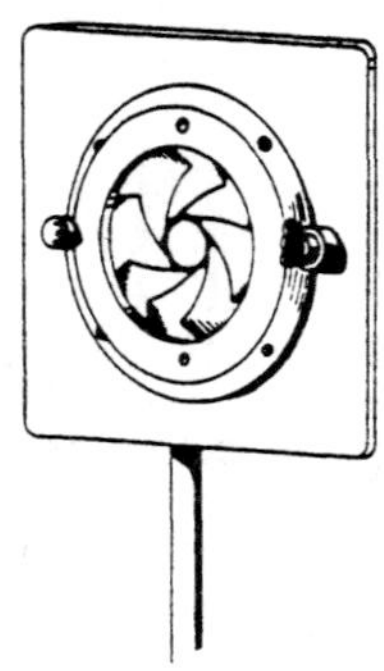

10. Iris diaphram
0° freedom

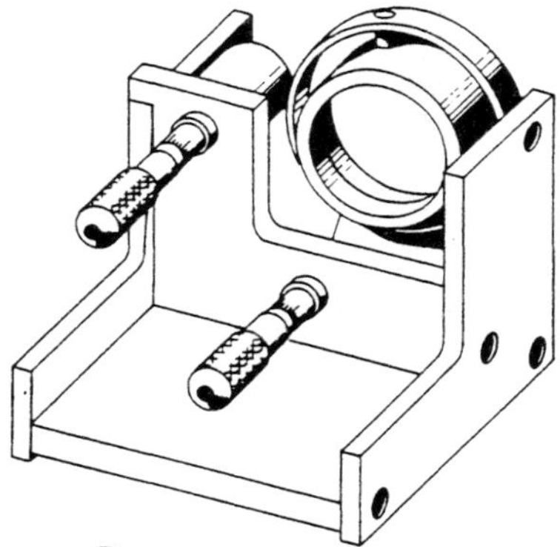

11. Two axis mirror mount
2° freedom

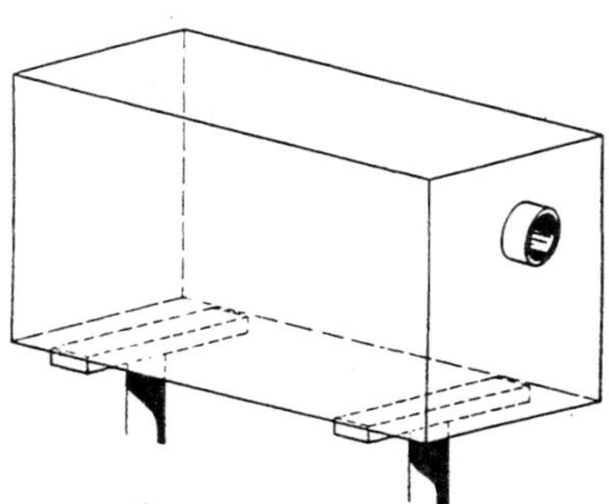

12. T Bar laser mount
0° freedom

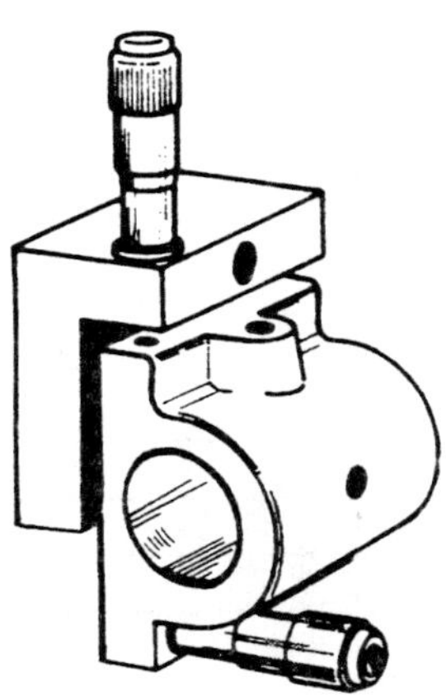

13. Typical cylendricall Laser mount
2° freedom

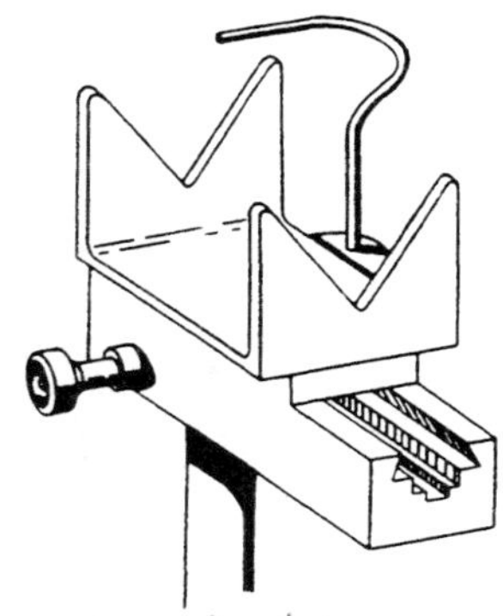

14. focusing holder
1° freedom
z axis

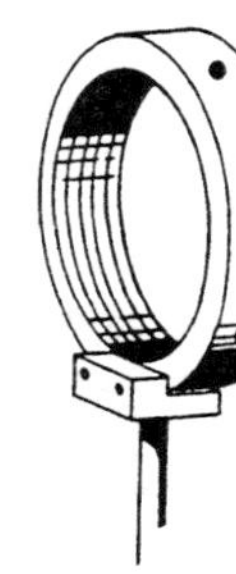

15. Threaded leas holder
0° freedom

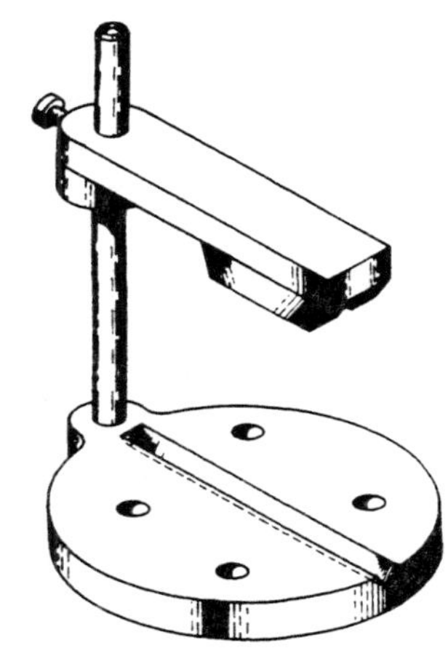

16. Hold down Clamp
0° freedom

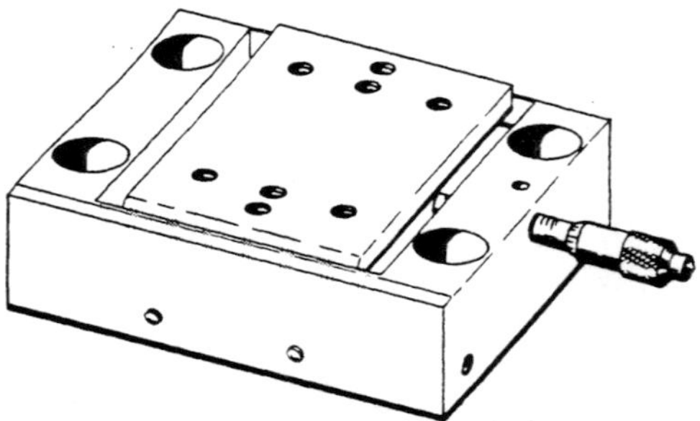

17. Translator table
1° freedom

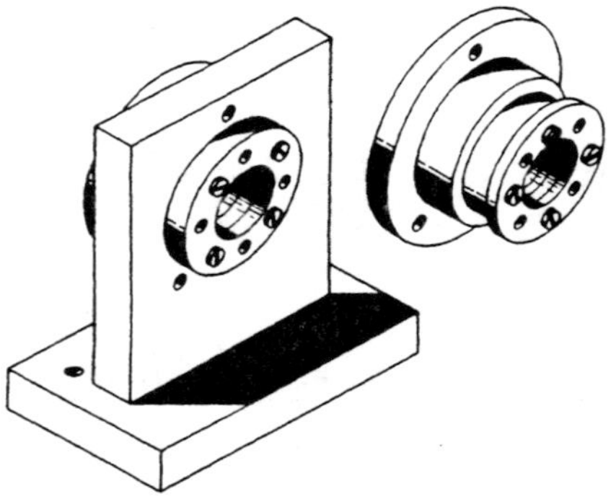

18. lens cavity mirror mount
2° freedom

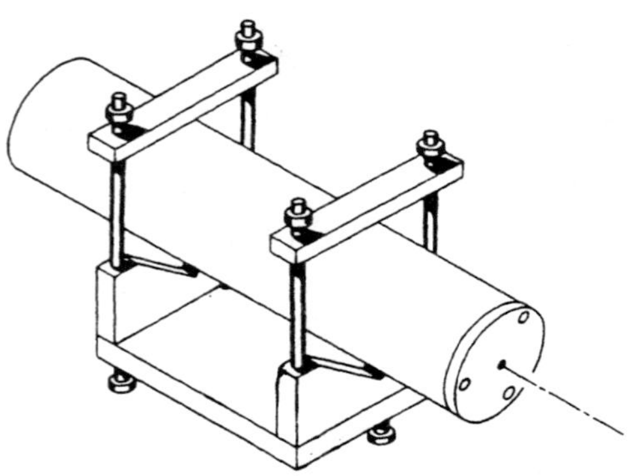

19. V Block laser mount
0° freedom

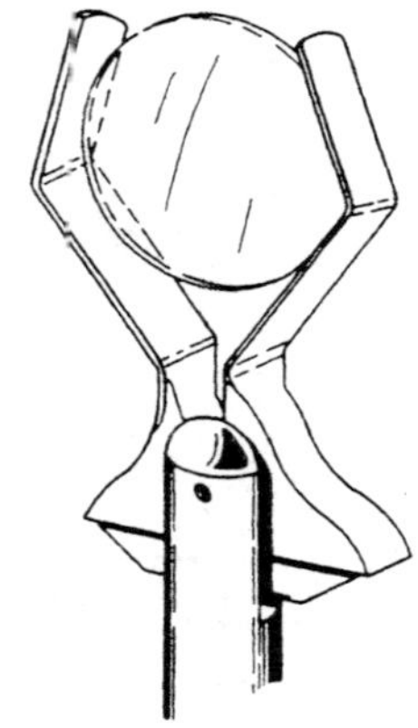

20. spring grip lens holder
0° freedom

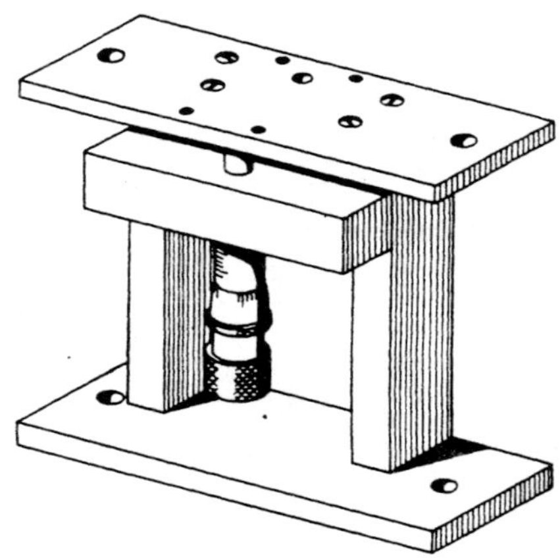

21. High resolution vertical
translator
1° freedom
y axis

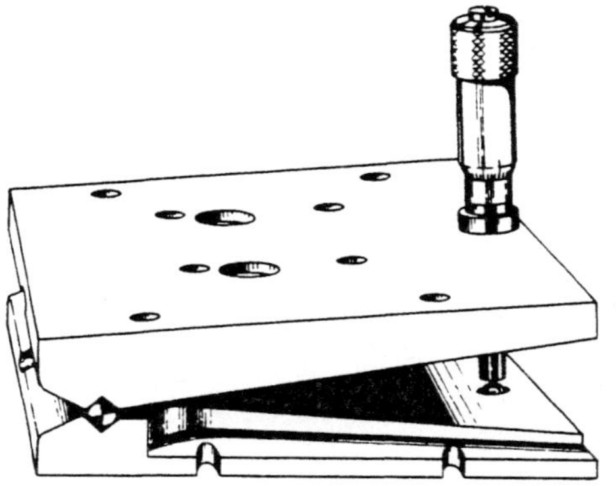

22. Sine table
1° freedom

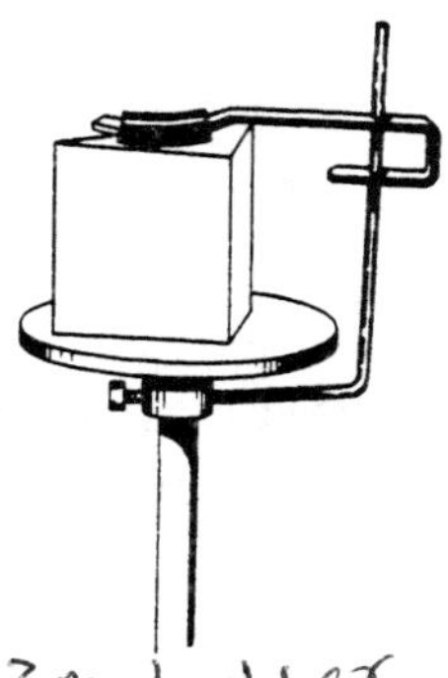

23. Prizm holder
0° freedom

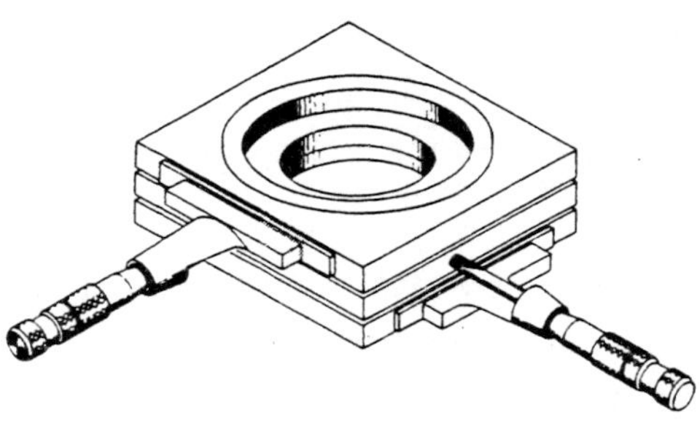

24. 2 axis translator or xy translator
2° freedom

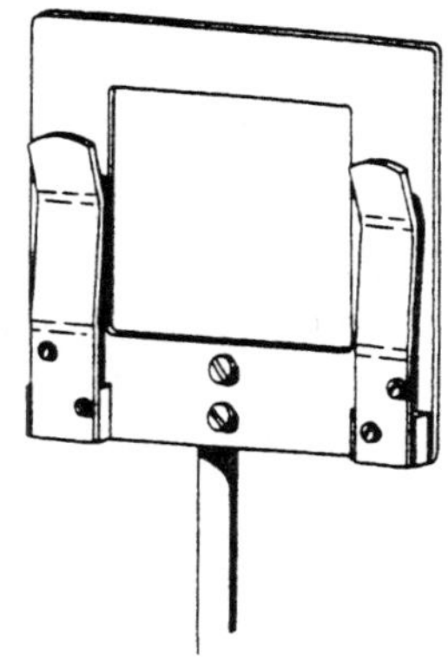

25. clip filter holder
0° freedom

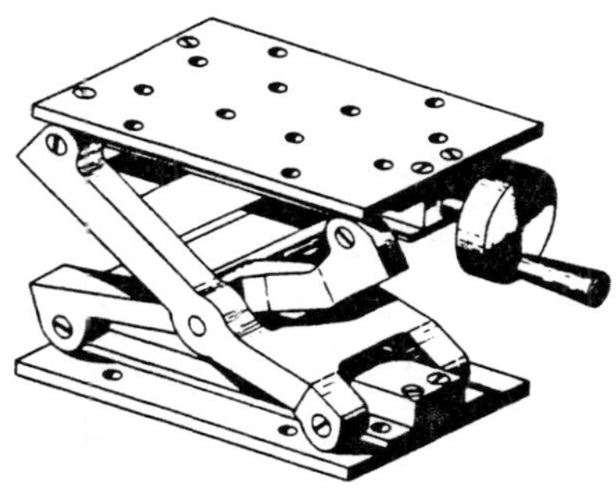

26. lab jack
1° freedom

MATERIALS

HeNe laser
Diverging lens, –10 to –20-mm FL
2 positive lenses, 10 to 20-cm FL
90° prism, 3/4-inch or larger focus
or first-surface reflector 3/4-inch x 1-inch or larger
Chromatic dispersion prism, 3/4-inch focus, or
diffraction grating, transmission type,
plastic, 10 to 15,000 lines per inch
Crosshair target (drawn on index card)
3 lens holders, compatible with lenses used
2 rotary tables, with holders for reflector and grating (or prisms)
Fixed mount and support for target (wood block)
2 optical rails, 1-meter length
5 carriages, compatible with rails and component supports
1 laser mount, preferably with 3 adjustable leg screws

PROCEDURES

1. Set up the system as shown in Figure 37, including the individual component supports mounted on the two rails. Don't mount the components at this time.

2. Mount the laser on its support. Turn it on and check its operation. **Observe Laser Safety!** Adjust laser height so that the beam will pass through the *approximate* center of the component mounting openings. Adjust the beam to be parallel with the centerline of the first rail, and level the beam throughout its length.

3. Place a 90° prism (or reflector) on a rotary table, rotated to a position so that the reflected beam is at 90° from its original path. (If you use a prism, use the first-surface reflection from the prism on the side facing the laser to establish normal incidence.)

4. Put a dispersive prism (or grating) on a second rotary table mounted on the second rail. Arrange the rail so that the beam enters the dispersive component at an angle allowing the dispersed beam to travel parallel to the rail's centerline. Use the first diffracted order from the grating. If you use a prism, rotate it to put the base (ground face) parallel with the rail.

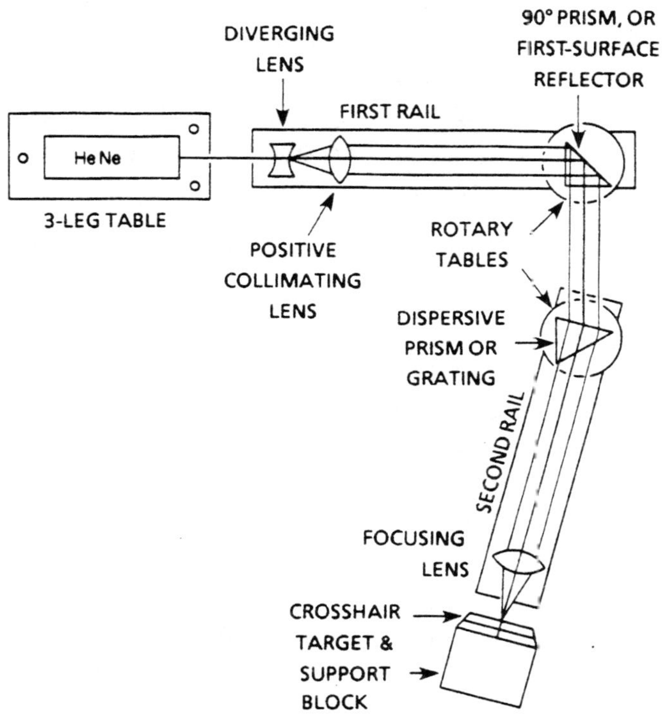

Fig. 37 Experimental arrangement.

5. Mount the target on its support block, so that the horizontal line is at the same height as the beam leaving the laser. Place it at a predesignated position near the end of the second rail and on a line coincident with the rail's centerline. Note the position and size of the laser spot on the target.

6. Temporarily remove the reflector and prism (or their alternates). Mount the diverging lens and accurately position it so that the beam is incident normally and passes through the center of the lens. (Check positions of the first-surface reflection and the transmitted beam with respect to the reflector mounts.)

7. Mount the positive lens to collimate the expanding beam. Adjust its position along the rail to establish a uniform beam diameter (zero divergence). Adjust the lens in a vertical plane to position the beam accurately onto the center of the reflector (or prism) at end of the first rail. (Replace the reflector to verify optimum alignment.)

8. Using the focusing lens, focus the beam to the spot size and position that were defined before the beam was expanded.

9. Consider the rotational stage in Example A in the Discussion. Assume that a small flat mirror is mounted vertically at the center of the stage and that a HeNe beam is directed at the center of the mirror. If you rotate the knob through an angle of 8.50°, how far will the laser beam move across a white card held 5.00 meters away in the reflected beam? Express you answer in cm.

10. Name the proper component support(s) to perform the following functions:

    a. Alter the height and horizontal direction of a light beam.
    b. Move a component along the optic axis.
    c. Adjust a prism to its minimum deviation angle.
    d. Adjust an aperture in two directions, vertically and horizontally perpendicular to the optic axis.
    e. Support unmounted (having no metal cell) lenses of several different diameters.
    f. Align a reflector to return a beam exactly back to its source.
    g. Support two unmounted sheet polaroids to form a polarizer and analyzer.

11. Calculate the linear resolution of a piezoelectric translator that moves at a rate of 5.00×10^{-7} meters with a voltage change of 200 volts, if the minimum possible voltage change from the power supply is 0.500 volt.

12. Calculate the linear resolution of a differential-screw micrometer that has a right-hand thread of 50 turns per inch and a left-hand thread of 40 TPI, if the minimum repeatable known rotation is 2.00°.

13. Use catalogs from two suppliers of optical component supports to compare and contrast the quality and price of five component supports discussed in this module.

REFERENCES

Ealing Electro-Optics Product Guide, Ealing Electro-Optics, Inc., 22 Pleasant Street, South Natick, MA 01760, (617) 655-6029.

Laser Focus Buyer's Guide, published annually by: Advanced Technology Group, PennWell Publishing Co., 119 Russell Street, Littleton, MA 01460, (617) 486-9501.

Melles Griot Optics Guide 3, Melles Griot, 1770 Kettering Street, Irving, CA 92714, (714) 261-5600.

The Newport Corporation Catalog of Precision Laser/Optics Products, No. 100 Newport Corporation (formerly NRC), 18235 Mt. Baldy Circle, Fountain Valley, CA 92728-8020, (714) 963-9811.

Oriel Corporation Catalog Vol. I, of Tables, Benches, Micropositioners, OpticalMounts, Oriel Corporation, 250 Long Beach Blvd., Stratford, CT 06497-0872, (203) 377-8282.

LASER/
ELECTRO-OPTICS

MODULE 6-3
PHOTOGRAPHIC RECORDING MEDIUMS

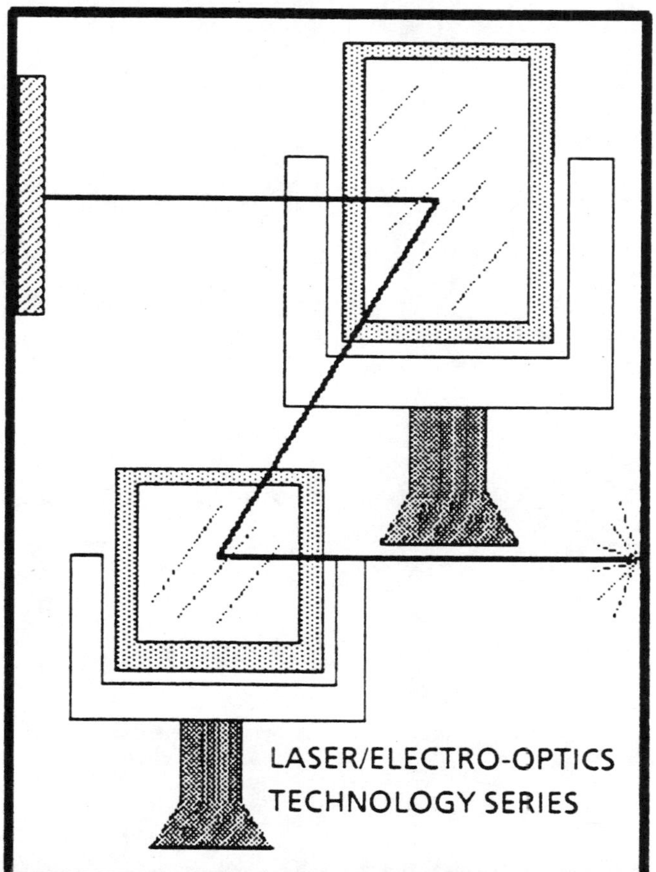

CORD

The Center for Occupational Research and Development
601 C Lake Air Drive
Waco, Texas 76710

Printed in the U.S.A.

ISBN 1-55502-024-0

PHOTOGRAPHIC RECORDING MEDIUMS

MODULE 6-3

Many experimental phenomena must be permanently recorded for future analysis and documentation purposes. When either high speed and/or high resolution is desired, such recordings frequently make use of photographic techniques. The field of photographic science is so broad that it won't be feasible to cover the many facets of this interesting technology. We'll concentrate on the basic principles involved, and address important photographic material characteristics that must be considered for the optimum selection from a large range of candidate recording mediums.

This module will cover the basic concepts and techniques of the black and white photographic process with a brief introduction to color photography. We'll discuss properties of photosensitive materials such as sheet, roll, and plate film, as well as Polaroid emulsions.

Details of film processing procedures—from chemical changes that take place during exposure, through developing, washing, fixing, and various other photographic processing techniques—will be presented. You'll learn about sensitometry as a way to determine film performance and suitability, and work a lab experiment on sensitometry.

You should know standard precautions for dealing with high-quality optics and hazardous chemicals. You should be aware of the principles and procedures of introduction to laser safety, helium-neon gas lasers, F stops and apertures, laser power and energy measurements, the cleaning of optical components, sources and their characteristics, and radiometry and photometry.

OBJECTIVES

When you complete this module, you should be able to do the following.

1. Describe in your own words the process of latent image formation in photographic emulsions.
2. Assume an appropriate exposure has been made. Describe in functional detail the necessary processing steps to arrive at a black and white negative. The steps to be described will include development, stop bath, fixing, washing, and drying.
3. Sketch a typical H-and-D curve that defines the ordinates and identifies the following regions on the H-and-D curve:
   a. Inertia point
   b. Background fog level
   c. Gamma (γ)
   d. Saturation
   e. Latitude
   f. Speed
   g. Overexposure (reversal, solarization)
   h. Region of normal exposure
   i. Toe
   j. Shoulder
4. Define reciprocity and reciprocity failure as regards film behavior.
5. Exhibit—by means of a graph—the spectral sensitivity of a typical photographic emulsion. Describe its importance in the concept of spectral matching.
6. Through the use of given technical characteristics, select an appropriate emulsion for a given application to include spectral sensitivity, speed and resolution.
7. Given a reproducible light source and appropriate photographic materials, cameras and chemicals, expose and develop a photographic emulsion to determine a relative H-and-D curve for the emulsion under a given set of conditions.

DISCUSSION

Photographic images produced by various photosensitive emulsions are dependent on the formation of a *latent* silver halide (silver bromide, silver chloride, etc) image through photochemical reaction. This latent silver halide image can be reduced to a permanent image consisting of metallic silver through a process known as developing, whereupon the unexposed silver halides are removed by a solvent known as a fixing agent. Suspended particles of elemental silver remain behind in the emulsion. They reproduce the intensity distribution that was incident upon the film during the time of exposure. Figure 1 shows a typical film.

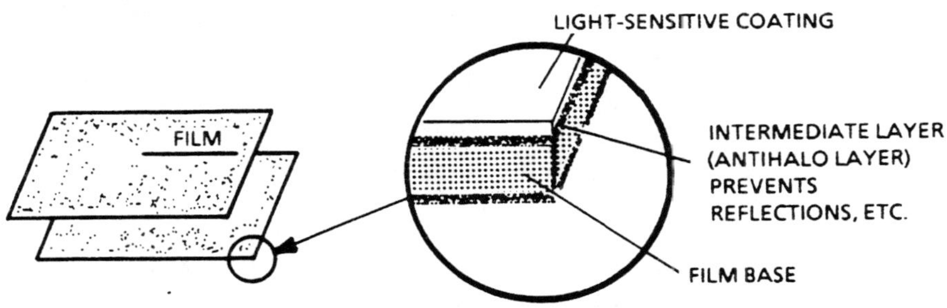

Fig. 1 Cross section of a typical black and white photographic film.

The final result of this process is called a *negative.* That's because the areas of greatest intensity in the photographed scene seem black, and the areas that are dim appear as clear (transparent) areas when viewed with back illumination. Of course, this negative can be inverted to produce a positive through an additional photographic cycle. This photographic process is shown in Figure 2. The above description briefly describes the process of image formation.

Let's look at a more detailed description of specific aspects of the process. Photographic film generally consists of a thin material or film base made of glass, cellulose acetate, nitrocellulose or other organic compounds. Frequently, film bases made of polyolefins and polycarbonate are used for scientific purposes because of their greater strength and dimensional stability. A light-sensitive emulsion is applied to this base.

Frequently, there's a thin, dark-color intermediate layer called the antihalo layer. It acts as an absorber for scattered radiation. The antihalo layer also decreases reflection from the film base. Reflection might lead to unwanted additional activation of silver halide particles. A typical emulsion

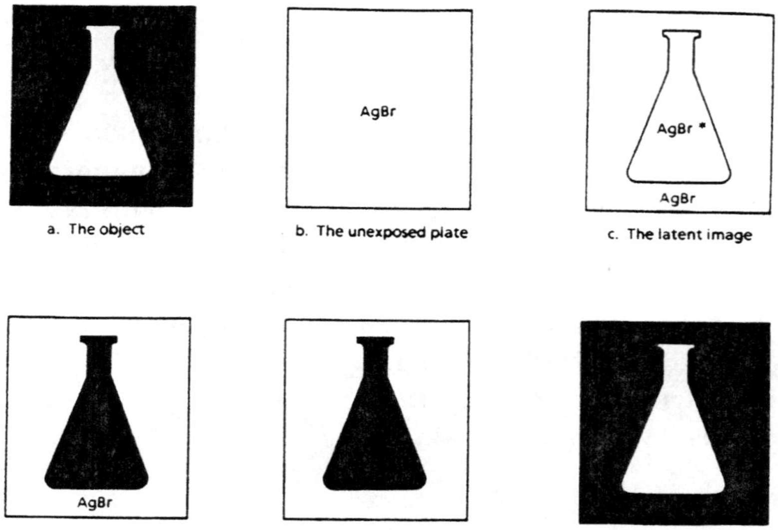

a. The object b. The unexposed plate c. The latent image

d. The developed plate e. The negative after fixation and washing f. The print

Fig. 2 The basic steps in the black and white photographic process.

may consist of 40% silver bromide crystals, 50% gelatin (as a binder), and 10% water. We'll see later that the size of the active silver halide crystals determines two important characteristics of photographic emulsions. These characteristics are speed and the resolving power.

The silver bromide emulsion normally used in many black and white films isn't equally sensitive to all frequencies of light. However, you can change the sensitivity dramatically by adding small quantities of additives such as gold, mercury, etc. By the use of other photosensitive materials, the spectral utility of films can be extended from the infrared down through the visible, to the ultraviolet, and into the X-ray region.

So, it's important to be able to select—from the numerous photographic emulsions available—that particular one that's best for recording the technical data you need in an experiment. You must therefore be familiar with the properties of photographic emulsions that will allow this optimum selection.

The best way to determine a photographic emulsion's technical characteristics is by sensitometry. **Sensitometry** is the science of evaluating the response of photographic materials after exposure and development under controlled and specified conditions. Sensitometric properties are closely related to the way the film was manufactured, exposed, processed, and so on. Factors that affect the darkening or exposure of photographic emulsions are the spectral content of the exposing radiation, the type and degree of development, the procedures used to measure the resultant density, etc. All are related through sensitometry. They will accurately describe the density or darkening obtained under a given set of conditions.

Let's define some useful terms. The first and most important is *exposure H* which is formally defined in sensitometry as the product of the irradiance E on the film and the time (Δt) for which the film is exposed to this irradiance.

$$H = E(\Delta t) \qquad \text{Equation 1}$$

Recall that optical density (D) of a material was defined as the negative log of the transmission.

$$D = -\log T \qquad \text{Equation 2}$$

Density is defined logarithmically since the response of the eye is logarithmic and film normally is evaluated by human observance. This is analogous to the decibel (dB) scale used to measure relative sound levels.

So, a density of 1 corresponds to a transmission of 10%. A transmission of 0.1% would correspond to an optical density of 3.

The Characteristic Curve

The *characteristic curve* normally is used to describe the sensitometric properties of main interest to technical photographers. This curve in reality plots the photographic effect as output versus input, that is, the density produced in an emulsion as a function of the exposure. Most frequently this curve is plotted as density versus the log of the exposure. So it's frequently known as the D, log H curve or the H-and-D curve, after Hurter and Driffield. Hurter and Driffield proposed that the log of 1/T (density) should be proportional to the mass of silver per unit area in the developed layer. Their experiment consisted of exposing photographic plates to a light of constant intensity and varying the time of exposure Δt. The result leads to a curve shown in Figure 3.

The H-and-D curve is probably the most simple and most-used way to describe the performance of a given photographic emulsion. This curve will be unique for each film, developing process, and exposure conditions such as

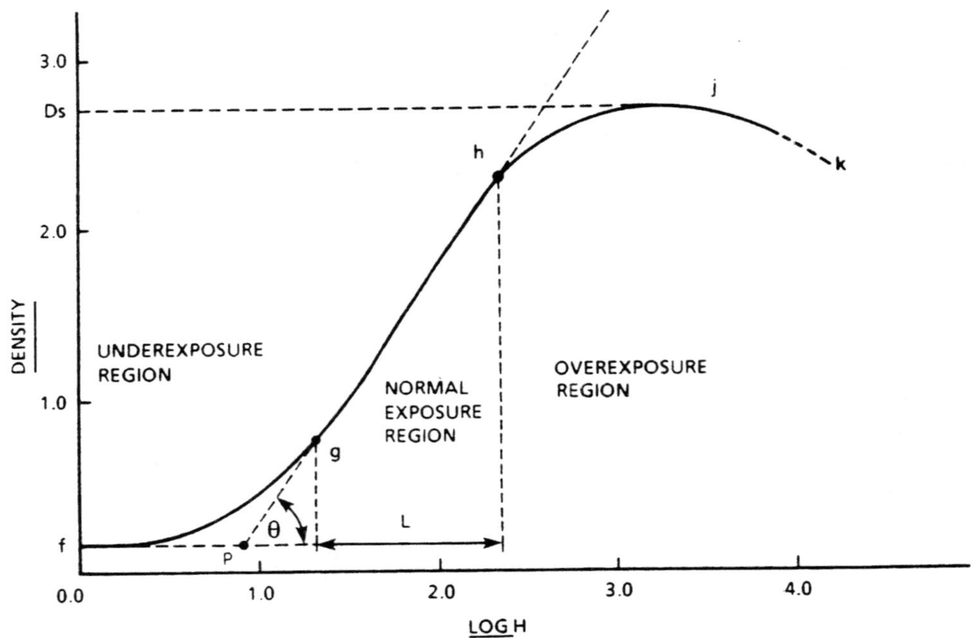

Fig. 3 Typical H-and-D curve.

spectral characteristics of the exposing source. Let's examine this curve in detail.

The first characteristic of the curve is the *toe or inflection* region $\overline{fg}$. This is followed by a region of *nearly constant slope* $\overline{gh}$, a region known as the *shoulder* $\overline{hj}$, and a *reversal* region $\overline{jk}$. Normally speaking, the region from f to g corresponds to underexposure, g to h the normal exposure region, and beyond h the overexposure region.

The density level f corresponds to the density of a developed photographic emulsion that hasn't been exposed to light. This is often called the *background fog density*. This density level can change depending upon specific processing procedures used. The extrapolation of the linear portion $\overline{gh}$ to the background fog density level produces an intersection at point p. Point p is called the *film inertia*. Generally speaking, you can think that this point shows the need for some measurable exposure before you get any evidence of photographic darkening. The specific value of the inertia and its reciprocal is a measure of the speed of the film. That is, the greater the value of p, the slower the film process combination (the more exposure it will take to produce an observable density change from the background).

For most applications, the linear portion $\overline{gh}$ corresponds to the most desirable region of photographic response, for in this region the density is a simple relation to the log of exposure. That is, over the range of exposure covered by this linear portion, there's a faithful correlation between photographic density and log of the exposure. The steepness or slope of this region is characteristic of the film's contrast. Quantitatively this contrast is defined as the "γ" (gamma) denoted by

$$\gamma = \tan\theta \qquad \text{Equation 3}$$

A high-contrast film has a high gamma.

The exposure range between g and h is called the latitude. Again, it's the part of the curve that's most suitable for recording photographic scenes that will yield the most satisfactory reproduction of various shades or tones.

It's obvious that under conditions of heavy exposure there will be fewer and fewer silver halide molecules left to be reduced to elemental silver. So a saturation takes place under these conditions. The onset of this process takes place at point h. It continues through the shoulder region to point j. The density at j is denoted d_{max} or d_s and is called the maximum or saturation density with additional exposure. This process is called *solarization* or *reversal*, since a greater exposure reduces the final density. The detailed reasons for this response are discussed in books and journal articles on photochemistry.

It's important to remember that a given H-and-D curve isn't only characteristic of the film, but can be modified through processing changes and changes in the spectral content of the exposing source. H-and-D curves for all commercially available films, together with the recommended processing criteria, are readily available in handbooks or pamphlets from various film manufacturers.

Sensitometry

Frequently you want to find the H-and-D curve for your particular experimental arrangement and processing procedures. Generally, you can get these experimentally with a sensitometer. A *sensitometer* is an instrument that produces, on a photographic material, a series of precisely known graduated exposures.

The correct performance of a sensitometric measurement depends on five determinable conditions. The conditions are:

1. a light source that emits radiation of known intensity and spectral composition,
2. a means of producing graded exposures of known relative values,

3. standardized development conditions,
4. a way to quantitatively measure the developed image,
5. a way to interpret the results by suitable quantities or expressions.

The essential components of a sensitometer are shown in Figure 4.

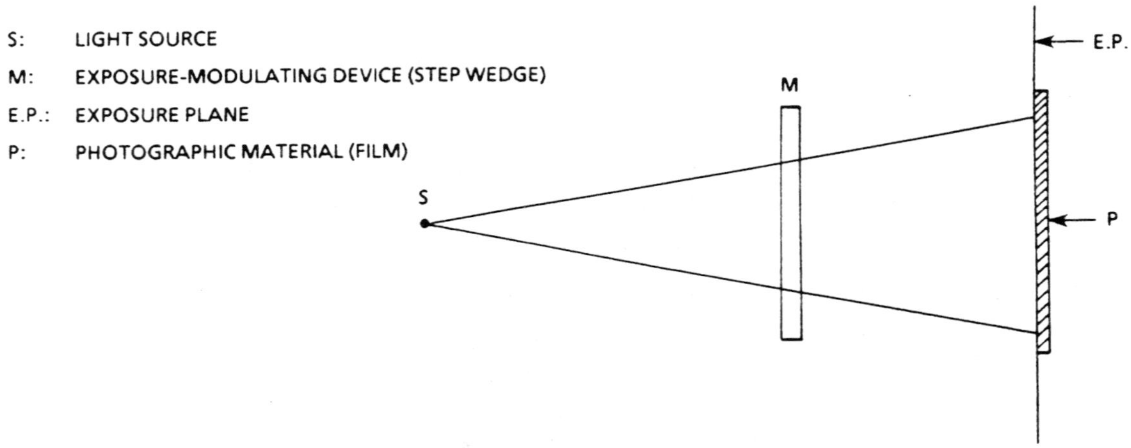

Fig. 4 Essential components of a sensitometer.

This simple arrangement makes use of a step wedge. A *step wedge* is usually a film that has known density gradations. The density gradations are usually sufficiently small. But, because of their large number (>20), they can vary the recorded density from a single exposure to cover the whole H-and-D response curve. Of course, each portion of the wedge must be illuminated under uniform conditions (intensity, etc.). Numerous types of sensitometers are used in the photographic field.

The various sensitometers can be designed to address specific measurements in the sensitometry field such as the very-short-duration response of emulsions, the response of photographic materials to repetitive exposures, multicolored sensitometers for color photography, and so on. We'll do a very simple sensitometry experiment in the lab for this module.

Reciprocity and Reciprocity Failure

Equation 1 indicates that, for a given exposure H, intensity and exposure time vary inversely. In other words, if one increases, the other must be decreased a corresponding amount to produce the same film density. This is *reciprocity*. It means that a source of unit intensity exposing a film for a given duration produces an image density the same as if the film were exposed by a source twice as bright for only half the duration as the original source.

However, if you investigate this reciprocity at the extremes of exposure, you'll find that the relation fails. For example, a Q-switched laser with a pulse length of 10 nsec illuminating a photographic emulsion at 10^6 W/cm^2 corresponds to an exposure of 10^{-2} joules/cm^2. Reciprocity states that a source that illuminates the film at a level of 1 milliwatt/cm^2 for 10 seconds—which can be produced by a CW gas laser—is also an exposure of 10^{-2} joules/cm^2. You'd expect the density to be the same from these two exposures. However, this won't be the case. This is not a failure of the film, but rather a characteristic of it.

This *reciprocity failure* comes about because we've assumed that the photochemical action is dependent only on the amount of light energy absorbed and not on the rate of absorption (power). For this reason, in many high-speed, time-resolved photoinstrumentation observations, you have to determine the H-and-D curve as a function of the temporal characteristics of the exposure. You can't rely on commercial curves that primarily govern the normal reciprocal response of photographic emulsions.

Spectral Sensitivity

Spectral sensitivity is another important characteristic of photographic emulsions. It's of particular concern to the scientist. The spectral sensitivity describes the degree of photographic response to radiation of various wavelengths.

Technically, sensitivity is the reciprocal of exposure required to produce a specified density above background fog when the material is processed as recommended. Normally, two graphical methods are used to describe spectral sensitivity—the sensitivity of an emulsion as a function of wavelength. The first, frequently used when dealing with a monochromatic source such as a laser, is the evaluation of an H-and-D curve at a specific wavelength. However—more generally as an aid in selection of emulsions—plots such as in Figure 5 often are used.

A summary of Kodak photographic emulsions frequently used in scientific photography is shown in Figure 6. As you can see, type Z emulsions are used often to photograph lasers operating at 1.06 µm. A variety of emulsions are suitable at the 0.6328-µm wavelength of the helium-neon laser.

The way that emulsions vary in spectral sensitivity comes from a process known as sensitization. *Sensitization* generally is changing or altering the spectral sensitivity by adding certain dyes to the emulsion. The variety of dyes used is very numerous. We won't cover them here. However, film

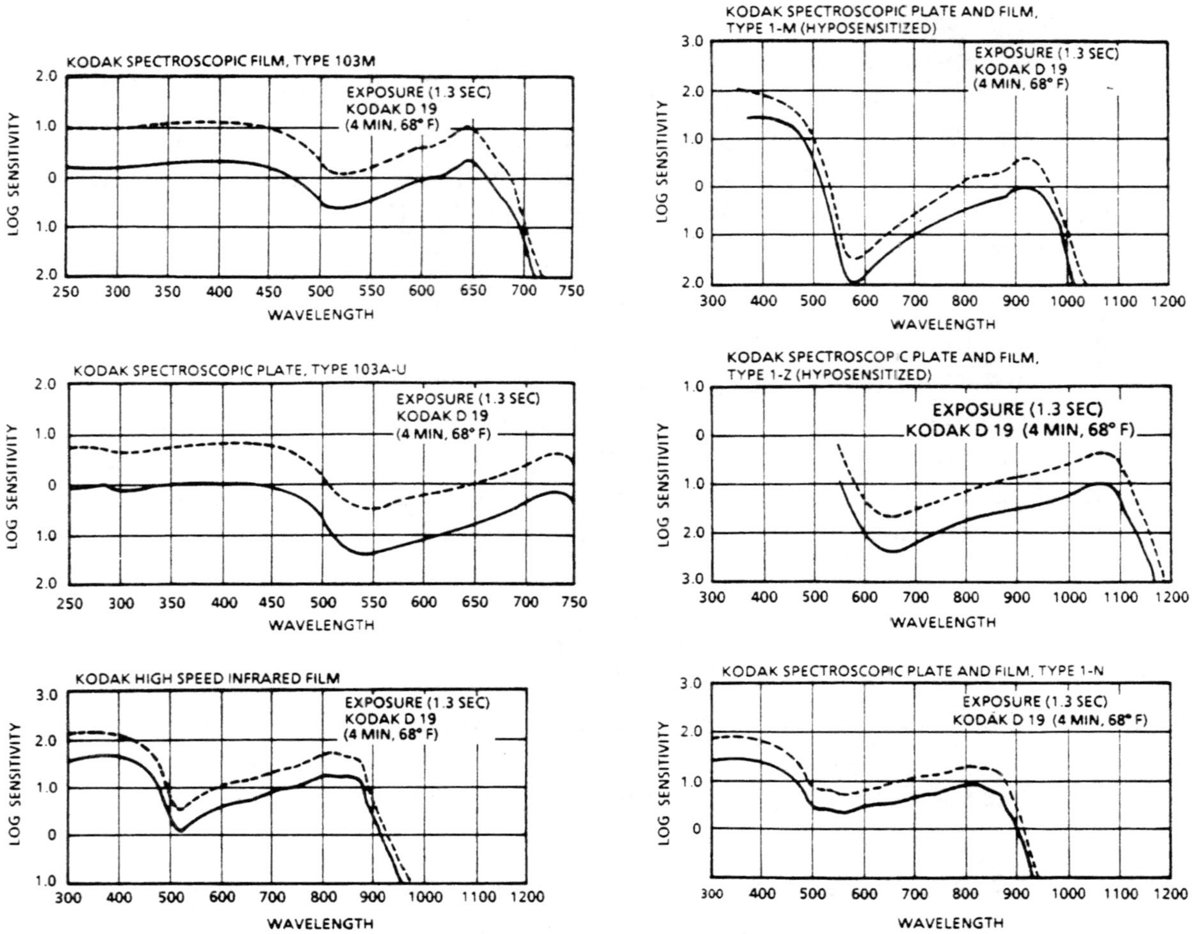

Fig. 5 Spectral sensitivity of selected Kodak scientific films.

manufacturers offer spectral sensitivity data on all photographic materials sold.

Another aspect of sensitization can be achieved in a similar way. That's the increase of a film's speed from its normal sensitivity or speed. Most sensitizations that increase a film speed come from changes in the processing procedure—either by the addition of heat or by chemical treatment of the film emulsion before exposure.

Hypersensitization is a frequently used process to increase the infrared sensitivity of films suitable for use in that portion of the spectrum. Here the emulsions are immersed in a caustic such as sodium hydroxide or ammonia immediately before exposure. This process greatly increases the response of

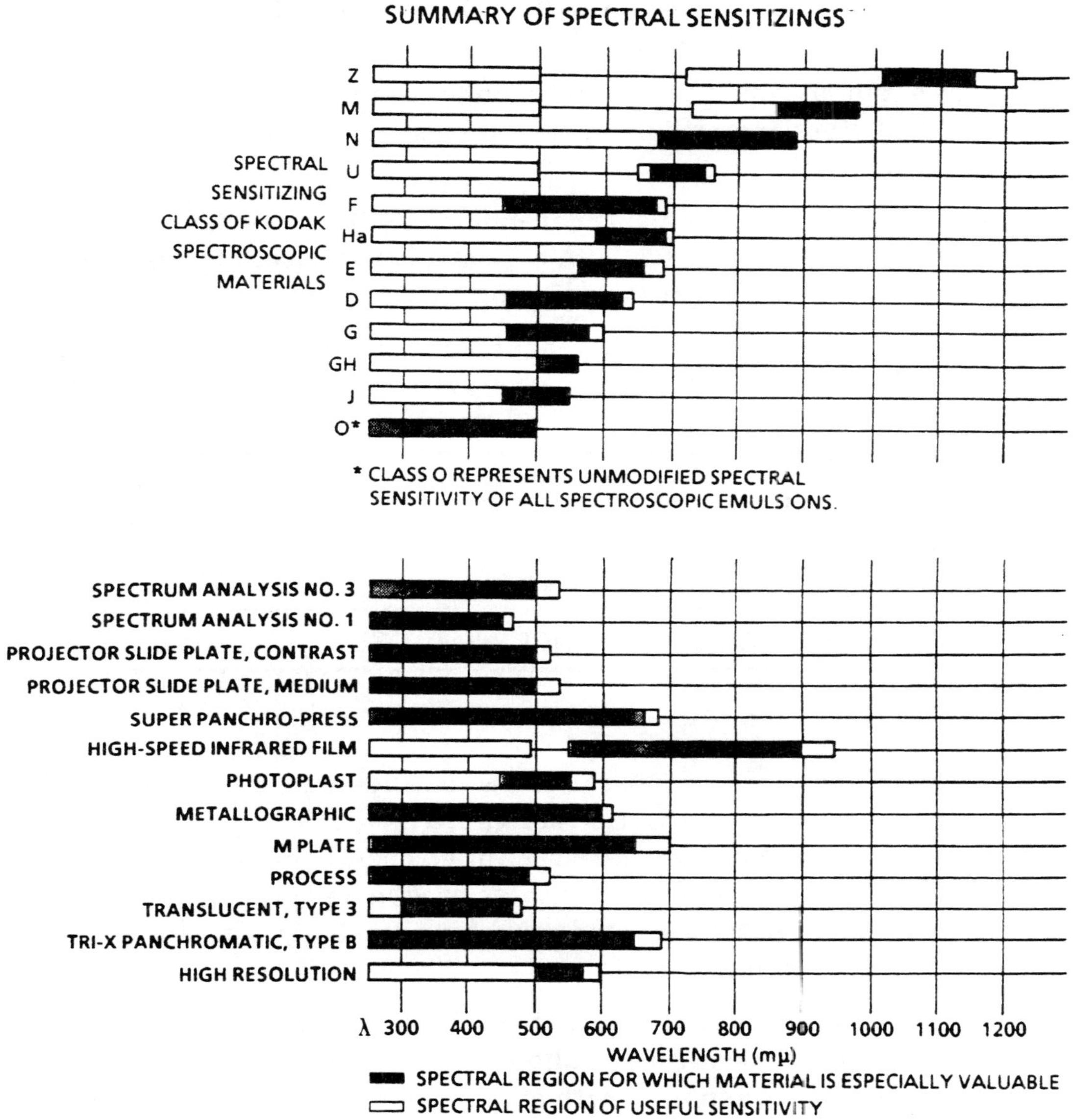

Fig. 6 Summary of spectral sensitizing for Kodak scientific and technical emulsions.

the emulsion. But it must be done carefully to ensure a uniform increase in sensitivity over the entire film area.

Resolution

A final characteristic of photographic emulsions that frequently must be specified is resolution. *Resolution* or *resolving power* is the ability of emulsions to record fine details. To find the resolution of an emulsion or complete photographic system, you use a bar chart that's photographed at a greatly reduced scale. This aspect will be covered in later modules.

Suffice it to say, upon examination under a microscope, you can use the number of dark lines per millimeter that are just recognizable in the image as an indication of resolving power.

The resolution of an emulsion depends only slightly on development. But there are great variations at both high and low exposure levels. The maximum resolution usually occurs at some intermediate exposure.

Resolving power is determined in large part by the physical construction of the film. (See Figure 7.) Film resolution can be altered (deteriorated) by

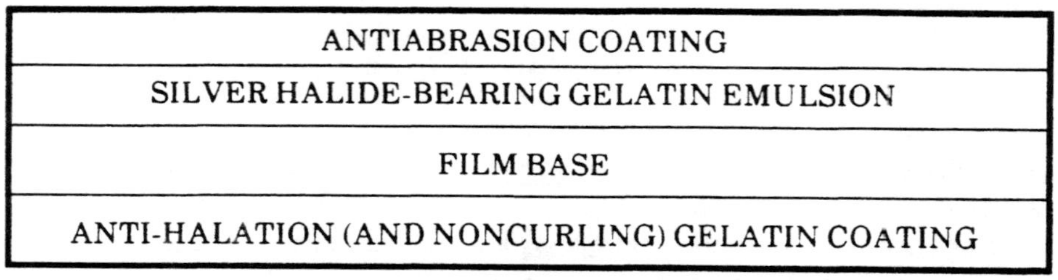

Fig. 7 Schematic view of film construction.

the various coatings of the film base. These coatings are essentially optical components that can cause deviation of light rays—generally from scattering. Low-resolving-power materials have resolutions below 56 lines per millimeter. Emulsions—for example, those used in holographic applications—have been produced with extremely high resolving power of greater than 2,000 lines per millimeter.

Some specialized emulsions have been developed that are essentially "grainless." One example is the so-called "DCG" or dichromated gelatin holographic plate. It's often used with argon lasers to produce bright "white-light" reflection holograms. These plates function on a principle that's entirely different from conventional silver halide film.

In the DCG plate, intense green light from the argon laser is absorbed by ammonium dichromate dissolved in the gelatin. The absorption of the light, coupled with a dehydration process, causes a change in the index of refraction of the emulsion. Therefore, the holographic image is stored as phase information in the plate, rather than areas of high and low transmission as is the case in traditional photography. In the DCG plate, the limit of resolution is down near the molecular scale rather than being dependent on grain size. Holography is discussed in greater detail in Course 7.

The ultimate practical limit to the resolution is the grain size of the photosensitive material in the emulsion. Generally speaking, emulsions used for high-speed films have very thick emulsions that lead to large grain sizes. The grain is actually a group of reduced silver atoms to form a detectable area. For thinner emulsions, you generally have smaller grain sizes, thus higher resolving power and reduced speed. (See Figure 1.)

The specific technical objectives will dictate the required characteristics. For example, in electron microscopy or optical microscopy at high magnification, you use a long exposure with a low-speed, high-resolution film. Dim transient events will need a high-speed film for recording, even though the resolution may suffer. We'll see in later modules how we can use other techniques—such as electronic image intensification—to compensate for some of the undesirable features of photographic emulsions.

Film Processing

Let's talk about—in some detail—the development process often used with black and white emulsions. (Refer to Figure 2.)

The first step in the process is called *development*. Development is achieved by chemical treatment of the emulsion to produce an observable darkening of the film corresponding to the object being reproduced or imaged. As we've said, this darkening comes from the reduction of activated silver halide photosensitive materials to pure elemental silver. The chemicals used in this process are called developing agents. A mixture of these chemicals in water is called the developer.

Most commonly used developer solutions include metol and hydroquinone. Specific developers are recommended for each type of emulsion. This information is readily available through many sources. You can find it in the Photo Lab Index and scientific and technical data available from specific film manufacturers.

The length of film immersion in the developer is a critical factor in the development process. If you leave the emulsion in the developer too long, too many silver halides—even those not activated—would be reduced to pure silver. On the other hand, the development process isn't immediate. So a rapid immersion would reproduce an essentially undiscernable darkening. At the end of the right development time, the "negative" is fully developed. As we've indicated, it's a reversed representation of the photographed subject.

Another important variable is the development temperature. The development process is a chemical reaction, so you can expect elevated temperatures to speed up the reaction. If the temperature is too cold, no reaction at all may take place. For normal processing, the preferred temperature is usually around 68°F. Many developing procedures, such as immersion times, are specified assuming the solutions are at 68°F.

When the emulsion has been developed according to proper procedure, the film still contains all the silver. It has both the reduced silver and the silver that's still combined in the halide form. To remove the unreacted silver halides, you must perform chemical reaction on the emulsion.

Before this next step, you should wash the emulsion with pure water to (1) stop the development process and (2) avoid contamination of the chemicals used in the next step by the developer chemicals. This is frequently called a *stop bath*. It's usually made weakly acidic to neutralize the alkaline developer. Acetic acid is the most commonly used chemical. This reaction actually dissolves the unexposed silver halides.

The next chemical step is *fixing*. The chemical commonly used is called hypo. Hypo dissolves the unexposed silver halide. This process is also selective. The hypo dissolves the silver halide, but not the reduced silver. Again if this process is too long, the hypo eventually will dissolve even the reduced silver. So this step is also critical in length of time and temperature that the emulsion is in contact with the fixing bath.

Fixers also contain chemicals that help to preserve the film by hardening the gelatin. Sodium sulphide is the chemical commonly used for this purpose.

At the end of the fixing process, the emulsion is washed in pure water to remove all traces of chemicals. Then it's carefully dried under controlled conditions. The total development process is summarized in Figure 8.

We've just discussed those aspects commonly found in routine development. But, many variations can be used to modify the sequence of steps. We won't delve into the variations to any extent.

It's important to realize that modifications to the sequence can make photographic emulsions more able to record luminous events. One such modification is the forced development of films.

Forcing can greatly increase the normal speed of typical emulsion. For example, Royal-X Pan film has a rating of ASA 1250. It can be "pushed" to an equivalent ASA speed of approximately 1600 by changes in the development process. You'd immerse it in the developer longer, with the developer at a much higher temperature. This change from normal processing gives a higher-contrast film. You'd get degraded resolution, however, due to the very grainy structure of the final developed film emulsion.

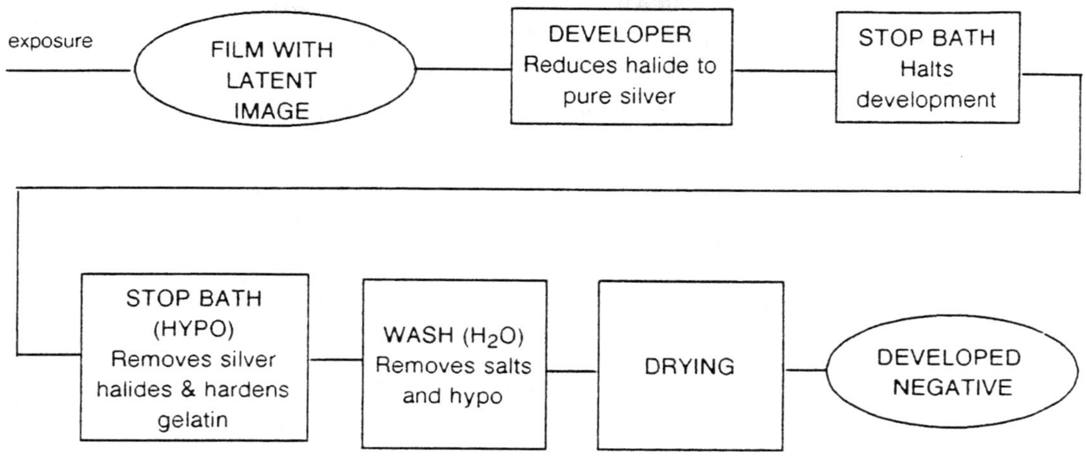

a. Schematic

Exposure	AgBr + light = AgBr* (activated)
Development	AgBr* = reducing agent (e^-) = Ag + Br^-
	(Unactivated AgBr is much harder to reduce.)
Fixation	$AgBr + 2S_2O_3^{--} = Ag(S_2O_3)_2^{--} + Br^-$
	(Unactivated AgBr is dissolved in "hypo" solution [sodium thiosulfate $Na_2S_2O_3$], developed Ag [metallic form] is unaffected.)
	At this point in the process, a negative has been produced.

b. Summary of the basic chemistry. After Fundamentals of General Chemistry.

Fig. 8 The photographic development process.

Printing

If you put a negative in close contact with another piece of unexposed and undeveloped sensitive material, the new material may be exposed by passing light through the negative under controlled conditions.

Each grain of silver will cast its shadow on the new material, and inhibit exposure of the silver halides at that point. Where the negative is clear, light will pass to expose the new material. Where the negative is dense and black, little or no light passes to the new material. The new material then is a negative exposure of the original negative. The final result, after development, is an accurate reproduction of the originally photographed object. The new negative is called the *positive.* The second sensitive material used determines the form of the positive. If the material is photographic paper, the positive is a *print.* When the positive is produced on a film, it's called a *transparency.* The transparency can be projected, like motion-picture film. From one negative, you can make many prints or transparencies.

For amateur, and some scientific work, such as oscilloscope recordings, a single print may be all that you need. In this case, you can eliminate the film negative through a special process called *reversal process.* Here, the originally exposed film contains the finished positive. The process differs in the steps that follow development.

Remember that after development, the negative silver image *and* all the unexposed silver halides are in the emulsion. You can bleach this silver image (make it soluble) with a selective chemical reaction that affects only the silver. It doesn't affect the undeveloped silver salts. Now the emulsion consists of soluble silver in the negative image and the silver halide left over from the development process.

A *clearing bath* removes the pure silver and leaves only the unexposed silver halides in the emulsion. A deliberate second exposure of the halides will render them reducible by any development that transforms all silver halides to pure silver.

The new image is precisely the opposite to the negative that would have been produced by normal processing. It's a positive. This reversal or "direct positive" system is used in some automatic cameras that use sensitized paper and produce a print directly (Polaroid film camera systems).

Color Photography

You'll often need to use color photography to maximize the information recorded in technical photography. There are many different processes of color photography, but generally they can be divided in two main groups:

- Additive processes used primarily for transparencies.
- Subtractive processes used for both transparencies and paper prints.

In all cases, the principles of color photography are essentially the same in that images are formed in three primary colors. Color film has three separate emulsions. One emulsion is sensitive to the yellow, one to the green, and the last to the red portion of the spectrum. This is done by using dyes that are sensitive to these portions of the spectra.

The principles in which these single-color images are combined to produce the final "essentially" continuous spectrum result varies from one type of process to another.

One method is the reversal process. It's a subtractive process, and it's the one most commonly used today. Because color photography development normally is done by a special processing laboratory, we won't go into these details.

A particular type of multiple-emulsion film—called extended-range film—is used often in scientific photography. If you're investigating a source that has dim as well as bright regions of emission, you can't record many important aspects of the event under study on normal film. That's because of their unlimited latitude. With extended-range emulsion, you can use three different layers. The layers correspond to three separate colors, and each has a different sensitivity. Events that have been investigated with this type of film include high-altitude nuclear detonations, exploding wires, laser-produced plasmas, and so on. Some of these films have a latitude of response over six orders of magnitude in exposure.

To evaluate and analyze the documented event, you simply put a filter in front of the projector. The filter is one that selects from the stack the individual emulsion whose sensitivity corresponds to the level of illumination of interest. Again, this film normally is processed under special development conditions not normally used with limited darkroom capabilities.

Polaroid Emulsions

Because of the popularity and frequent use of Polaroid photography in technical work, let's describe the characteristics of these emulsions. To help you to understand both Polaroid films and color photography, we'll address specifically the Polaroid Land color film process.

Processing of ordinary color films takes many complex operations. So the rapid one-step production of colored positives by Polaroid colored film is used a lot in photography. This large use is in no small measure due to the fact that you can reproduce a colored positive within one minute of the exposure.

Figure 9 shows Polaroid color film. It's coated with three photosensitive layers that have the silver halide crystals, three layers that have dyes and developers, and two intermediate layers. Each pair of silver halide and spectrally selective dye layers responds to the blue, green, and red regions of the spectrum.

As shown in Figure 9, blue light is primarily stopped in the first layer, green will be stopped in the second silver halide layer, and red will pass through to the third silver halide layer. Obviously, light sources that have all three colors will cause an exposure in each silver halide layer as shown.

After exposure, the film or a strip is pulled from the camera. This releases the developer-activating paste. Through the action of smooth precision-spaced steel rollers, the negative layer is pressed firmly onto the so-called positive layer. The positive layer has the color and intermediate layers and an acid layer for neutralization and stabilization of the color image.

Paste rapidly diffuses into the negative. It first enters the blue-sensitive layer and liberates a yellow dye in the adjacent layer. It then diffuses by similar process through the other pairs of layers. In the green-sensitive pair, it liberates a purple dye, and in the red-sensitive layer, a blue-green dye is liberated.

The dyed molecules that have been liberated by the paste are a reproduction of the latent image. They now diffuse in all directions. When they encounter a grain of reduced silver they're retained and held by it. The dye molecules that don't encounter silver grains keep migrating until they reach the positive layer. They're retained there.

At the end of the process, water is formed in an alkali neutralizing process. It serves to wash the positive.

This explanation didn't just describe the way Polaroid prints are obtained. It also told you how one-step rapid processing emulsions can lead to positive prints that faithfully reproduce the color tones in the photographed scene.

Between the extremes of the Polaroid Land process and the one originally described for black and white film negatives, there are a lot of slight variations in photographic development processes. If you're interested in more detail, you should look at the references cited at the end of this module.

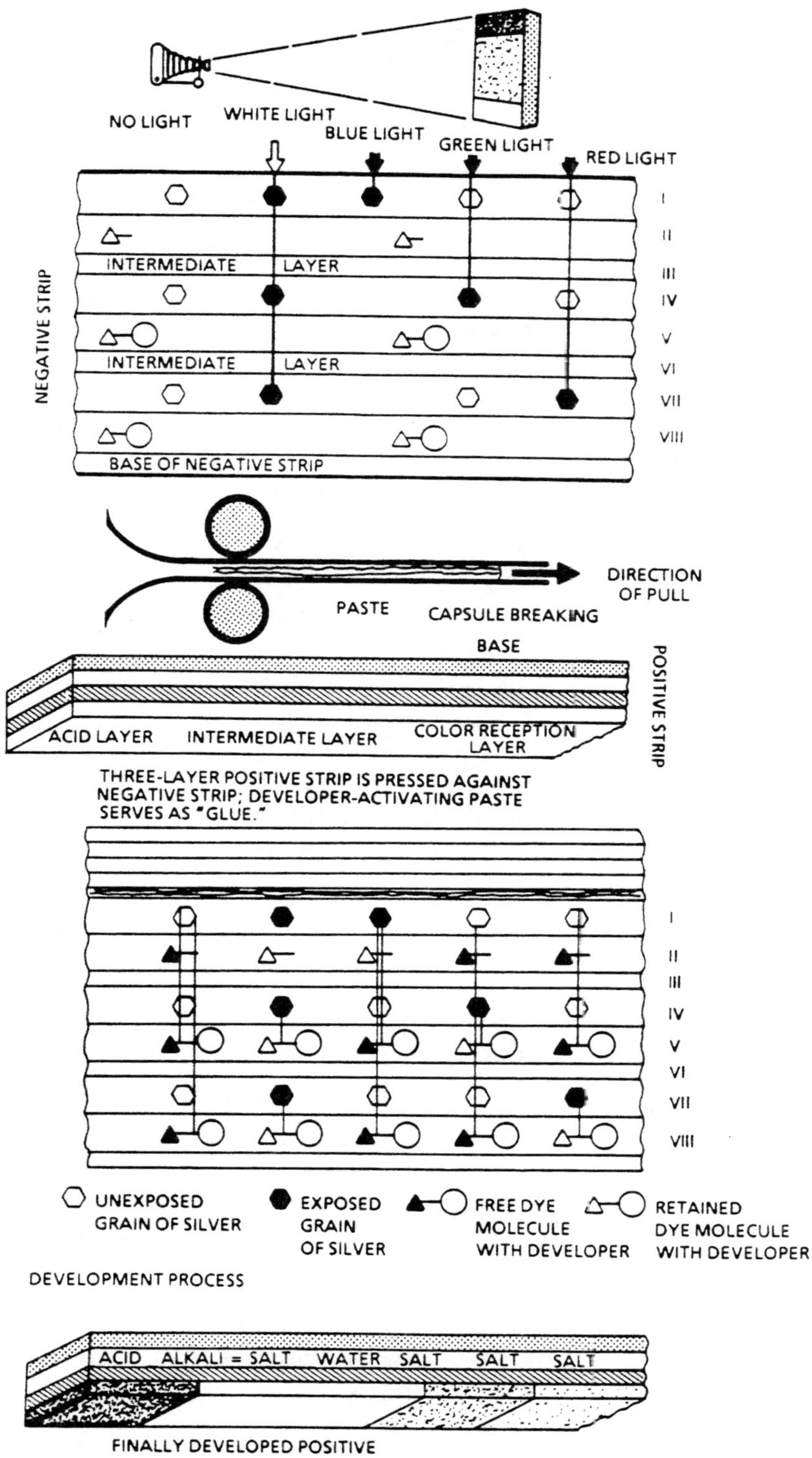

Fig. 9 Polaroid Land colorfilm.

Spectral Matching

Spectral Matching is an important concept in technical photography. To get the best performance inherent in the photographic process, you choose the specific photographic emulsion for optimal response in recording the event under study.

For example, Polaroid emulsions often are used to record oscilloscope traces. To get the maximum resolution and performance of the oscilloscope-recording process, you need to use as small a spot as is detectable on the oscilloscope face.

As you know, the oscilloscope face is coated with a phosphor that fluoresces from impact of the electron beam that's translated across the face plate as a result of the various sweep circuits. There's a wide selection of phosphors available on oscilloscopes. They have different emission characteristics.

To get the maximum resolution in the recording process, you should use an emulsion of the required resolution. It should be one that's maximized in spectral sensitivity at the peak of the wavelength emitted by the fluorescent screen. For example, you'd use a blue-sensitive film when you record images on an oscilloscope with a P-11 phosphor that emits primarily in the blue.

From the technical data in this module, you should see that there's limited infrared response of normal photographic emulsions. So, when you record weak emissions from 1.06-μm-operating neodymium YAG lasers, you'll often use type-1Z plates that have the greatest response in the region. (See Figure 6.)

You can see that, to select the best photographic emulsion for recording any technical event, you don't just consider the characteristics of the film. You also try to match the film to the emission characteristics of the object to be recorded.

Characteristics of Photosensitive Materials

To help you evaluate the suitability selection of photographic emulsions for specific technical applications, you should get information that's readily available from film manufacturers. You can find an excellent compilation of commercially available films and plates normally used within the United States for various scientific and industrial photography. It's Kodak Booklet No. 9. The title is "Kodak Plates and Films for Science and Industry." An example is shown in Figure 10, which is shown on two pages as Figures 10a and Figure 10b.

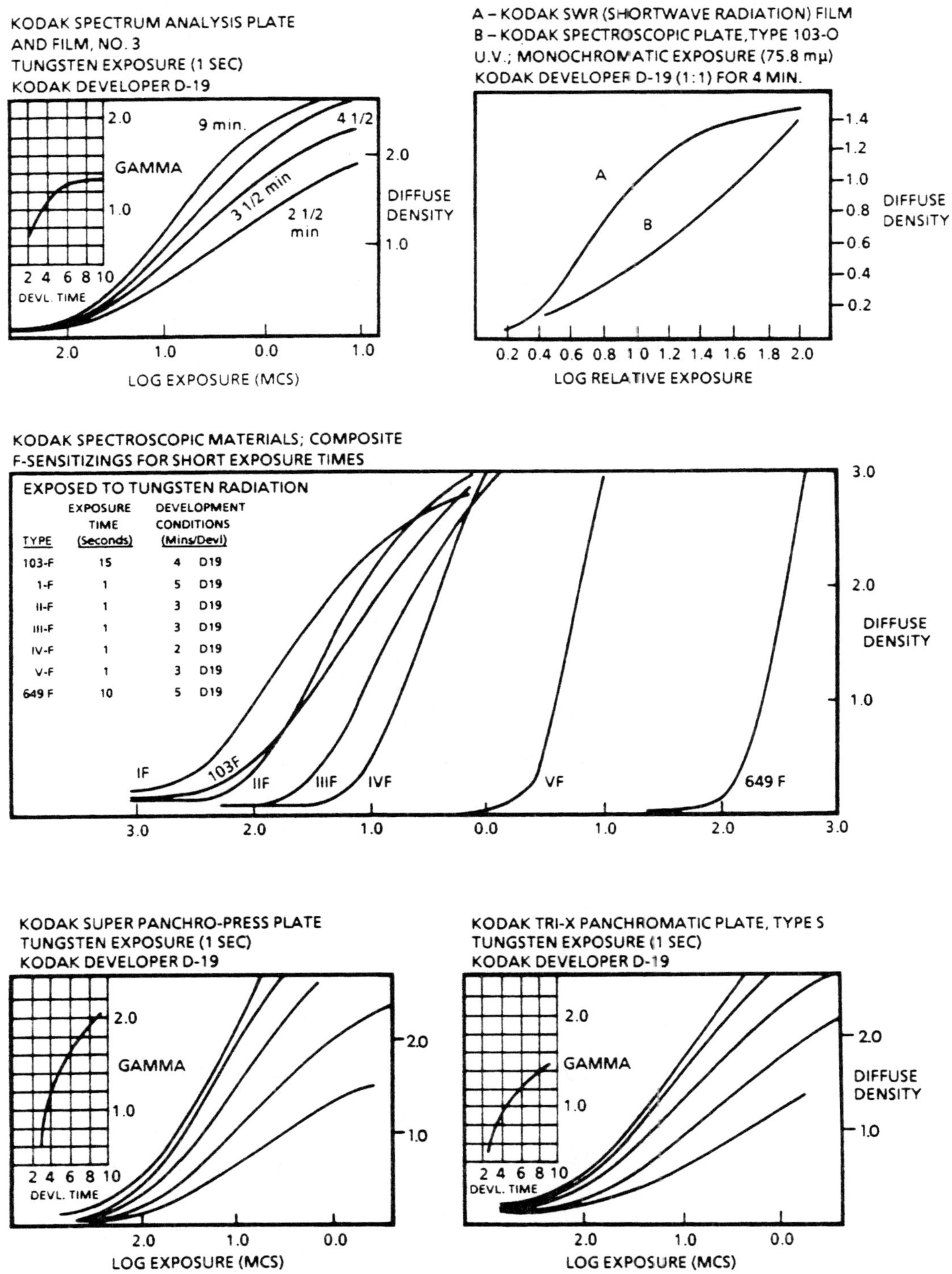

Fig. 10a Some typical Kodak Film characteristics.

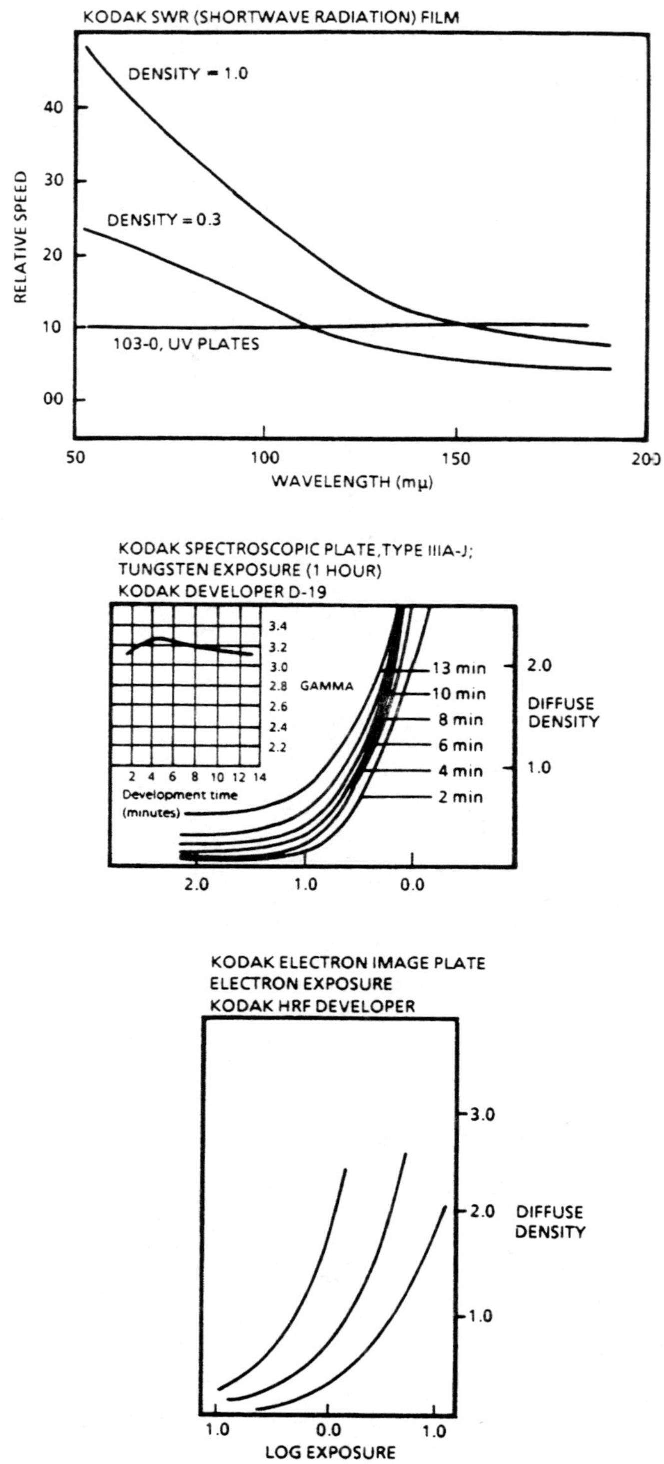

Fig. 10b Some typical Kodak Film characteristics.

You can find similar information for other films made within the United States, such as DuPont X-ray films, or from film manufacturers in other countries, such as Ilford in England, Agfa-Gevaert in West Germany, Fuji in Japan. In many cases films made in other countries have unique characteristics that make them suitable for specific applications such as high-resolution holographic recording.

Characteristics of available Polaroid films are given in Table 1. The spectral response and density curves for the seven more commonly used Polaroid films are shown in Figure 11. This information has been taken from the Photo Lab Index.

Table 1: Speed of Polaroid Films*

Film Type	Format	Picture Obtained	Picture Size (inches)	Speed (ASA Equiv)	Devel Time at 75°F (24°C)
Roll Films					
20	8-exp roll	B & W print	$2\frac{1}{2} \times 3\frac{1}{4}$	3000	15 sec
32	8-exp roll	B & W print	$2\frac{1}{2} \times 3\frac{1}{4}$	400	15 sec
38	6-exp roll	Color print	$2\frac{1}{2} \times 3\frac{1}{4}$	75	60 sec
42	8-exp roll	B & W print	$3\frac{1}{4} \times 4\frac{1}{4}$	200	15 sec
48	6-exp roll	Color print	$3\frac{1}{4} \times 4\frac{1}{4}$	75	60 sec
Pack Films					
107	8-exp pack	B & W print	$3\frac{1}{4} \times 4\frac{1}{4}$	3000	15 sec
108	8-exp pack	Color	$3\frac{1}{4} \times 4\frac{1}{4}$	75	60 sec
4 x 5 Film Packets					
51	20 single-exp packets	High-contrast B & W print	1 x 5	200	15 sec
52	20 single-exp packets	B & W print	4 x 5	400	15 sec
57	20 single-exp packets	B & W print	4 x 5	3000	15 sec
58	10 single-exp packets	Color print	4 x 5	75	60 sec
55P/N	20 single-exp packets	B & W and a negative	4 x 5	50	20 sec
Special-purpose Films					
46-L	8-exp roll	Cont-tone B & W transparency	$3\frac{1}{4} \times 4$	800	2 min
146-L	8-exp roll	High-contrast B & W transparency	$3\frac{1}{4} \times 4$	200	15 sec
410	8-exp roll	High-contrast B & W print	$3\frac{1}{4} \times 4\frac{1}{4}$	10,000	15 sec**

* Processing times on this table may need some modification dependent on the temperature conditions. You can find more detailed information on the instruction sheet packed with each particular type of film.

** To facilitate rapid-sequence recording, Type 410 Land film can be advanced at 2-second intervals and prints developed 13 seconds outside the camera.

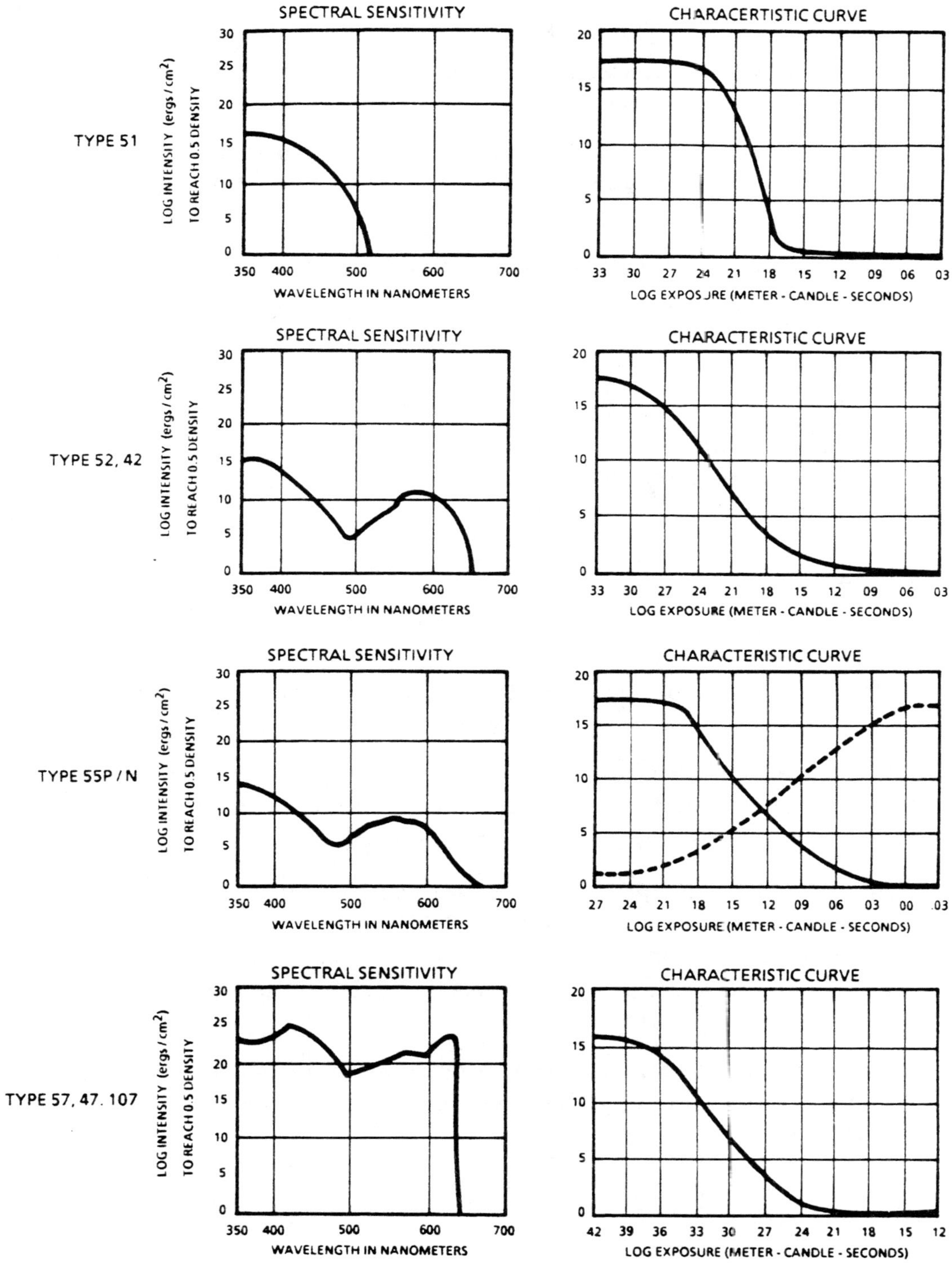

Fig. 11 Spectral response and density curves for Polaroid films.

EXERCISES

1. Describe in your own words the process of latent image formation in photographic emulsions.
2. List the required steps, and name the various bath solutions and times required for processing the film used in the laboratory (in proper sequence).
3. What are the two variables plotted on an H-D curve?
4. Draw a typical H-D curve and identify the following:
   a. Toe
   b. Shoulder
   c. Latitude
   d. Inertia point
   e. Solarization region
   f. Contrast
   g. Background fog level
   h. γ
   i. Saturation
   j. Speed
5. Define reciprocity and reciprocity failure as regards film behavior. Convey the same meaning as given in the text. Use examples.
6. Draw a graph of the spectral sensitivity of a typical photographic emulsion similar to the graphs given in the text. Relate the importance of this graph to the concept of spectral matching.
7. A technical photographer wants to photographically record a weak 1.06-micrometer laser radiation pattern in a holographic investigation. What two prime characteristics should she look for in selecting an emulsion? Check the most appropriate emulsion from the technical information given in the text.
8. If, after initial experiments, she determines that the chosen emulsion doesn't have enough sensitivity, what alternatives does she have for overcoming this deficiency?

For the next two problems, you'll need a copy of the Kodak B&W Darkroom Dataguide.

9. You are about to develop Kodak Tri-X Pan film using Polydol developer in a darkroom that's at a temperature of 24.5°C. Assume that you will develop the film "normally" (no "push"). How long should you keep the film in the developer?

10. Repeat the above problem. But now consider that you'll be "pushing" the film. (Push processing compensates for underexposure, but at the expense of increased graininess.)

MATERIALS

Roll of twenty-exposure black and white film (Kodak Tri-X Pan)
Camera (35 mm) with bulb ("B") or time ("T") shutter settings available
Electronic flash unit that can be single-flash triggered (Strobotac)
Darkroom facility and chemicals for black and white processing
Optical bench with fixtures (holders)
Graph paper
Kodak Black and White Darkroom Dataguide
HeNe Laser (about 2 mW)
Collimator/beam expander
Iris
Optical power meter

PROCEDURES

1. Set up on a 1-meter optical bench, the large frosted glass plate and camera, so that the camera is focused on the glass plate and the plate covers the field of view. See Figure 12.
2. Set up the Strobotac behind the glass plate (approximately 120 cm) so that illumination on the plate will be fairly uniform. Place an ND 3.0 filter in front of the lamp to attenuate the bright flash (ND value will depend on brightness of strobe).
3. Set the Strobotac setting for single-flash operation.

   Caution: If you decide to use the "auto-fire" mode of your flash for the longer exposures (128, 256, etc), make sure that the flash intensity is the same in the single-shot and auto modes. The old-style General Radio strobes decreased flash intensity significantly in the "auto-fire" mode.
4. Set the camera on time ("T") setting, after advancing the film to the first frame. Then set the camera aperture to the smallest setting: f/22 or f/16.

   Note: Some cameras do not have "T" setting. So you must use the bulb ("B") setting.

   On "T" one click opens the shutter and another closes it (two clicks per picture required).

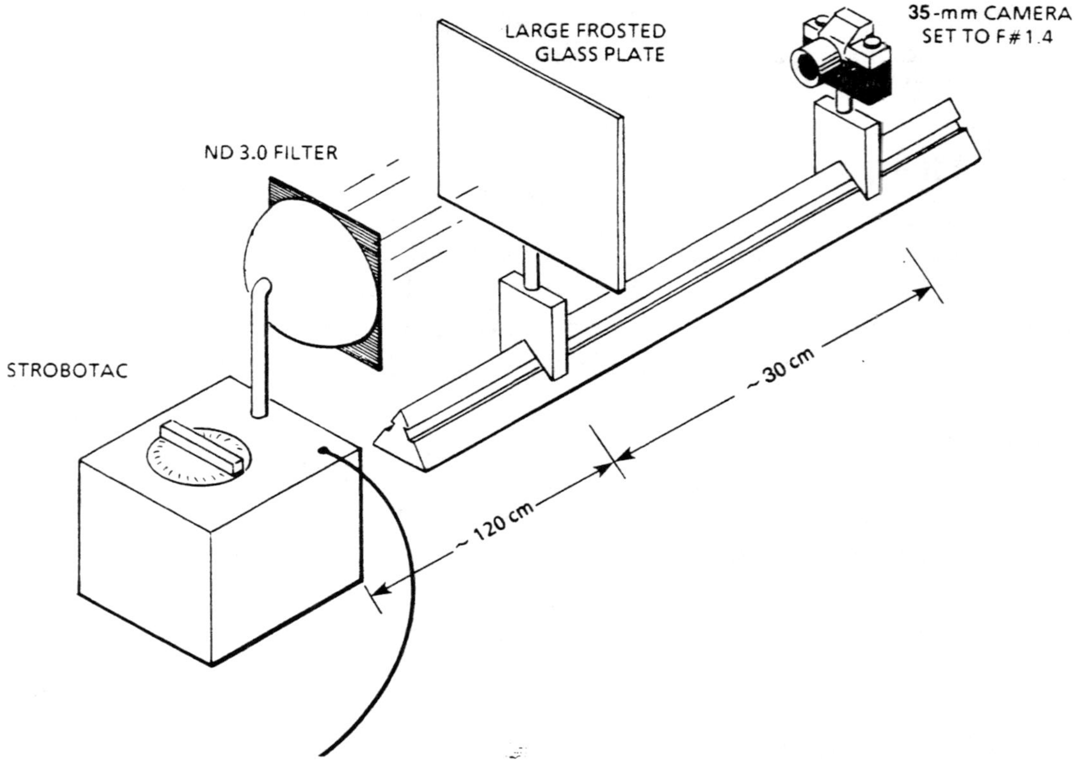

Fig. 12 Exposure setup for production of H-D curve.

On "B" the shutter stays open as long as the shutter button is depressed. It closes when you release the button. In this case, you should use a shutter extension cable.

5. Darken the laboratory, open the shutter and flash a single flash. Then close the shutter and advance the film.
6. Repeat Step 5, but flash the unit two times.
7. Repeat Step 5, but flash the unit four times.
8. Repeat Step 5, doubling the number of flashes for each succeeding frame.
9. After you've made about 9 exposures, remove the film from the camera.
10. Develop the film according to manufacturer's instructions (see Appendix A). Cut film into 35-mm slides and mount in cardboard or plastic slide holders. Number the slides by order of exposure.
11. Note visually the variation in the film density as a function of the number of flashes. If all of the exposures look approximately the same density you'll have to repeat the experiment. Change either the value of

the ND filter or the F# on the camera lens to either increase or decrease the exposure until you get a good "dynamic range" of exposures.

12. Set up an optical arrangement as shown in Figure 13.

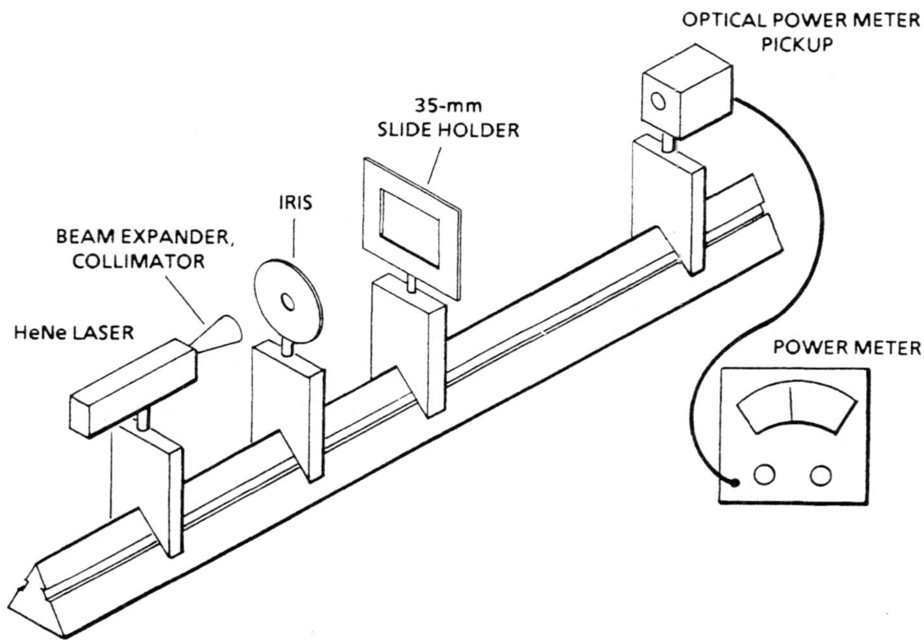

Fig. 13 Setup for making relative H-D curve.

13. Adjust the collimator so that the beam is about 1 cm in diameter. Set iris to about 1/2 the collimated beam diameter so that only the center of the beam is used. Align the beam with the center of the exposed region on the film.

14. Measure the power, with meter, for each exposure made. Tabulate "number of flashes." Plot the data obtained in this step with density (103 1/T) on the Y-axis and log to the base 2 of the number of flashes on the X-axis. Use the following conversion table:

No. flashes	Log2 (No. flashes)
0	–
1	0
2	1
4	2
8	3
16	4
32	5
64	6
128	7
256	8
512	9

15. Plot density (log 1/T) versus $\log_2$ (# of shots) on graph paper. On your graph show the linear portion of the H-and-D curve, saturation (if you achieved it), the film's gamma and the film's inertia. Note that your values for gamma and inertia are only "relative" values because the horizontal axis of the graph isn't given in actual exposure units (joules or ergs per cm^2).

DATA TABLE

Number of Flashes	$\log_2$ (Number of Flashes)	Power Reading (mW) (transmitted)	Power Reading ÷ Incident Power
0			
1			
2			
4			
8			
16			
32			
64			
128			
256			
512			

P_o = incident power = ___________mW transmission T

REFERENCES

Baines, H. The Science of Photography. New York: John Wiley & Sons, Inc.

Brode, Wallace R. Chemical Spectroscopy. New York: John Wiley & Sons, Inc., London: Chapman and Hall, Limited.

Eastman Kodak Co., Rochester, NY.
Basic Developing, Printing, Enlarging, No. J-2.
Kodak Black-and-White Darkroom Dataguide, No. R-20.
Kodak Films for CATHODE-Ray Tube Recording. No. P-37.
Kodak Films in Rolls for Black and White Photogaphy, No. F-13.
Kodak High Resolution Plates, No. P-47.
Kodak Plates and Films for Science and Industry, No. P-9.
Processing Chemicals and Formulas for B&W Photography, No. J-1.
Ultraviolet Sensitization of Kodak Plates, No. M-13.

__________. "A Guide to Polaroid Film," Industrial Photography, Vol. 22, No. 8, August 1973, p 53.

Mees, J. The Theory of the Photogaphic Process. New York: MacMillan Company.

Neblette, C. B. Fundamentals of Photography. Van Nostrand Reinhold Co.

Photo Lab Index. Morgan and Morgan, Inc. Publishers, 101 Park Avenue, New York City, N.Y. 10017.

Sorum, C. H. Fundamentals of General Chemistry. Second edition, Prentice-Hall, 1963.

Appendix A
Instructions for Developing Kodak Tri-X Pan Film

Step	Time (min)
Developer (D76)	See development dial
Rinse (DI water)	1.0
Stop bath	0.50
Fixer	3.0
Wash (DI water)	1.0
Hypoclear (optional)	2.0
Wash (DI water)	5.0
Photoflo (optional)	1.0
Dry	30 (depends on temp)

Development Dial

Set for development number 33 (Tri-X Pan in D-76)
See next page.

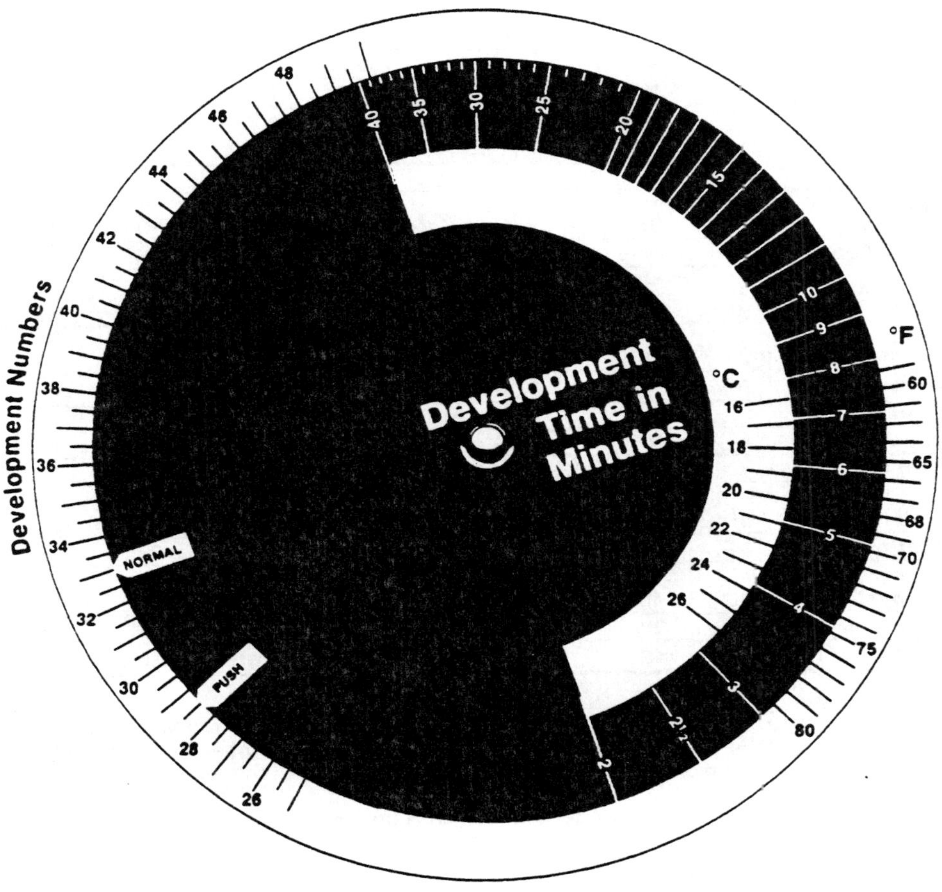

Development Number Adjustments	If your negatives are consistently:			
	Too thin (lacking shadow detail)	Too dense	Too flat	Too contrasty (especially when using a condenser enlarger)
	Use a lower film speed	Use a higher film speed	Add 3 to the development number	Subtract 3 from the development number

This and the following page are taken from: Kodak Black-and-White Darkroom Dataguide, 7th edition, publication # R-20, Eastman Kodak Company, 1980.

DEVELOPMENT NUMBERS

KODAK Black-and-White Films	Development Numbers for KODAK Developers						
35 mm Films	D-76	D-76 (1:1) MICRODOL-X	MICRODOL-X (1:3)	DK-50 (1:1)	POLYDOL	HC-110· Dil. A	HC-110 Dil. B
PANATOMIC-X	32	36	42	NR	33	NR	30½
PLUS-X Pan	33	36	42	NR	33	NR	32
TRI-X Pan	37½	40	46	34½	36	28	36½
Recording 2475	NR	NR	NR	NR	NR	31	38½
High Speed Infrared	41	NR	NR	NR	NR	NR	34
Technical Pan 2415	See Film Instructions						
Roll Films							
PANATOMIC-X Professional	32	36	42	NR	33½	NR	30½
VERICHROME Pan	36	38½	43½	NR	34	NR	32
TRI-X Pan	37½	40	46	34½	36	28	36½
TRI-X Pan Professional	37½	40	NR	37½	39	NR	39
PLUS-X Pan Professional	33	36	41	NR	33	NR	32
ROYAL-X Pan	NR	NR	NR	NR	36	32	38
Sheet Films			DK-50				
ROYAL-X Pan 4166	NR	NR	34	NR	38	34	40
ROYAL Pan 4141	40	41	32	37½	37½	29	37½
TRI-X Pan Professional 4164	36	39	NR	35	37½	NR	36½
SUPER-XX Pan 4142	NR	NR	32	38	38½	29½	35
PLUS-X Pan Professional 4147	37½	40	NR	34	37½	NR	36
TRI-X Ortho 4163	36½	39	NR	35	37½	NR	36½
EKTAPAN 4162	40	43	NR	34	40	28½	34

NR = not recommended

How to find the development time

Find the development number for your film and developer above. Set the NORMAL pointer at the development number and read the development time opposite the temperature of the developer.

Push Processing—Use the PUSH pointer instead of the NORMAL pointer to determine development times. See page 14 for details.

LASER/
ELECTRO-OPTICS

MODULE 6-4
WINDOWS

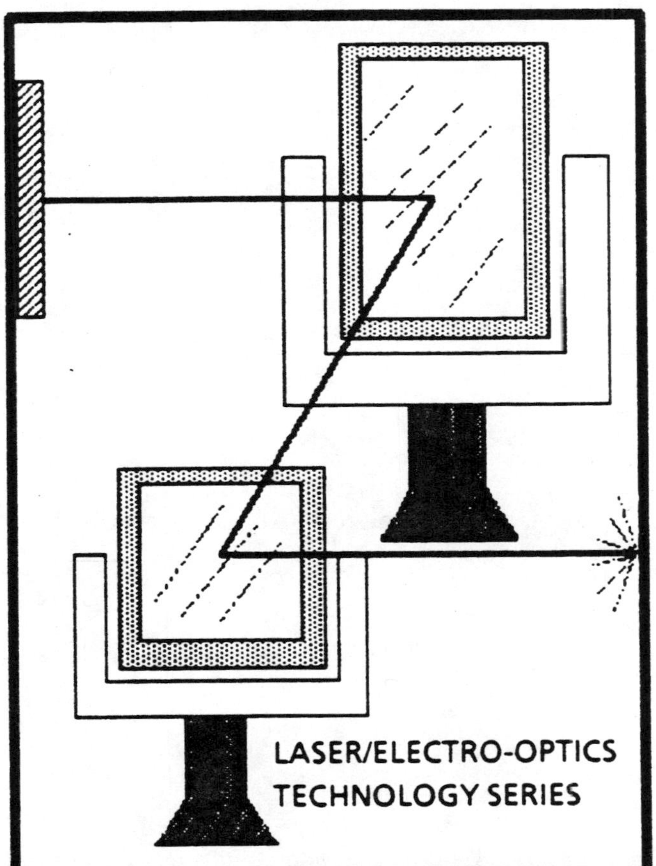

CORD

The Center for Occupational Research and Development
601 C Lake Air Drive
Waco, Texas 76710

Printed in the U.S.A.

ISBN 1-55502-024-0

Prion the Damned ~~m~~
Lord of the Sea
Fuzzums_007

WINDOWS

MODULE 6-4

During the Apollo (manned lunar exploration) space program, thousands of pictures were taken from inside the space vehicle of objects and phenomena on the moon and in outer space. These pictures constituted basic scientific data. From them, objects were identified, topography was mapped, courses were charted, phenomena were observed and records were maintained.

The accuracy and validity of the data were, among other things, dependent on the windows' ability to transmit images from outside the spacecraft to the inside for photographic recording without significant distortion. These windows also had to help to maintain a positive atmosphere inside the spacecraft. They had to prevent excessive solar radiation from entering and heating the vehicle, and withstand the intense heat and shock met on reentry to the earth's atmosphere.

This example of an optical window demonstrates that windows are precision optical components that must be selected, specified, and designed prior to construction and usage. Optically, a window should appear *as though it were not there.* Windows most commonly are used in these two classes of applications:

1. To retain a specific gas pressure or a partial vacuum in a tube or other enclosure.
2. To prevent dust, moisture or other particles from entering the system.

Windows often are constructed to serve additional purposes. They act as filters, beam splitters, output mirrors, polarizers, lenses or prisms.

This module describes the desirable characteristics of optical windows and the principles for selection, specification and use of these components. For you to fully understand this description, you also will learn the characteristics, quality and finish of optical materials used in windows.

Before you begin this module, you should have studied an introduction to lasers, especially laser safety. You also should understand geometric optics, types and properties of optical materials, limitations and testing of optical materials, and polarization.

OBJECTIVES

When you complete this module, you should be able to do the following:

1. Describe Fresnel reflection, and state the conditions under which it occurs. Using a sketch, show the incident and reflected rays, angles of incidence, reflection and refraction, and the interacting surface. Give the approximate values (in %) of the ratios of reflected to incident, and transmitted to incident light, for Fresnel reflection of a material with a refractive index of 1.5.
2. Explain, and illustrate with a sketch, Brewster's angle. Give its relationship to the refractive index of the material, and its effect on the polarization of the reflected portion of the incident light. Include in the sketch the reflecting interface, the incident and reflected rays, Brewster's angle, and the polarization of the reflected ray.
3. List five defects found within optical materials referred to in the Discussion.
4. List two defects found on optical component surfaces.
5. List five important characteristics of an optical window.
6. Using an intense, coherent light beam, visually examine optical surfaces of seven different materials. Evaluate their finish by observing scattered light.
7. Use a shop microscope and grazing incidence illumination to examine two surfaces to determine the nature and severity of surface defects.
8. Use an expanding beam of coherent light to test five windows for their effect on the intensity pattern of that beam.
9. Evaluate the surface flatness of five windows, using interference fringes produced between them and a master flat.
10. Determine which of two lenses has an antireflective coating.
11. Calculate the pressure differential that can be applied to a window before significant distortion occurs.

DISCUSSION

A *window* could be defined as *a pane of glass or other transparent material used to transmit light with negligible distortion and to isolate the atmospheric and thermal environments on either side.*

One common example of windows used in a laser is shown in Figure 1. Here, the windows are the ends of a plasma tube for a helium-neon laser. The intracavity laser beam must pass through each of these windows twice during each round trip through the laser cavity. This is a very low-gain laser. That means that, if each window caused as much as one percent loss per pass to the laser beam, the laser would not operate.

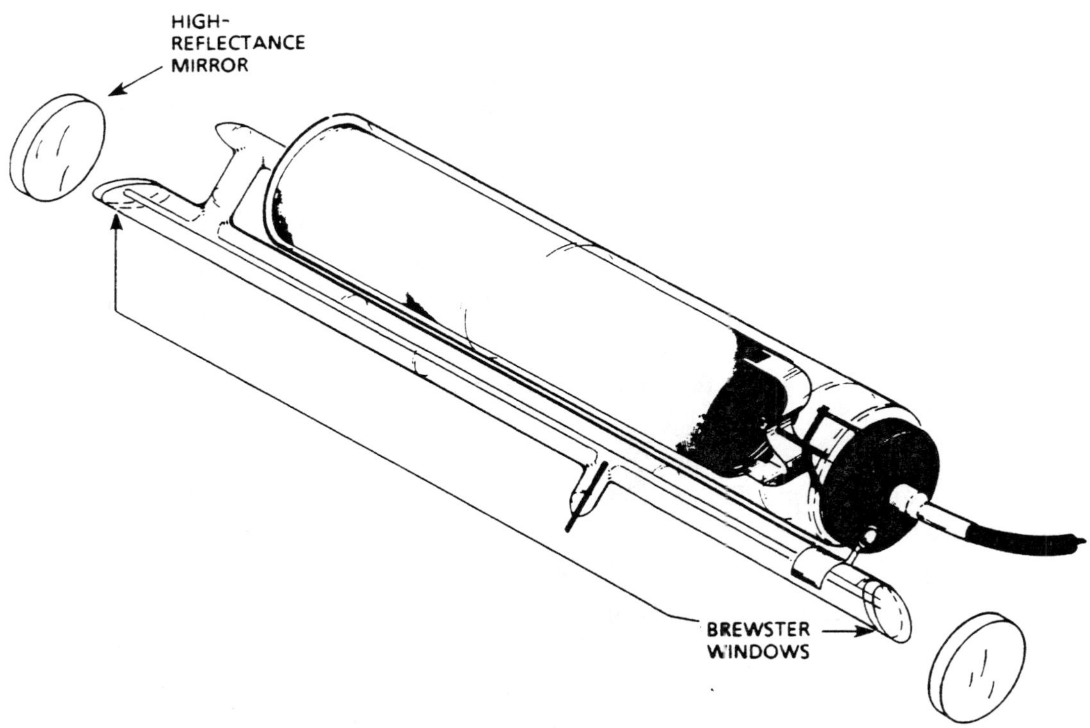

Fig. 1 Windows on plasma tube of helium-neon laser.

Another application of windows in an electro-optical system is shown in Figure 2. Here, the window protects the output optics of a laser metal-cutting instrument from splattered, molten metal. The window is cleaned or replaced periodically.

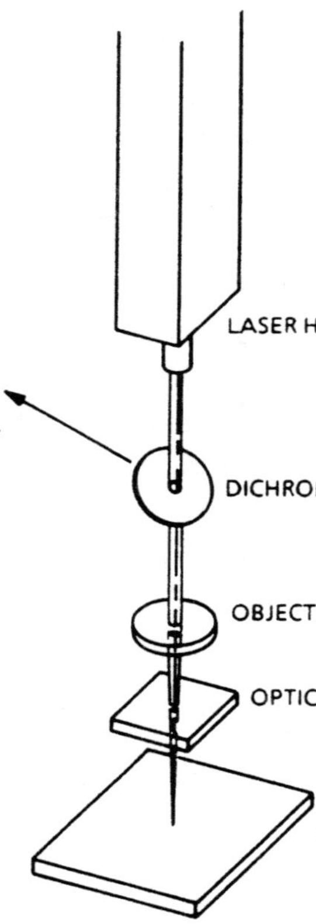

Fig. 2 Window used to protect output optics in a laser metal-cutting operation.

The optical quality of a "perfect" window would be such that it would have *no effect* on a parallel, plane wavefront of an optical beam passing through it. Based on this "as if it weren't there" concept, we can list the desirable characteristics of an optical window.

1. Low absorption of transmitted light
2. Low reflection of light incident on the surfaces of the window
3. Low refraction (or bending) of the transmitted light rays
4. Minimum distortion of the transmitted beam due to imperfections in the optical material or the surface finish.
5. Durable, not susceptible to damage by the laser, and an ability to withstand degradation from environmental causes. This may include strength, resistance to water vapor or temperature, and so on.

The rest of this discussion will describe how to get these desirable characteristics in optical windows by proper material selection, fabrication, and component use.

Low Absorption

You can achieve low absorption in a window by selecting the window material properly and by minimizing the window thickness. Light absorption in an optical material is dependent on wavelength of the light intensity, absorption characteristics of the particular material at that wavelength, and thickness of the material. Look at the window model in Figure 3. The relationship between incident light irradiance, E_0, and transmitted light, E, is given by:

$$E = E_o e^{-\alpha x} \qquad \text{Equation 1}$$

where: α = Absorption coefficient of the material at a particular wavelength.
x = Thickness of the material.

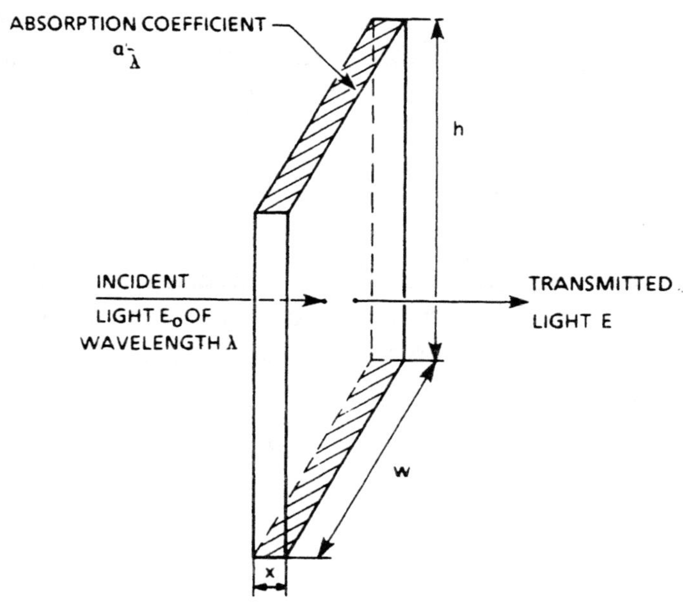

Fig. 3 Model of an optical window.

The ratio of transmitted to incident light is the transmissivity of the window (or percent transmission if it's multiplied by 100%). You can rearrange Equation 1 to calculate percent transmission in the following way:

$$\text{Percent Transmission} = \frac{E}{E_o} \times 100\% = \left(e^{-\alpha x}\right) \times 100\% \qquad \text{Equation 2}$$

The curves in Figure 4 (a-d) show the percent transmission over a broad wavelength range of a variety of optical materials in common use. Inspection of this curve shows that families of materials are suitable for a UV window (quartz, sapphire); other materials for visible light and near infrared (various oxide glass and plastic); for intermediate infrared wavelengths (special oxide glasses and nonoxide glasses such as aluminate, germanate, arsenic-sulfur, germanium and silicon); and far infrared or the "thermal region" (various salt crystals, sulfides, selenides and tellurides). For a given application (perhaps knowing the wavelength of light to be transmitted) you can select an optical material for minimum absorption (maximum transmission).

The values of transmission shown in Figure 4 can be improved by various coatings that decrease the reflection at the interface between the window material and air. Note that the curves in Figure 4 are for a window of specific thickness X. Study Equation 2 to see that the transmission is higher if the thickness X is smaller. It's a logical conclusion that the window should be made as thin as possible. This is a valid assumption as long as the window is thick enough to be rigid and maintain its flatness. If there is no pressure differential on the window (the pressure is the same on both sides), the ratio of the diameter D to the window thickness t can be as high as 100:1 (for a round window).

Mechanical distortion of windows is discussed in greater detail later in this module.

Many infrared optical window materials are available for a wide variety of applications. Transmission curves for several of these materials are provided in Figure 4. Wavelength ranges for some IR window materials are given in Figure 5.

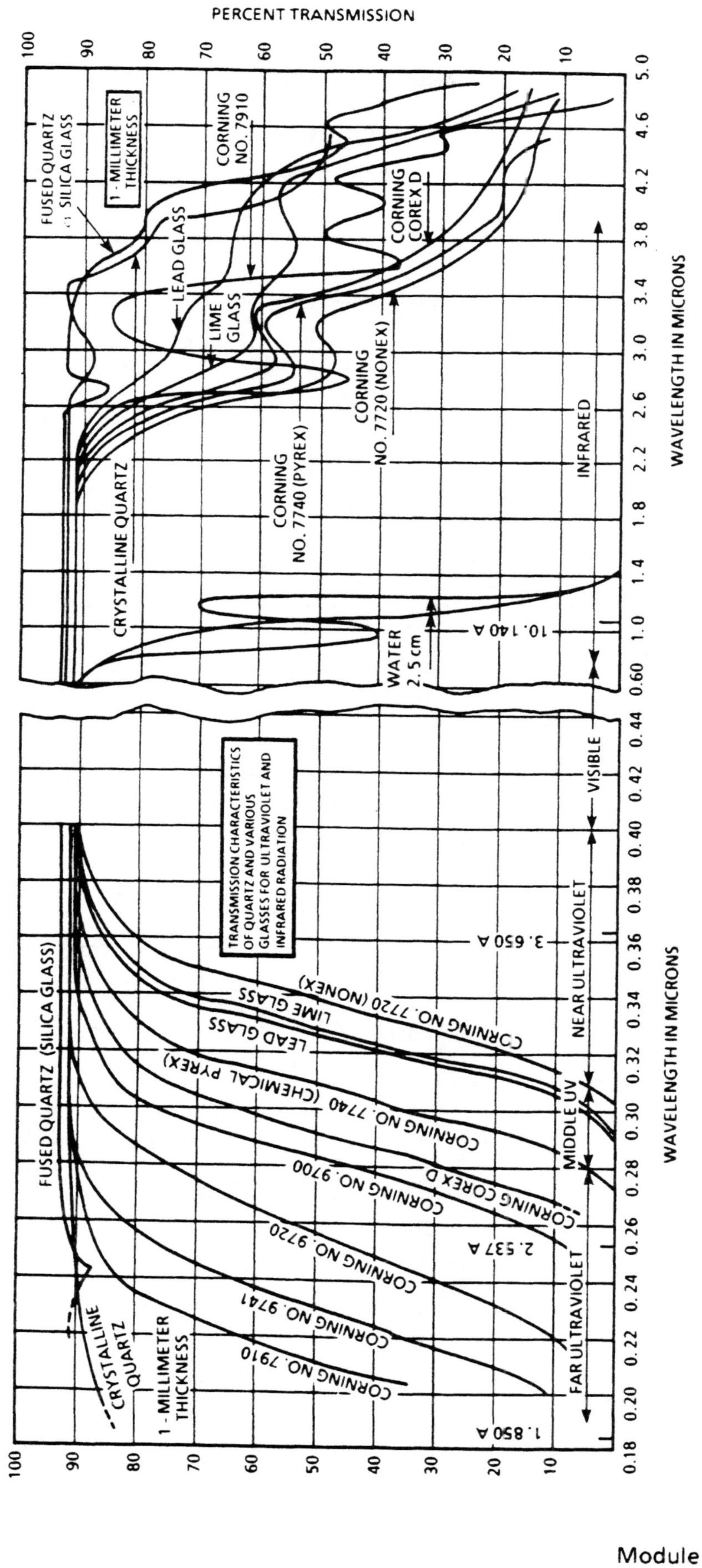

a. UV visible, near-IR window materials.

Fig. 4 Transmission curves for various optical materials.

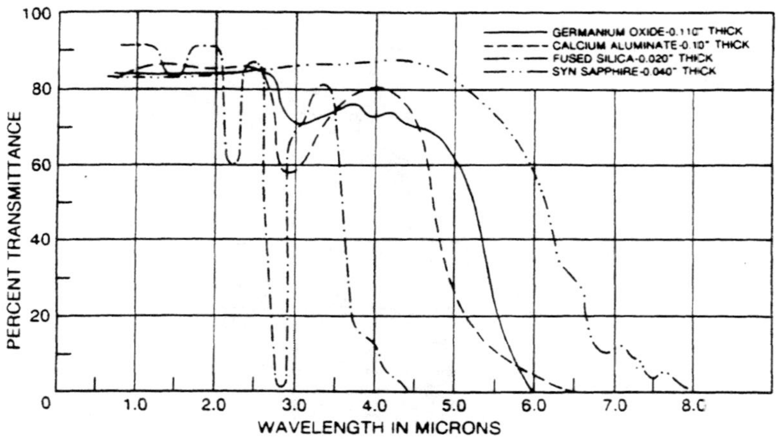

b. Various infrared window materials.

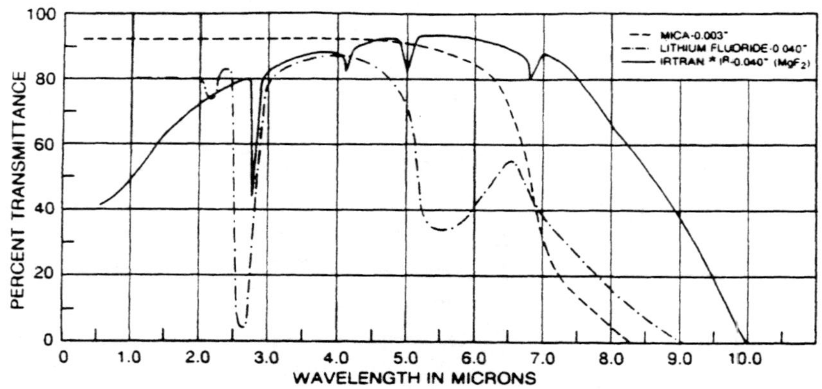

c. Various medium IR window materials.

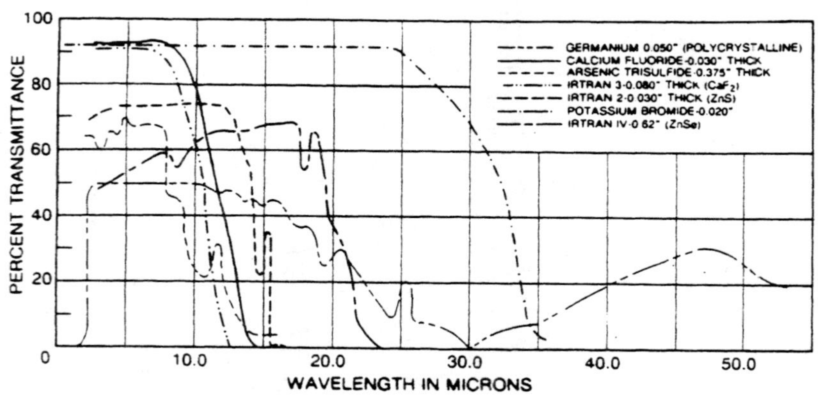

d. UV visible far-IR window materials.

Fig. 4 continued

---

* Irtran is a registered trademark of the Eastman Kodak Company.

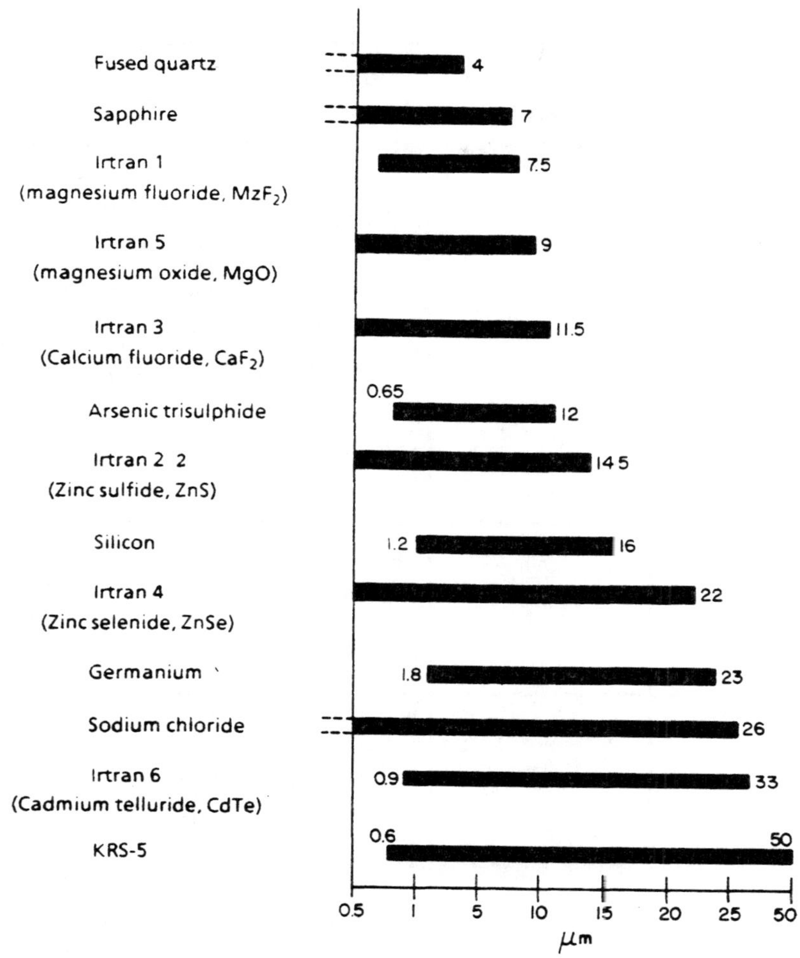

Fig. 5 Useful wavelength range of selected materials.

You should note the usefulness of certain IR optical materials— including sodium chloride, germanium, gallium arsenide, cadmium telluride, zinc selenide, Irtran 2, Irtran 4, and some of the alkali halides—for use as optical windows in CO_2 and other infrared lasers. Some of the properties of IR window materials are shown in Table 1.

Table 1: Characteristics of Selected Infrared Transmitting Materials

Material	Wavelength Range	Refractive Index (@ 2 μ)	Notes
Sodium chloride ($NaCl$)	0.25 to 16 μ	1.52	Useful because of range and low cost, stands thermal and mechanical shocks well; water-soluble; fogs slowly.
Potassium bromide (KBr)	0.25 to 25 μ	1.53	Useful with extended-range spectrophotometers; fogging does not seriously affect transmission; softer than NaCl; stands thermal and mechanical shock fairly well; water-soluble; fogs only a little faster than NaCl.
Calcium fluoride (CaF_2)	0.15 to 8 μ	1.42	Insoluble in water; resists most acids and alkalides; one of the strongest crystal materials and particularly useful in high-pressure cells, doesn't fog.
Barium fluoride (BaF_2)	0.13 to 13.5 μ	1.46	Insoluble in water; good resistance to fluorine and fluorides; must be handled with care because it is easily cleaved by thermal or mechanical shock; doesn't fog.
Cesium bromide ($CsBr$)	1 to 37 μ	1.67	Water-soluble; a light fog doesn't affect transmission seriously; not resistant to oxidizing agents; soft and easily deformed; fogs slowly.
Cesium iodide (CsI)	1 to 50 μ	1.74	Water-soluble; generally easier to handle than cesium bromide; fogs slowly.
Irtran - 2	2 to 14 μ	2.26	Especially useful for water solutions; attacked by strong oxidizing agents; stands thermal and mechanical shock well; high refractive index induces persistent fringes.
Polyethylene (high density)	16 to 300 μ	1.54	Needed for far-infrared; attacked by few solvents; difficult to clean thoroughly; very inexpensive.
KRS - 5	0.5 to 40 μ	2.2	Not hygroscopic, wide wavelength range, toxic. Do not grind or polish.
Silver chloride ($AgCl$)	0.4 to 28 μ	2.07	Soft material, will cold-flow, darkens under intensive UV radiation; insoluble in water.

Table 1: Characteristics of Selected Infrared Transmitting Materials

Material	Wavelength Range	Refractive Index (@ 2 μ)	Notes
Silver bromide (AgBr)	0.4 to 35 μ	2.31	Attacks base metals; soft; will cold-flow; insoluble in water.
Germanium (Ge)	1.8 to 23.0 μ	4.00	Low absorption coefficient @ 10.6 μ; takes good optical polish; can be coated with hard, durable dielectric films; good for CO_2 laser window or mirror at low power densities; relatively inexpensive.
Gallium arsenide (GaAs)	1.0 to 11.0 μ	3.12	Lower absorption coefficient than Ge @ 10.6 μ; much better thermal runaway characteristics than Ge; good for CO_2 laser windows and mirrors at high power densities ($\geq$ 150 W/cm^2); expensive
Zinc selenide (ZnSe)	0.5 to 22.0 μ	2.40	Transmission region facilitates inspection for internal flaws, allows system alignment with HeNe laser; lower absorption coefficient than GaAs; non-hydroscopic; good for high-power IR laser applications; expensive
Cadmium telluride (CdTe)	2.0 to 25.0 μ	2.80	Low absorption coefficient; high thermal runaway temperature; good for high-power IR laser windows and mirrors; expensive.

Low Reflection

If a ray of light of any polarization strikes normal (perpendicular) to an interface (junction) between two optical materials of dissimilar indices of refraction (n_1 and n_2), a fraction of the incident ray will be reflected according to the relationship:

$$\%\,Reflection = \left(\frac{n_1 - n_2}{n_1 + n_2}\right)^2 \times 100\% \qquad \text{Equation 3}$$

This type of reflection sometimes is called "Fresnel reflection." (See Figure 6.)

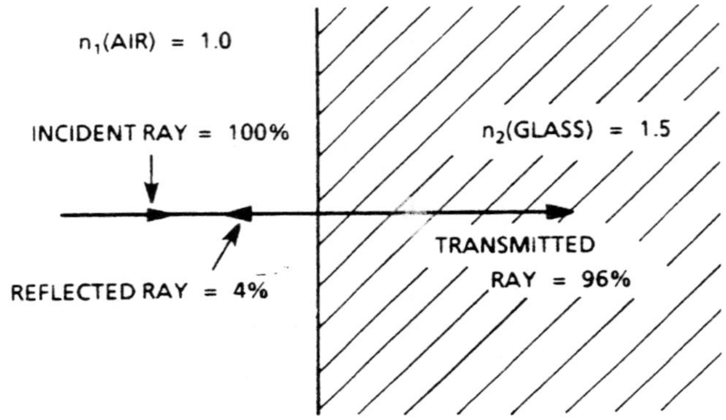

Fig. 6 Diagram of Fresnel reflection.

For parallel rays of visible light passing through a window, one medium probably will be air ($n_1 = 1.0$). The other probably will be some other glass ($n_2 \sim 1.5$). Using Equation 3, you can calculate a reflection of $\sim$ 4%. This calculation is for one side of the window. When the rays emerge from the other side, another 4% reflection loss is encountered. In summary, we realize that a common window pane (example, glass) with air on either side would reflect approximately 8% of the light incident normal to the surface. For example, you see a faint image of your face when you look out a window.

Optical theory shows that, if multiple reflections within a window are taken into account—as well as the first-surface reflection—the total theoretical external transmittance is given by:

$$T_{tot} = \frac{2n}{n^2 + 1} \qquad \text{Equation 4}$$

where:

$$n = \frac{n_2}{n_1}$$

So the total reflectance is:

$$R_{tot} = 1 - \frac{2n}{n^2 + 1} \qquad \text{Equation 5}$$

where:

$$n = \frac{n_2}{n_1}$$

In other words, $T_{total} + R_{total} = 1$. In Equations 4 and 5, zero absorption is assumed. Table 2 shows the total reflective loss theoretically expected from a variety of infrared window materials.

Table 2: Reflective Losses from Selected IR Materials

Material	Approximate Index (n) Over Major Region of Use	Percent Theoretical Reflection Losses From Two Surfaces
Kodak Irtran 1 (MfF_2)	1.35	4.3
Kodak Irtran 3 (CaF_2)	1.40	5.4
Barium fluoride	1.45	6.5
Quartz	1.45	6.5
Fused silica	1.46	6.7
Sodium chloride	1.50	7.7
Potassium bromide	1.52	8.2
Kodak Irtran 5 (MgO)	1.65	11.3
Sapphire	1.70	12.6
Spinel	1.70	12.6
Cesium iodide	1.70	12.6
Silver chloride	1.98	19.5
Strontium titanate	2.20	25.1
Kodak Irtran 2 (ZnS)	2.22	25.1
Cadmium sulphide	2.25	25.8
KRS-5	2.36	28.2
Arsenic trisulphide	2.40	29.0
Kodak Irtran 4 (ZnSe)	2.41	29.2
Chalcogenide glass	2.60	33.0
Kodak Irtran 6 (CdTe)	2.67	34.3
Gallium arsenide	3.25	43.8
Silicon	3.42	46.1
Germanium	4.00	52.9

As an example, use $n = 1.35$ for the refractive index of a Kodak Irtran 1 window. Using Equation 5, we get for the percent total reflective loss:

$$\% R = \left[1 - \frac{2(1.35)}{(1.35)^2 + 1}\right] \times 100\%$$

$$\% R = (1 - 0.9566) \times 100\%$$

$$\% R = 4.34\% .$$

These reflections can be substantially reduced by coating the surfaces with an antireflection layer or by positioning the windows at the Brewster angle (for TM polarized light). These two techniques are discussed in the following paragraphs.

1. Surface reflections can be reduced to less than 1% by applying vacuum-deposited antireflection coatings to each surface of the window. The coatings can be simply a thin film of a dielectric material (such as magnesium fluoride) selected because its index of refraction (n_3) is close to the square root of the product of the indices of refraction of the two materials (n_1 and n_2).

   In other words,

$$n_3 \approx \sqrt{n_1 \bullet n_2} \quad and \ t = \frac{\lambda_o}{4n_3} \qquad \text{Equation 6}$$

   where: λ_o = Wavelength of the light in a vacuum
   t = Thickness of the film (assuming normal incidence).

   You can determine by visual observation if a window is antireflection (AR) coated. If you look at the reflection from a light source at a low grazing angle to the window, the light will appear to be somewhat blue-violet. Antireflection coatings are usually very "hard." They can be cleaned with alcohol and lens tissue without danger of scratching or otherwise damaging the coating.

2. Brewster's angle window—In the previous discussion, the angle between the incident light rays and the window was considered to be 90° (normal incidence). When the window is positioned at a different angle, the percentage of reflected light changes.

   Alignment of the window to the incident ray or optic axis usually is expressed by the angle, ϕ. This angle is measured between the incident ray and a line normal to the plane of the window. (See Figure 7.) For normal incidence, the incidence angle is therefore $\phi = 0°$.

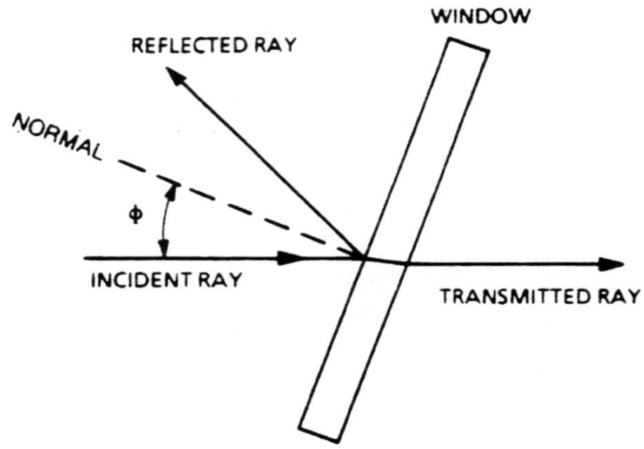

Fig. 7 Diagram of angle to incidence for a window.

A light beam (which is an electromagnetic wave) is said to be *linearly polarized* if the amplitude of the corresponding electric field (called the "E-vector") varies in only one plane along the optic axis. If a linearly polarized light beam is transmitted through an optical window, the percentage of the beam that's reflected at each surface varies with the incidence angle, ϕ, according to the curves shown in Figure 8. Note

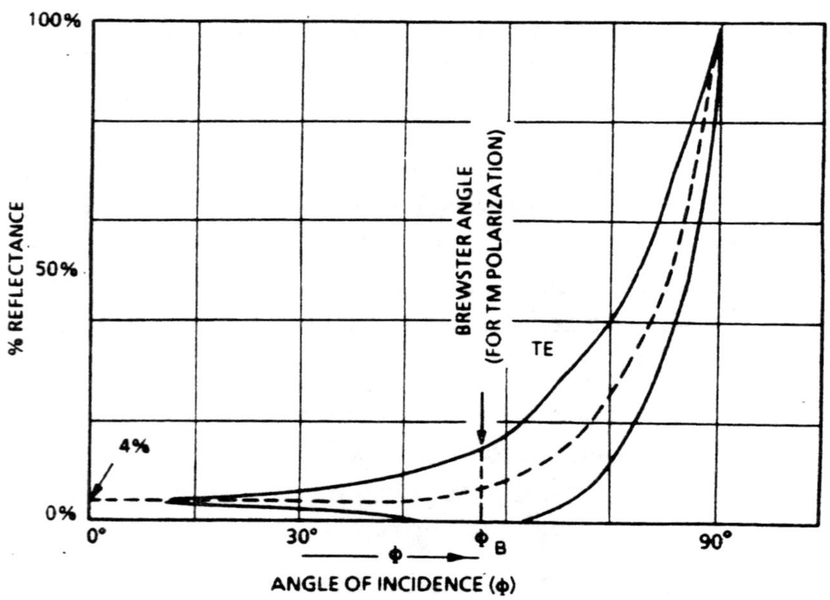

Fig. 8 Reflection of polarized light by an air/glass (n = 1.5) interface.

that there are two curves. One is designated TM (transverse magnetic) polarization, and one is designated TE (transverse electric) polarization. (See Figure 9.)

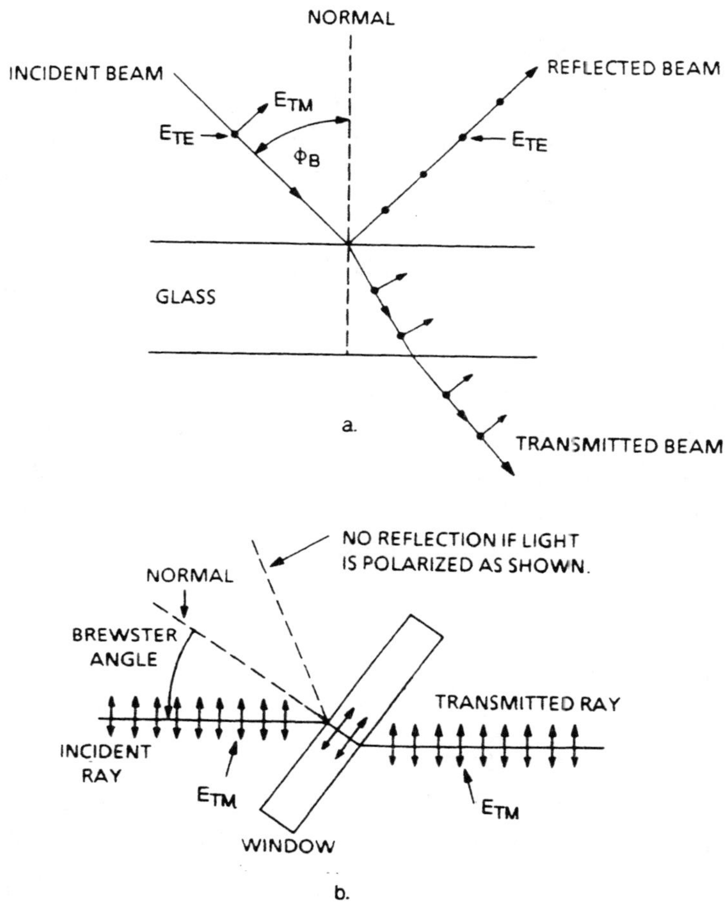

Fig. 9 Reflection and refraction of light at Brewster's angle.

TM polarization means that the "B-vector" lies perpendicular to the plane of incidence or "transverse" to it. Hence the name "transverse magnetic" polarization. The plane defined by the incident ray and the normal is called the plane of incidence. E_{TM}—the polarized E-vector is normal to the incident beam but in or "parallel" to the plane of incidence. TE—the "perpendicular" polarized E-vector is also normal to the incident ray, but it's perpendicular—or "transverse"—to the plane of incidence (shown as a black dot in Figure 9a). At an angle of incidence equal to Brewster's angle ϕ_B, the reflected ray contains no E_{TM} vector, only E_{TE}. If the incident beam were totally linearly polarized with E_{TM} only, at Brewster's angle of incidence, there would be no reflected light. All of the E_{TM} vibration would be refracted at the first surface.

TE polarization means that the E-vector is oriented perpendicular to the plane of incidence. Incident light has only E_{TM} polarization and is incident on the window at Brewster's angle. So no light is reflected from the first surface. All of the light of E_{TM} polarization is refracted. The transmitted beam is totally parallel polarized.

Keep in mind that the electric and magnetic fields in a light wave are always at right angles to each other. So, for example, in TM polarization the magnetic field is "transverse" to the plane of incidence, and the electric field must lie *in* the plane of incidence.

Note that the parallel polarization curve in Figure 7 approaches zero reflection for an incidence angle of approximately 57°. This is called "Brewster's angle."

Many laser outputs are linearly polarized. So Brewster-angle windows often are used to eliminate losses due to air-to-glass surface reflections. Brewster angle is not always 57°. It will vary according to the ratio of the index of refraction of the window material to the index of refraction of the surrounding medium according to

$$\phi_B = \tan^{-1}\left(\frac{n_{window}}{n_{medium}}\right) \qquad \text{Equation 7}$$

where: ϕ_B = Brewster angle
n_{window} = Refractive index of the window
n_{medium} = Refractive index of the surrounding medium

If the window is used in air, ϕ_B simply equals $\tan^{-1} n_{window}$.

Reducing Distortion Due to Refractive Effects

This section describes techniques and specifications on gross dimensions of optical windows to minimize transmitted wavefront distortion due to refractive effects caused by the window. Refractive effects from material inhomogenieties or finish will be discussed in the last section.

To prevent distortion of a transmitted wave, the two plane surfaces of the window must be reasonably flat and parallel. We can achieve flatness of 1/4 to 1/10 micrometer (μ) for a surface of several square inches without too much difficulty. This degree of flatness is probably enough to prevent the window from acting as a lens or from causing irregular bending of the beam due to refraction. Parallelism of the two surfaces to within 1-2 arc minutes is reasonable and probably adequate for several square inches of surface. If the surfaces are not parallel, the window will be a small wedge, bending the beam slightly. If you know the wedge angle accurately (and it can be measured with an autocollimator or laser), the bending can be compensated for. So it probably won't cause a problem. If the surfaces of the window are allowed to cause reflections, a wedge in the window will cause multiple beams from the multiple reflections between the two surfaces. (The intensity of the first reflected beam will be less than 0.2% of the intensity of the primary beam.)

You also should be aware of distortions induced in a window because of thermal gradients across the surface. Here's an example. Assume that a round glass window five inches in diameter and 0.5 inch thick is being heated from absorbed sunlight or a transmitted laser beam. The primary source of heat loss to the window will be through the edges of the window to the frame. This will make the window hotter in the center and cooler on the edges. The hotter area of the window will expand more, causing the center to be thicker than the edges. In a thick window like this, the expansion can cause a serious distortion in the window. The window will look and act like a positive lens, focusing the rays of the transmitted beam.

There's also a distortion that can be introduced by a change in refractive index with temperature. This distortion can cause similar problems.

Window Distortion Due to Imperfections in Optical Material or Surface Finish

The undesirable properties of optical glass are as follows:

1. *Striae*—a series of fine, hair-like lines visible because of sharp variations in the index of refraction. These imperfections are caused by improper mixing or insufficient annealing of the glass during manufacturing.
2. *Stones, seeds or inclusions*—tiny, solid, undissolved particles, not chemically united with the rest of the ingredients in a batch of glass. They are caused by an improper melting process or foreign particles contaminating the batch (from the handling equipment or the crucible). Point inclusions are common in glass.
3. *Bubbles*—air or gas pockets in the glass caused by improper stirring or insufficient annealing of the ingredients.
4. *Cloudiness or milkiness*—a lack of, or poor, transparency due to improper cooling (annealing) methods.
5. *Strain*—appears as light and dark streaks, not seen by the naked eye but revealed by polarized light. It is induced during manufacture by

improper annealing. Strain also can be caused by mechanical stress when an optical element is improperly mounted in an instrument or holder.

The defects in optical material surface finish are:

1. *Scratches*—any marking or tearing of the optical surface. Scratch types are identified as the following:
   - Block reek—chain-like scratch produced in polishing.
   - Runner cut—curved scratch caused by grinding.
   - Sleek—hairline scratch.
   - Crush or rub—surface scratch or a series of small scratches generally caused by mishandling.
2. *Digs*—small rough spots on the polished surface similar to pits in appearance, generally caused by mishandling.

All of the defects in optical materials described in this section, with the exception of "strain," generally will tend to scatter light from the transmitting beam. In laser beams, this type of scattering will cause an inhomogeneous field, diffraction and streaks—effects similar to those caused by dirty or dusty optics. They also often act as initiating sites for laser-induced damage. Strain will cause more gross distortion in the beam due to bending of the light rays.

It is beyond the scope of this module to try to train a technician to produce quality specifications for optical components. Moreover, it's even more difficult to describe inspection tests necessary to determine if rigid manufacturing standards have been followed. It would be a good idea for the student to acquire and study two commonly used military specifications:

- MIL-0-13830A – General Specifications Covering the Manufacture, Assembly and Inspection of Optical Components.
- MIL-G-174 – Optical Glass.

Mechanical Distortion of Windows*

A pressure differential applied to an optical window may cause distortion that is unacceptable in many applications. In extreme cases, the stress even could cause catastrophic failure. The maximum stress in a uniformly loaded window is given by,

* This discussion is based on an analysis presented in Catalog OP 101, Harshaw/Filtrone, Solon, Ohio, (216) 248-7400.

$$S_{max} = \frac{KD^2 p}{4t^2} \qquad \text{Equation 8}$$

where:

S_{max}	=	Maximum stress, psi
D	=	Diameter of support
t	=	Thickness (same units as D)
K	=	A constant
p	=	Load per unit area, psi

To avoid plastic deformation of the window, which in turn could cause significant optical distortion, we must keep the maximum load on the window below a threshold number called the "flexural apparent elastic limit" F_a. This limit is defined as the load point on the load versus deformation curve at which the material will deform at a rate 1.50 times the initial rate. So we can write

$$S_{max} < F_a \qquad \text{Equation 9}$$

For additional assurance that our window will not significantly deflect under an applied pressure differential, we can introduce an additional safety factor, SF, such that,

$$S_{max} = \frac{F_a}{SF} \qquad \text{Equation 10}$$

Substituting into Equation 8 gives us a design equation based on the dimensionless ratio of window thickness to diameter,

$$\frac{t}{D} = \left(\frac{SF \times K}{4}\right)^{1/2} \left(\frac{p}{F_a}\right)^{1/2} \qquad \text{Equation 11}$$

The value of the constant K in Equation 11 depends on the method of supporting the window. Two popular methods of support are shown in Figure 10. In Figure 10a, no motion of the edge of the window is allowed. If the window is supported by rubber O-rings or other flexible seals, the value of K in method 10b should be used.

Typical values of the elastic limit F_a are given in Table 3. A summary of the equations for the deformation of a circular window and a rectangular window is given in Table 4.

CLAMPED, OR FIXED EDGE
(S_{max} AT EDGE)

UNCLAMPED OR FREELY SUPPORTED EDGE
(S_{max} AT CENTER)

a. $K_c = 0.75$

b. $K_u = 1.50\, K_c = 1.125$

Fig. 10 Two methods of supporting a window.

Table 3: Elastic Limit F_a for Some Common Window Materials

Material	F_a, psi
CaF_2 (1)	5300
Polytran® CaF_2 (5)	8000
BaF_2 (1)	3900
KRS-5 (2)	3800
LiF (3)	1580
NaCl (6)	350
Polytran® NaCl (5)	1400
KCl (6)	330
Polytran® KCl (5)	1600
KBr (6)	160
CsI (6)	810
MgF_2 (5)	7200

These design equations are for circular and rectangular windows that are either clamped or free at the edge. The equations assume a safety factor (SF) of 4.00.

Table 4: Design Equations for Circular and Rectangular Windows

	Clamped	Unclamped
Safety Factor	4	4
K	$K_c = 0.75$	$K_u = 3/2K_c = 1.125$
Circular	$t/D = 0.866\left(\frac{p}{F_a}\right)^{\frac{1}{2}}$	$t/D = 1.06\left(\frac{p}{F_a}\right)^{\frac{1}{2}}$
Rectangular	$t = 1.23b\left[\frac{p}{F_a(1+c^2)}\right]^{\frac{1}{2}}$	$t = 1.5\,b\left[\frac{p}{F_a(1+c^2)}\right]^{\frac{1}{2}}$

where:
- t = Thickness
- p = Load per unit area, psi
- F_a = Apparent elastic limit, psi
- D = Unsupported diameter, for circular window
- b, a = Length and width of rectangular window
- c = b/a

Example: Window Thickness to Prevent Optical Distortion

Given: You are constructing a large test chamber for laser-induced fusion. The pressure inside the chamber is essentially zero. Pressure outside is one atmosphere (14.7 psi). The windows will have a diameter of 12.0 inches and will be made from NaCl, held with rubber O-rings at the edge.

Find: Minimum thickness window that will prevent significant optical distortion.

Solution: *Assuming a safety factor of 4.0 and a free (unclamped) window,*

$$t = D\left(1.06\right)\left(\frac{D}{F_a}\right)^{\frac{1}{2}}$$

$$t = \left(12.0"\right)\left(1.06\right)\left(\frac{14.7\ psi}{350\ psi}\right)^{\frac{1}{2}}$$

$$t = 2.60" .$$

Note: *In these equations, the applied pressure P is the* pressure difference *between the two sides of the window.*

> *For example, if one side of the window were at 1 atmosphere and the other at 5 atmospheres, P would equal the difference, which is 4 atmospheres.*

Maintenance and Durability of Optical Windows

To select a material for an optical window and to determine its proper maintenance, you must consider the environmental conditions to which it may be subjected:

1. *Thermal*—Extreme heat or cold, or rapid changes in temperature, can cause temporary or permanent changes in the physical characteristics of the window. The critical material characteristics are melting temperature, coefficient of thermal expansion, and variation in thermal expansion along two orthogonal coordinates in the material.
2. *Surface abrasion*—The surface of the window can be altered by external abrasions in the form of scratches from other materials (even cleaning cloths), or from particles in the air (sand, grit, dust). The critical material characteristic in determining a window's vulnerability to surface abrasion is *hardness*.
3. *Chemical damage*—The damage to window surfaces from a chemical stems primarily from the chemical's ability to partially dissolve the window material. Examples are acetone on plastic windows, and water on infrared "salt" windows.
4. *Laser damage*—Very high-power lasers can damage windows either at the surface or internally. The subject of laser damage to optical components is beyond the scope of this module. It's enough here for you to know that this damage mechanism exists and to seek additional information on the subject if the possibility of the condition exists.
5. *Contamination*—The surface of a window can become contaminated from its exposure to polluted atmospheres. This contamination can be in the form of dust, grease, chemical films, or moisture. You usually detect it by visual observations. Proper operation of an optical system depends on the technician recognizing conditions of contamination and performing periodic maintenance in the form of surface cleaning.

Flats

An *optical flat* is a window of high-grade quartz glass with one or both surfaces flat to at least one tenth of the wavelength of light. It's generally used in one of two ways.

Due to their flatness, flats make excellent substrate material on which to put a desired coating. They can be used as the substrate material for first-surface reflectors, dielectric-coated beam splitters, or for coated windows.

A more common use for an optical flat, though, is to test the flatness of other optical components. The flat is placed on the surface to be tested, then illuminated with monochromatic light. Interference banks appear, produced by variation in the optical path length of light going through different thicknesses of glass. The curvature of the interference bands with respect to the distance of the bands indicates the amount of flatness error. Study the interference patterns shown in Figure 15. Straight, parallel, and evenly spaced bands indicate that the work surface is accurately flat with respect to the master flat.

Flats are versatile devices for measuring surface flatness of polished areas such as the surfaces of prisms, filters, or optical windows.

EXERCISES

1. Describe Fresnel reflection. State the conditions under which it occurs. Use a sketch to show the incident and reflected rays, angles of incidence and refraction, and the reflection surface. State the approximate values (in %) of the ratios of reflected to incident, and transmitted to incident light, for Fresnel reflection of a material with a refractive index of 1.5.
2. Explain Brewster's angle and illustrate it with a sketch. State its relationship to the refractive index of the reflecting material and its effect on the polarization of the reflected portion of the incident light. Include in the sketch the reflecting interface, the incident and reflected rays, Brewster's angle, and the polarization of the reflected ray.
3. List five defects found within optical materials referred to in the discussion.
4. List two defects found on optical component surfaces.
5. List five important characteristics of an optical window.

6. The following is an interference pattern obtained by comparing a sheet of glass to a master flat. What is your interpretation of the flatness of the glass tested?

7. Describe the visual difference between a lens with an antireflective coating and a lens without a coating.

8. A flat glass plate has an index of refraction of 1.52. If light is to strike the plate at normal incidence, what percent of the light will transmit through the plate?

9. Consider the same plate as in Exercise 8. Assume now that we will apply a single-layer AR coating to one surface. Ideally, what will be the index of refraction of the coating? If we're going to illuminate the plate with a HeNe laser, how thick should the single-layer AR coating be?

10. Find Brewster's angle for TM polarized light striking an interface from air into water. ($n_{water} = 1.33$.)

11. A 10.0-cm-diameter window, fabricated from LiF, is designed to withstand a 15.0 atmosphere pressure differential without suffering significant optical distortion. Assume that the window is clamped around its circumference and that we will allow a safety factor of 4.0. How thick should we make the window?

12. Repeat Exercise 11, but this time allow a safety factor of 2.50 and allow free movement of the window's edges.

13. A rectangular window, 4.00 x 5.00 cm on a side, is fabricated from Polytran KCl. The window is held with flexible rubber O-rings along its perimeter, and has a uniform thickness of 6.00 mm. Assuming a safety factor of 2.00, what pressure differential can be applied to the window before it begins to suffer significant optical distortion? Express your answer in atmospheres of pressure.

MATERIALS

Optical windows
- Borosilicate crown glass
- Quartz (fused)
- Sapphire
- Plexiglass or similar plastic
- Ordinary glass (window pane, microscope slide, etc.)
- Irtran
- Germanium

HeNe laser, 1 mW
Optical flat, $\lambda/10$ or better
Mercury vapor lamp, high intensity
Green filter (gelatine or glass)
Diffuser (ground glass, translucent paper, etc.)
Shop microscope
Diverging lens, focal length –2 to – 4 cm
Ordinary plate glass, 1/8" to 1/4" thick, about 3" square
Supports and mounts for lens, windows, filter and glass plates
Lens, antireflection coated, any size or focal length
Lens, without coating, any size or focal length

PROCEDURES

Surface Condition of Windows

Visually examine the windows for cleanliness, surface condition, and the presence of antireflection coating.

1. Clean all surfaces that are contaminated. Use high-purity ethanol, methanol or Kodak lens cleaning fluid, with clean lens tissue. Set cleaned windows on a fresh piece of lens tissue. Cover them with another piece to avoid further contamination.

2. Set up the HeNe laser, with a diverging lens in the beam. Examine each surface using this source of illumination.

   Use Extreme Care. Avoid directing the reflected beam into your own or others' eyes.

   Rotate the test piece as shown, to use various angles of incidence. The intense beam will readily reveal dirt and scratches, by scattering light. More intensity is available in the narrower part of the beam. Moving the

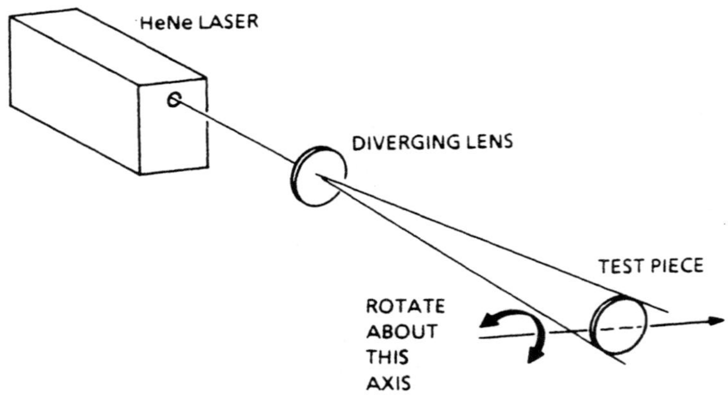

Fig. 11 Inspecting surface with divergent beam.

test piece closer to the diverging lens provides this, and makes the test more revealing. As a final check, remove the diverging lens and inspect the piece with the full intensity of the narrow beam. Hold the piece nearly perpendicular to the beam path. Again, be careful of the direction of the reflected beam!

3. Record your observations of each surface. Note scratches, hazy appearance, and any other defects. Identify the piece with the best surface polish.
4. Use the microscope to examine one or two of the worst surfaces. Use illumination at a large angle of incidence, as shown in Figure 12.

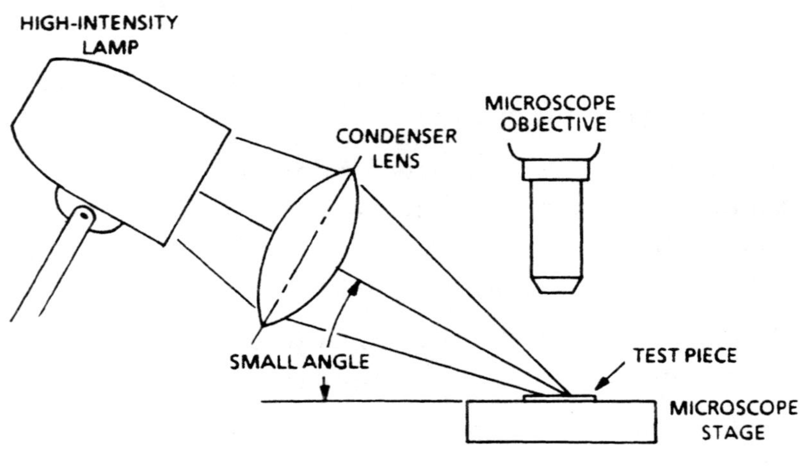

Fig. 12 Microscopic examination of surface.

Internal Defects in Window Materials

Visually examine the effect on an expanded HeNe beam that has passed through the window.

1. Mount window and diverging lens to project a spot several inches in diameter onto a sheet of white paper, as shown in Figure 13.

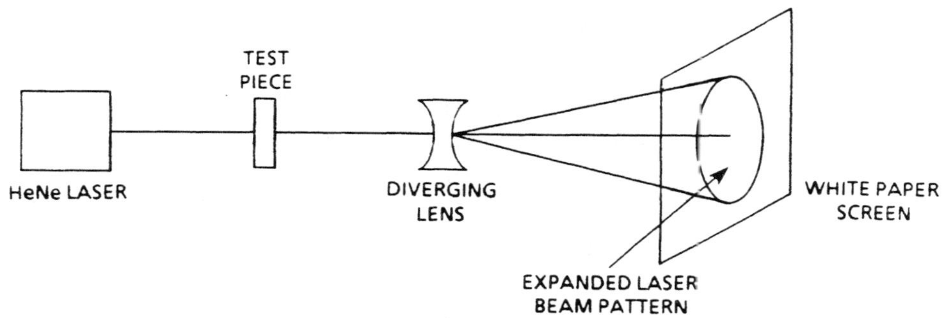

Fig. 13 Testing the quality of a window by examining the expanded transmitted beam.

2. Carefully examine the projected spot for any local variations in brightness. To confirm that any variations are not due to either the laser itself or the diverging lens:
   a. Remove the window and compare results.
   b. Rotate the lens, while observing spot variations. Those caused by the lens will follow lens movement.

Interference Test for Optical Flats

Evaluate the flatness of the surfaces by comparison with a master flat of known accuracy. Use the viewing arrangement shown in Figure 14 to display interference fringes.

1. Very gently lay a test piece onto the master flat. Use care not to slide it across the flat.
2. While you observe the pattern as shown above, gently slide the window around over the flat until 5 to 10 fringes are visible. A clean toothpick, tongue depressor, or a pointed kitchen match makes a good tool to manipulate the test piece. This will allow most of the air between test piece and flat to escape, broadening and reducing the number of fringes. (If this doesn't happen, carefully remove the piece and reclean both it and the flat.)

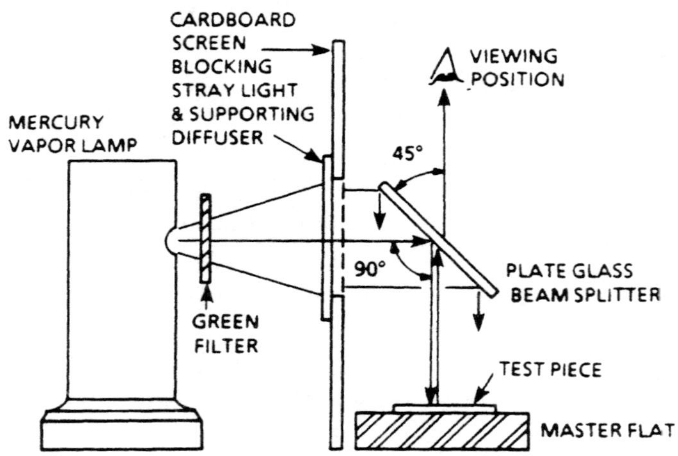

Fig. 14 Viewing arrangement for interference test.

3. When only two or three fringes are visible, compare one of them to a straightedge held over the center of each end of one fringe. See Figure 15.

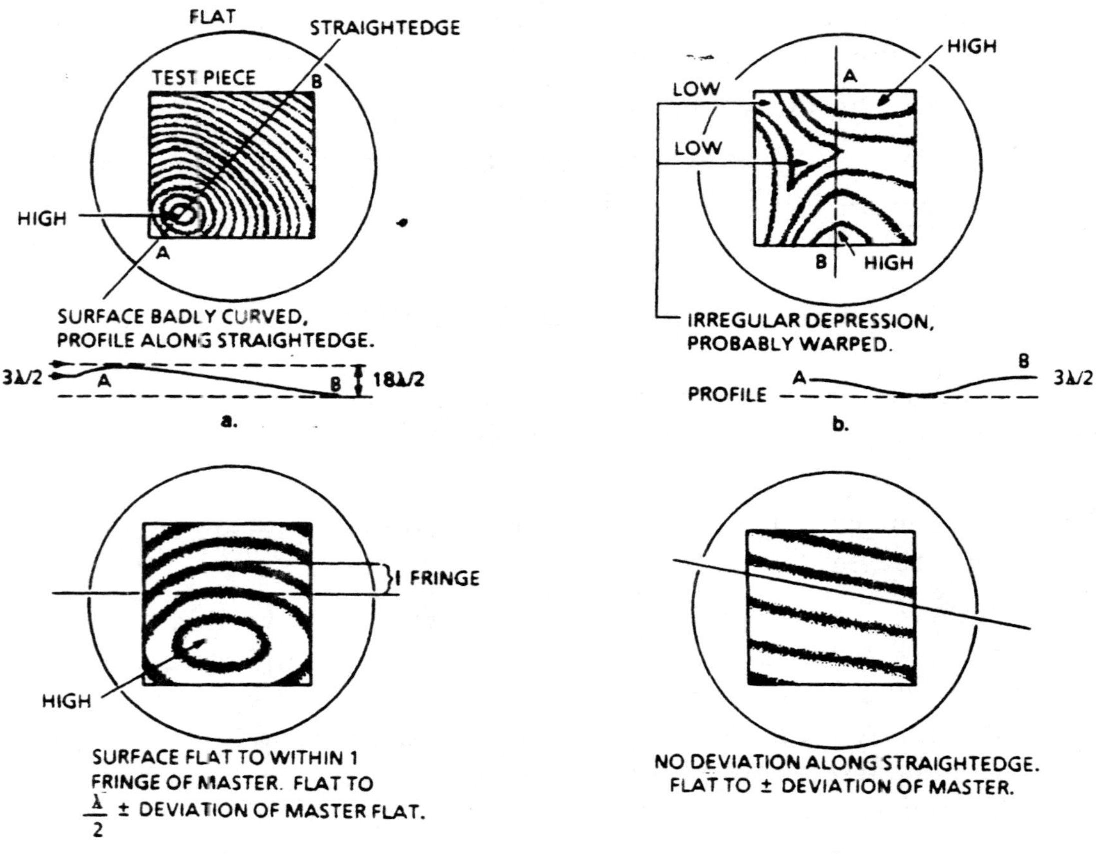

Fig. 15 Examples of fringe patterns in interference test.

Since two successive fringes mean that a variation of the optical thickness between the surfaces of the master flat and test piece equal to a half wavelength has occurred, the curvature of the fringe shown in Figure 15c indicates that the test piece varies by approximately $\lambda/2$ from whatever flatness the master plate possesses. To determine if it's concave or convex with respect to the flat, simply lower your eye toward the piece. If convex, the fringes will appear to spread out. If concave, they will move closer together.

4. Test and record the deviation of both surfaces of each window.

Identification of Antireflection Coating

Examine the two lenses and determine which has the antireflection coating.

REFERENCES

Catalog OP 101, Crystal Optics, Harshaw/Filtron, Solon, Ohio, 44139, (216) 248-7400.

Harshaw Optical Crystals, Harshaw Chemical Co., Crystal & Electronics Products Department, Cleveland, Ohio.

Kodak Irtran Infrared Optical Materials. Eastman Kodak Publication U-72, 1971.

Kruse, P. W., and L. D. McQuistan. Elements of Infrared Technology. Chapter 4, John Wiley, 1962.

Mauro, J. A. Optical Engineering Handbook, 1966, General Electric Company.

MIL-G-174 Glass, Optical,US Government Printing Office.

MIL-O-13830A, Optical Components for Fire Control Instruments, General Specification Governing the Manufacture, Assembly and Inspection of, 11 September 1963, US Government Printing Office.

SPIE, Basic Optics and Optical Instruments, Vol. 1, 1965, Nortronics Division of the Northrop Corporation.

Walker, Bruce. "Anti-Reflection Coatings," Optical Spectra, October 1972.

Waugh, John. "Component Testing by Interferometry," Optical Spectra, June 1973.

Wolfe, William L., editor. Handbook of Military Infrared Technology. Office of Naval Research, 1965. Department of Documents, US Government Printing Office, Washington, DC.

Zanoni, Carl A. "Specifying and Measuring Optical Glass Homogeneity," Electro-Optical Systems Design, February 1973.

LASER/
ELECTRO-OPTICS

MODULE 6-5
Mirrors and Etalons

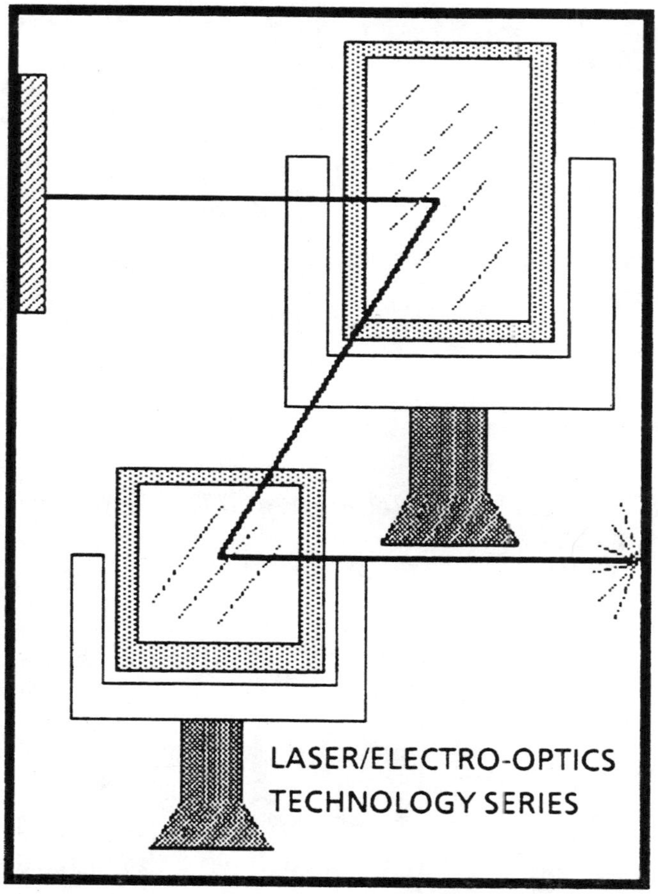

ORD

The Center for Occupational Research and Development
601 C Lake Air Drive
Waco, Texas 76710

Printed in the U.S.A.

ISBN 1-55502-024-0

MIRRORS AND ETALONS

MODULE 6-5

A mirror is a smooth surface that can reflect enough light to serve a useful purpose. Mirrors have been used since man first saw his image in a calm pool of water. They have progressed from the great silver-blackened hall mirrors of the eighteenth century to where, today, they are used to reflect light within a laser cavity.

This module will instruct you about the recognition and specifications of various mirror types, their limitations and their applications.

Before you begin this module, you should have studied optical cavities and modes of oscillation, light rays, reflection of light, reflection of light at a plane boundary, and reflection of light at a spherical surface.

OBJECTIVES

When you complete this module, you should be able to do the following:

1. Write a statement in your own words comparing the advantages and disadvantages of front- and rear-surface mirrors.
2. Explain and give one application for each of the following:
   a. Dielectric-coated metal mirror
   b. Hot mirror
   c. Cold mirror
   d. Etalon
   e. Autoreflection mirror
3. Using three manufacturers' catalogs, outline a comparison of the quality and price of mirrors that meet given specifications.

4. Construct a retroreflector that gives performance satisfactory to your instructor.
5. With the aid of an appropriate diagram, explain how "ghost images" are formed and how they can be eliminated.

DISCUSSION

When light strikes the surface of any object, some of the light is reflected, some is absorbed by the object, and some is transmitted through the object. The extent to which each of these processes occurs varies. Some objects transmit the major portion of the incident light, some objects absorb most of the incident light, and other objects reflect most of the incident light. Objects designed to reflect most of the light are called "mirrors."

The amount of light reflected by a mirror depends on:

1. The nature of the reflecting surface (composition, energy level, structure, density, color, and so on).
2. The texture of the reflecting surface (smooth or rough, regular or irregular, dull or polished, etc).
3. The wavelength and polarization of the light.
4. The angle at which the light strikes the surface.

We will discuss some of these factors in more detail in this module.

The reflection from the surface of a high-quality mirror is called "specular." This means that the rays striking the surface are reflected from the surface according to the law of reflection, $\theta_i = \theta_r$. Figure 1 shows specular reflection from a polished mirror surface.

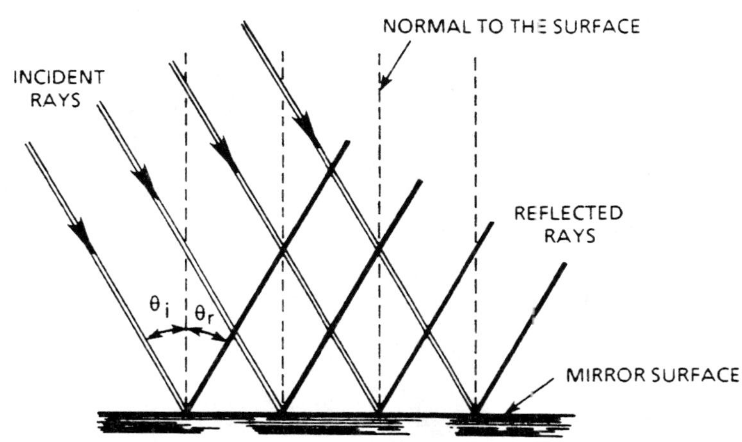

Fig. 1 Specular reflection.

Not all reflectors, however, are specular reflectors. Some are diffuse reflectors, which means that the rays are reflected in random directions. Figure 2 shows diffuse reflection. Common examples of diffuse reflection include the reflections from this paper and from projection screens.

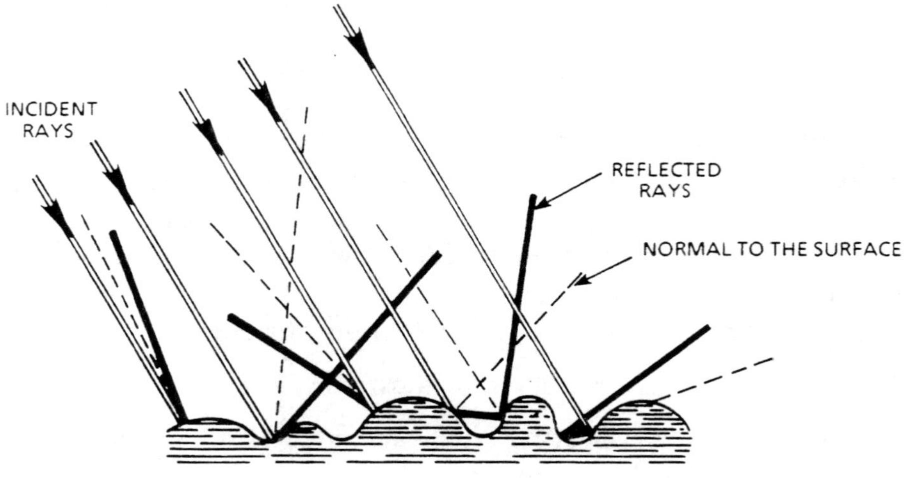

Fig. 2 Diffuse reflection.

Figure 3 illustrates a plane mirror used to form a virtual image by reflection. The virtual image formed by a plane mirror corresponds in size and shape to the object and is the same distance behind the mirror as the object is in front of the mirror. The image is correctly oriented from top to bottom, but the right and left parts are interchanged.

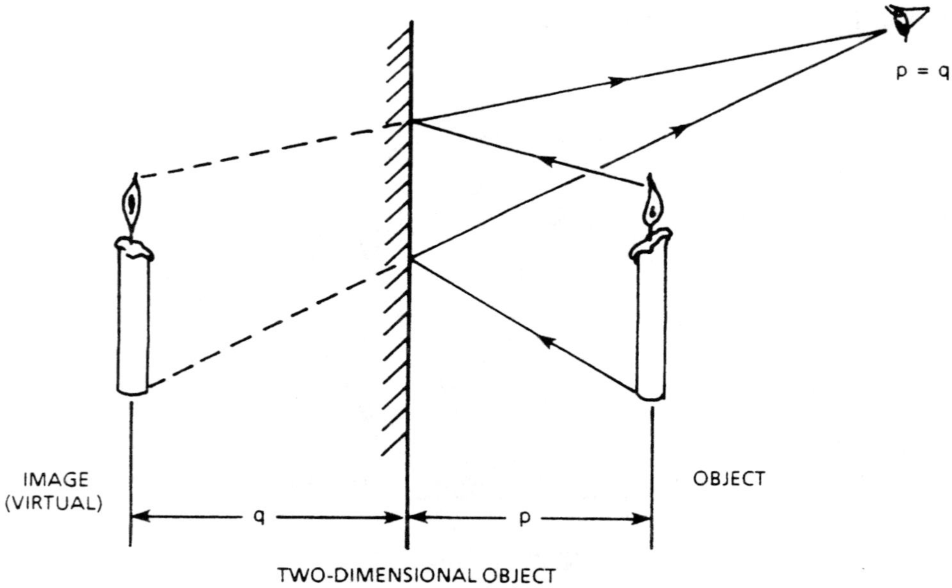

Fig. 3 Image formed by reflection.

To illustrate this point further, let's consider Figure 4. If the word READ is considered to be right-handed and is readable, then the mirror image of the word must be considered to be left-handed. It has been turned left for right. No matter how the word is turned, it will remain left-handed. To read it correctly, you must either use a second mirror to turn it or turn the paper over and look at it from the other side. The image from a single surface reflection is always left-to-right inverted.

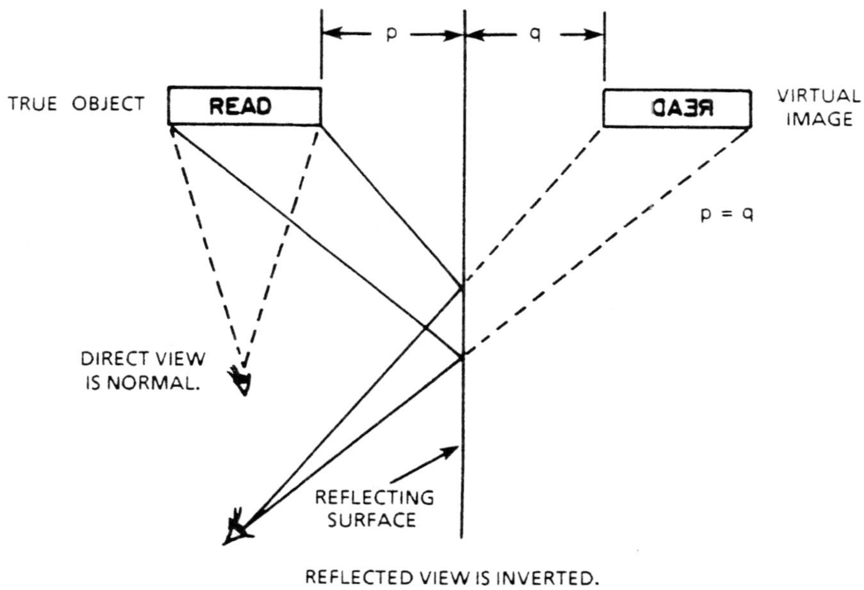

Fig. 4 Inversion of the image formed by a single surface reflection.

As you learned from an earlier module, curved mirrors form either real or virtual images by converging or diverging light. Inexpensive, curved, diverging mirrors can be used to enlarge a field of view. Such an application is the small "spot" mirror attached to the rear view mirror of an automobile. Low-quality converging mirrors often are used as shaving mirrors to enlarge an image.

The quality, size and reflectivity of mirrors largely determine their use. For instance, the searchlight shown in Figure 5 does not require a high-quality mirror since a focused image is not being formed by the system.

The reflecting telescopes shown in Figure 6 require mirrors of very high optical quality to produce usable images. The reflectivity of the telescope mirrors doesn't have to be extremely high since a reasonable amount of light absorption in the system can be tolerated.

Mirrors used for the end reflectors in laser cavities must be of high optical quality and have high reflectivity. Even small absorption or scattering losses in this application may render the system inoperable.

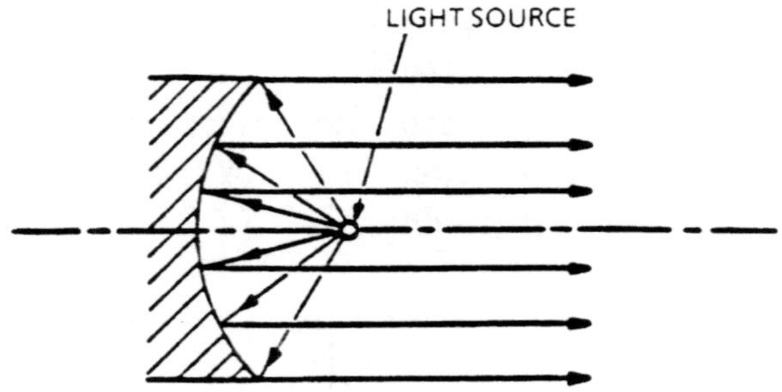

Fig. 5 Searchlight as an application of concave mirrors.

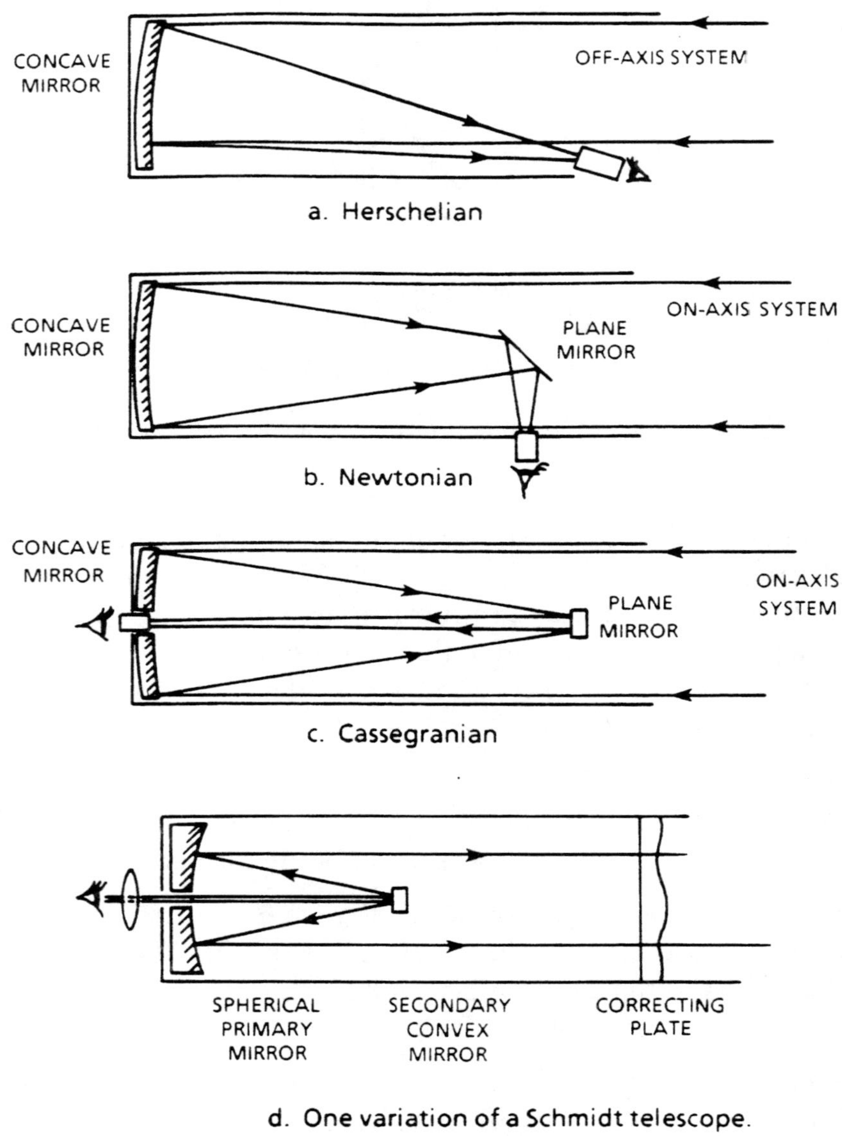

Fig. 6 Types of reflecting telescopes.

Mirror Limitations

Mirrors used as imaging systems suffer from many of the same defects as lenses. These defects include spherical aberration, coma, astigmatism, curvature of field, and distortion. You can find a more complete description of these defects in a lesson on lenses.

Rays of light striking the outer portions of a spherical reflecting surface don't focus at the same point as rays striking the center portion of the surface. This is known as spherical aberration. It will cause an out-of-focus image. Short-focal-length mirrors are plagued by spherical aberration more than are longer-focal-length mirrors. This is because the short-focal-length mirror has a much sharper curvature.

Spherical aberration can be controlled or eliminated in several ways. If the aperture of the mirror is limited, the outside rays of light will not strike the reflecting surface. This prevents them from causing the out-of-focus image. This is normally the least expensive way to reduce spherical aberration.

A more costly method is also the more perfect method. It consists of using a parabolic shape for the reflecting surface instead of the spherical shape. To understand the reason for using a parabolic shaped reflecting surface, refer to a good analytic geometry textbook.

Note that, when it's used off-axis a spherical mirror may produce a better image than a comparable parabolic mirror. Parabolic mirrors usually are used in optical systems, such as telescopes, where the incident rays are nearly parallel to the mirror's axis of symmetry.

A third way to minimize spherical aberration is a technique developed by Bernhardt Schmidt, the so-called "Schmidt telescope" or "Schmidt camera." In this optical system, a thin transparent "correcting plate" is placed at the center of curvature of the spherical mirror. (See Figure 6d.) The plate is essentially a specially shaped lens that corrects the spherical aberration, astigmatism and coma of the telescope's primary mirror. Schmidt telescopes are popular for applications such as small astronomical instruments where it's sometimes more cost-effective to fabricate a spherical mirror plus a correcting plate rather than a parabolic primary mirror.

Astigmatism is a tendency for a mirror to focus light entering the surface vertical, with respect to a particular axis, at a different point from light entering horizontal with respect to the same axis. This defect causes images of a point to be extended into a line. All optical elements display some degree of astigmatism although it can be greatly reduced by appropriate combinations of lens and/or mirror elements.

Because the quality of a mirror depends on the rigidity of the substrate material, it's usually made of some stiff material like glass. In less stringent applications plastic may suffice and in more severe applications, materials more rigid than glass may be necessary. The substrate material must be thick enough to minimize distortion of the mirror surface. It is also advisable to mount the mirror on three support points only, to avoid stress-induced distortion.

When glass is used as the mirror substrate, reflection from an uncoated front surface generally will produce a "ghost image." The "ghost image" is slightly displaced from the main image that arises from the highly reflective rear surface. For this reason most high-quality mirrors are made with front or first reflecting surfaces on top of the glass substrate.

Many optical systems can operate very well even with low-quality mirrors. Other systems require one or all the defects to be minimized. A mirror of too low quality may hinder the operation of your equipment or decrease its accuracy. Too good a mirror may represent a needless expense. As discussed later, mirror quality must be related to its application.

Reflecting Surfaces

To obtain high specular reflectance, the polished glass or plastic surface can be coated with a metallic layer. Silver was the classical material used for this purpose. A highly polished silver surface reflects ~ 95% of the incident light in the visible range. Silver, however, has to be protected from tarnishing. So it has been largely replaced by aluminum evaporated onto the polished substrate in a high vacuum. On contact with the atmosphere, a thin layer of aluminum oxide forms on the surface of the aluminum film and provides some protection. Other protective coatings or "overcoats" often are needed. These will be mentioned later.

The reflectance of metals depends on the wavelength, the surface preparation, the angle of incidence, and specific properties of the metal itself. Figure 7 shows the variation of the reflectance of a number of typical metals as a function of visible wavelength.

Silver and aluminum are of particular importance for general use because they maintain their high reflectance throughout the visible spectrum. It's now standard practice to coat the mirrors of large telescopes, such as the 200-inch Hale reflecting instrument at Mt. Palomar, with evaporated aluminum. For reflecting surfaces where the small advantage of silver over

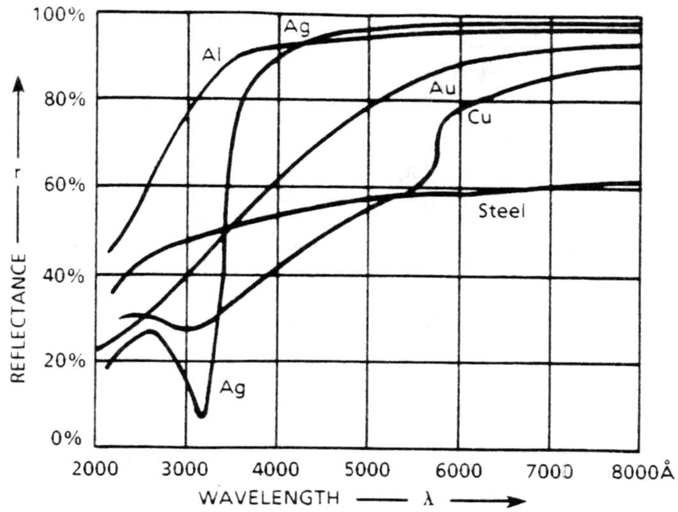

Fig. 7 Reflectances at normal incidence of aluminum, silver, gold, copper and steel.

aluminum is important, as in Fabry-Perot etalons, silver is preferred. For use with visible and infrared light, silver is also preferred. But for ultraviolet light aluminum is better, and gold may be preferred at longer wavelengths.

As was mentioned earlier, some metals (aluminum) form their own protective layer. This layer is very thin, however, and it is usually advisable to provide additional protective coatings. Silicon monoxide frequently is used for this purpose. It causes some reduction in reflectance, but this reduction is less than is experienced by aging processes. Magnesium fluoride also can be used as a protective coating. The effect of these two coatings on the reflectance of aluminum is shown in Figure 8.

You should note from Figure 8 that different protective coatings affect the reflectivity differently as a function of wavelength. For example, if you needed a mirror for reflecting light of 2-micrometers wavelength, silicon monoxide would be best. If 10 micrometers were the area of interest, magnesium fluoride would be the better choice.

In some applications that involve multiple reflections—such as a laser cavity—even the high reflectance of the metals shown in Figure 7 may not be enough. Surfaces that have a higher reflectance than any metal can be produced. These surfaces generally are composed of several dielectric materials and use the principle of interference to produce high reflectance values.

The dielectric or multilayer film mirrors sometimes are combined with a metal mirror to improve the reflectance. Figure 9 shows the reflection enhancement of a metal mirror due to multilayer coatings. The dielectric

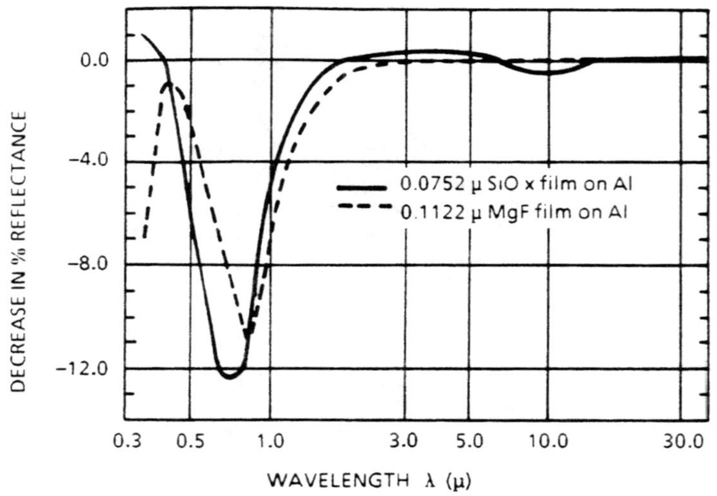

Fig. 8 Spectral reflectance of coated aluminum mirrors.

materials in Figure 9 are zinc sulfide and cryolite, both vacuum deposited on a metal mirror to a highly controlled thickness.

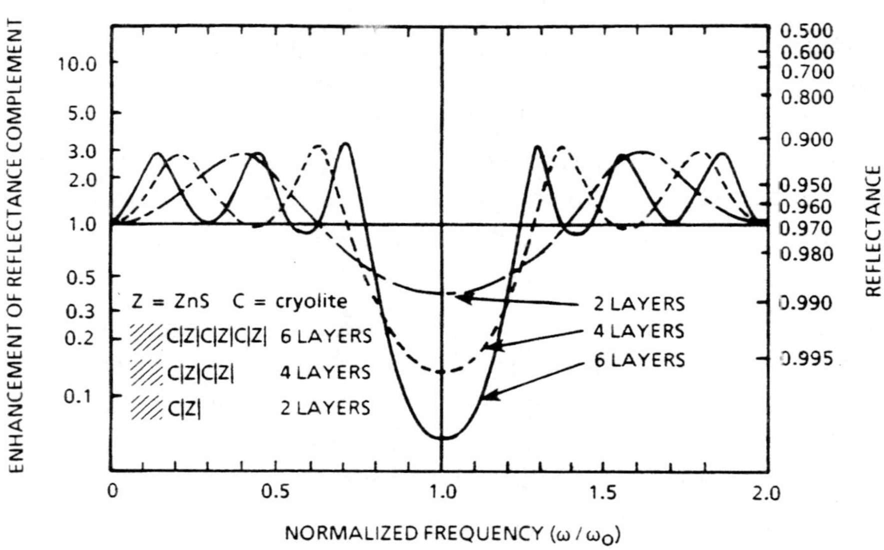

Fig. 9 Enhancement of spectral reflectance at a specific wavelength due to multilayer coatings.

Notice that reflectances very near to 100% can be achieved by this method. The high reflectance does not, however, extend over as great a wavelength range as does that of bare metals. The reflectance of a dielectric or dielectric coated metal is peaked very strongly around a particular wavelength of light. The reason for this should become apparent in the following discussion.

Consider Figure 10. The incident light is divided into two different beams by a partially reflecting surface. Each of these beams travels a different distance and interference is observed when these two rays recombine. As shown in Figure 10 the light from the source impinges upon surface S_1. Part of this beam is reflected at a point A. The remainder penetrates the medium between the two surfaces S_1 and S_2 and reflects back and forth between them. The thickness of the dielectric is h_1, and θ_1 is the angle of incidence in the medium of index of refraction n_1.

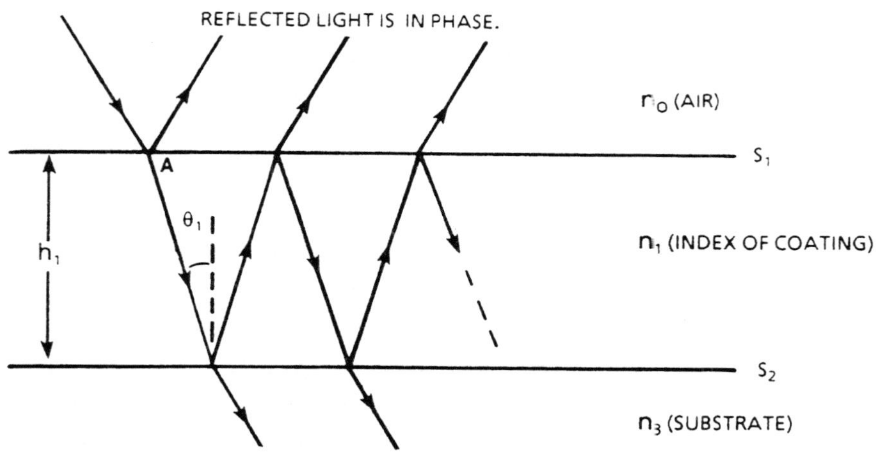

Fig. 10 Reflection from a single thin-film coating layer.

If the reflectance of the surfaces is low, the beam may reflect only once before it's greatly attenuated. If the reflectance is close to unity, it may reflect as many as 30 times. In either case, the beams that reflect from either side of the film have traveled different optical paths and interference is observed when they recombine.

If several dielectric films—each a quarter wave thick and with alternately high and low index of refraction—are stacked, the reflected beams from all the interfaces are in phase upon leaving the uppermost boundary. This situation is shown in Figure 11.

Note that light may reflect many times before returning to the upper surface. But when it does this, it's in phase and going the same direction as all the other reflected light rays. Modern coating technology produces reflectances in excess of 99.999% using approximately 20 layers in a dielectric "stack."

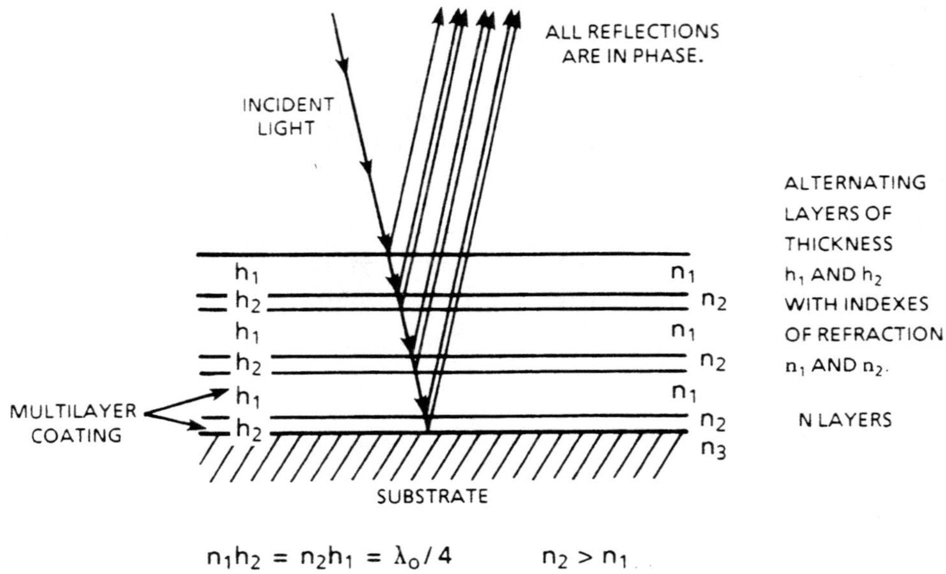

Fig. 11 Operation of a multilayer dielectric coating.

You can see in Figure 12 that the reflectance of a dielectric mirror is a strong function of the wavelength of the incident light. The angle of incidence, as well as the specific properties and construction of the dielectric coating are also important mirror parameters. If either of these parameters changes, the light may not travel the right distance through the coating to be in phase at the upper boundary.

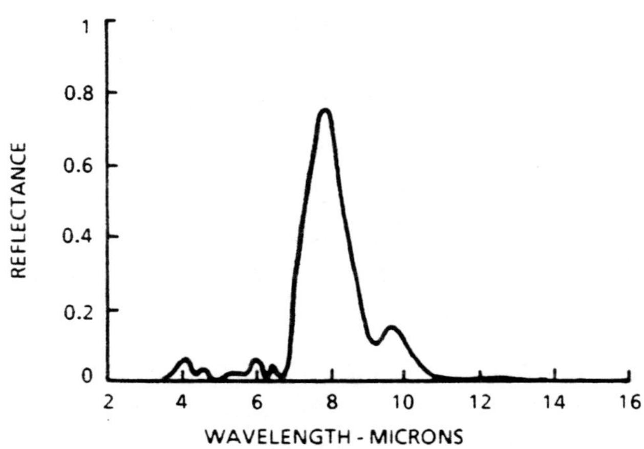

Fig. 12 Experimental curve of reflectance of a three-layer dielectric mirror as a function of wavelength.

Figure 12 shows the reflectance of a three-layer dielectric mirror as a function of wavelength. The curve would be even narrower and the peak higher if more dielectric layers were used.

Multilayer dielectric mirrors have been devised for use in the ultraviolet (220 nm) to the infrared (20 μm). A difficulty met in the ultraviolet is that most high-index materials, such as zinc sulfide and titanium dioxide, are absorbing. So you must use materials with a lower refractive index. Antimony trioxide and lead fluoride are good examples of low-refractive-index materials.

Laser end reflectors are very common examples of multilayer dielectric mirrors. When you specify such reflectors you must consider a number of items. The mirror must be chosen for use at the desired wavelength. Diameter and thickness must be chosen to coordinate with the size of the mirror mounts to be used. The curvature of both surfaces of the mirror must be specified to match the type of laser cavity and the cavity length. The reflectivity, scattering and absorption at the particular wavelength being used must be a major consideration.

Substrate material and surface quality also must be specified. Manufacturers specify surface quality as being flat to a fraction of a wavelength. This flatness is determined by observing the interference pattern produced when the surface is compared to a "perfect" master flat.

The flatness of a mirror (or other optical components such as a window's) surface should be matched to the intended application. For example, wavefront distortion can usually be tolerated in laser material-processing applications such as heat treating and welding. The beam-steering optics in such systems can be of relatively poor flatness—$^1/_4\,\lambda$ should suffice.

On the other hand, in many laser applications wave-front distortion must be minimized. As an example, the optics in holographic systems should be of high surface quality to preserve the phase information. Surface flatness of $^1/_{20}\,\lambda$ is often desirable. Similarly, the plane cavity mirror and Brewster windows used in many lasers usually are specified to at least $^1/_{10}\,\lambda$ flatness to ensure good output beam quality.

Note that, unless otherwise specified, optical flatness is measured with respect to the two yellow Sodium D lines (5890 Å and 5896 Å) that historically have been used as a convenient standard (almost) monochromatic light source for optical measurements.

A more complete discussion of flatness specifications is given in Module 6-4, "Windows and Flats."

Durability of the surface coating usually is specified rather vaguely. But you should consider it when a laser and reflector are to be chosen.

One application of multilayer mirrors of particular importance is heat control in projection systems. In most projection systems, a mirror is placed behind the projection lamp to direct the maximum amount of light toward the film. Formerly these mirrors were coated with silver, but this reflected the infrared heat as well as the visible light toward the film. This problem is solved if you use a "cold" mirror instead of the silver. The cold mirror is a multilayer coating that has a high reflectance in the visible spectral region and a high transmission in the near infrared. As shown in Figure 13, this allows the infrared to pass through the rear of the lamp housing.

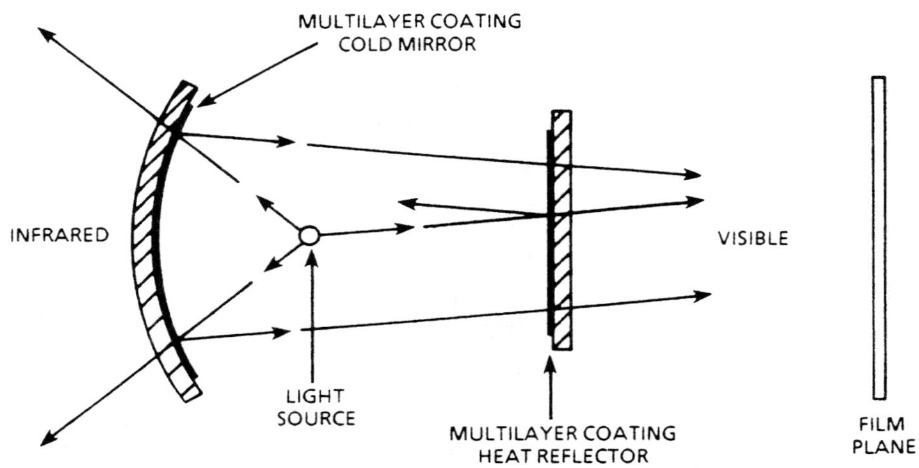

Fig. 13 Use of cold mirror and heat reflector in a projection light source.

The performance of such a system is increased even more if a heat reflector is inserted between the lamp and the film. The heat reflector is a mirror in the infrared and a high transmission window in the visible. This reflects even more of the heat from the film. In practice this combination of mirrors can result in a two-to-three times reduction in heat incident on the film. A similar arrangement could be used to pump a laser cavity at the desired wavelength and keep heat loading to a minimum.

Before we leave the subject of dielectric multilayer mirrors, we should mention another device that uses the same principle of reflective interference. The Fabry-Perot interferometer, or etalon, is basically a multilayer device on a larger size scale. As shown in Figure 14, the light used in an etalon is that which is transmitted in phase after multiple reflections from two mirrored surfaces.

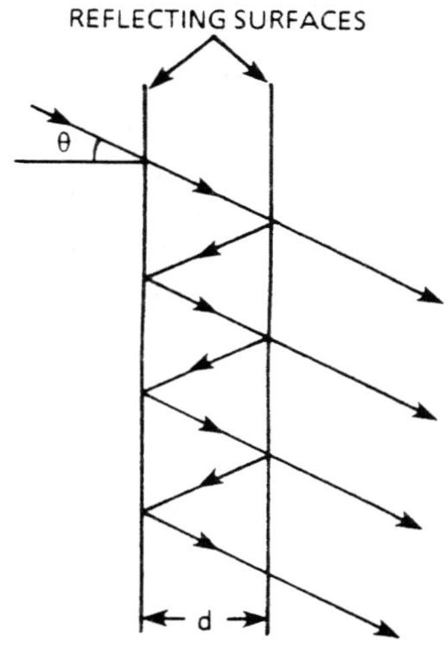

Fig. 14 The etalon is composed of two reflecting surfaces spaced to provide transmission only at a particular λ.

Etalons are used inside laser cavities to select a single wavelength (that wavelength that permits reflections to be in phase) from a multimode laser. The etalon serves as a mirror for all wavelengths except the desired one, which is transmitted. Figure 15 illustrates the use of an etalon in a laser cavity.

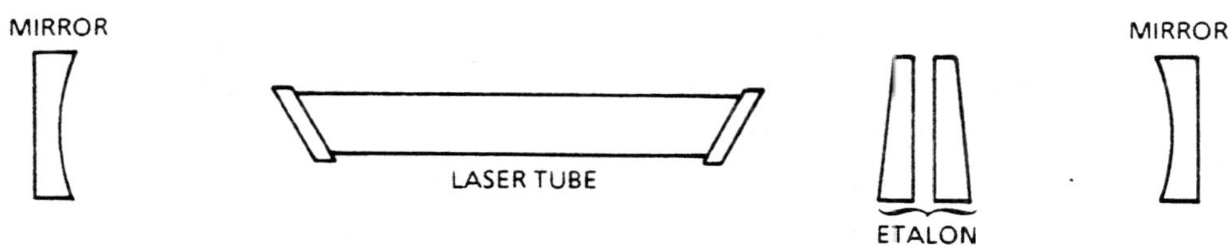

Fig. 15 The etalon forces the laser to operate only at wavelengths that are transmission maxima of the etalon.

Use of Mirrors

This section will describe some uses for mirrors that have not been mentioned previously.

The law of reflection states that the angle of incidence equals the angle of reflection. If the angle of incidence is increased by one degree (as would be caused by a mirror rotation of one degree) the angle of reflection which must be equal also is changed by one degree. So the angular difference in degrees between the incident and reflected ray is twice the mirror rotation in degrees. This principle often is used for angular magnification.

The simple periscope makes use of mirrors to change the direction of line of sight. One obvious use of such an instrument is on submarines. The periscope allows the crew to elevate their line of sight to one above the water. Figure 16 depicts a simple periscope.

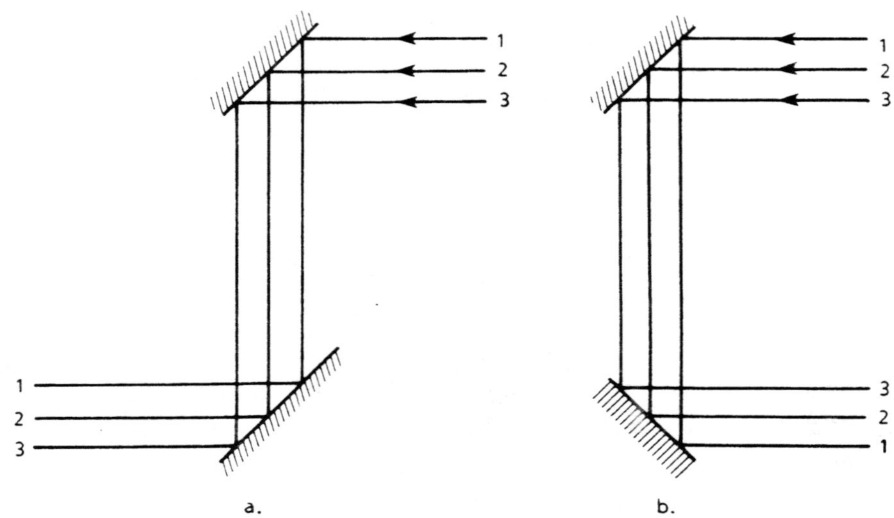

Fig. 16 The simple periscope.

This same instrument can be used, for example, to raise or lower a laser beam on an optical table. Note that the three rays on Figure 16a emerge from the second mirror unreversed and that you are looking in the same direction as the periscope. The reversal at each of the two reflections has the effect of canceling the image reversal.

However, in Figure 16b the image is inverted and you're looking in a direction 180° from that of a periscope. Such image reversal is of little importance in most laser applications.

If three mirrors are placed so that they are all at right angles to one another, they form a corner cube. Such an arrangement is shown in Figure 17. A corner cube also is known as a retroreflector because any light striking the device will be reflected back along an exactly parallel path, within the limits of accuracy of the cube's construction. A little geometric study will convince you that three reflections from orthogonal mirrors will send light back in the direction of its origin.

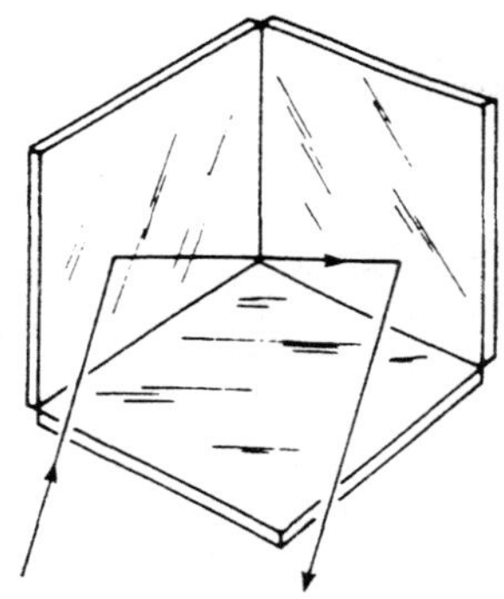

Fig. 17 A corner cube or retroreflector.

A device like this was left on the moon by the Apollo astronauts. A laser on the earth was aimed at the retroreflector and, by determining the round-trip travel time of the light pulse, the precise distance to the moon was determined.

Mirrors often are used to scan a beam of light along a surface. Figure 18 is an example of a rotating mirror being used to scan across a collimated beam. One use for such a scan system would put a detector behind an aperture to determine the variation in intensity across a light beam. This device with other optics is used as a flying spot scanner to send pictures for newspapers over telephone lines.

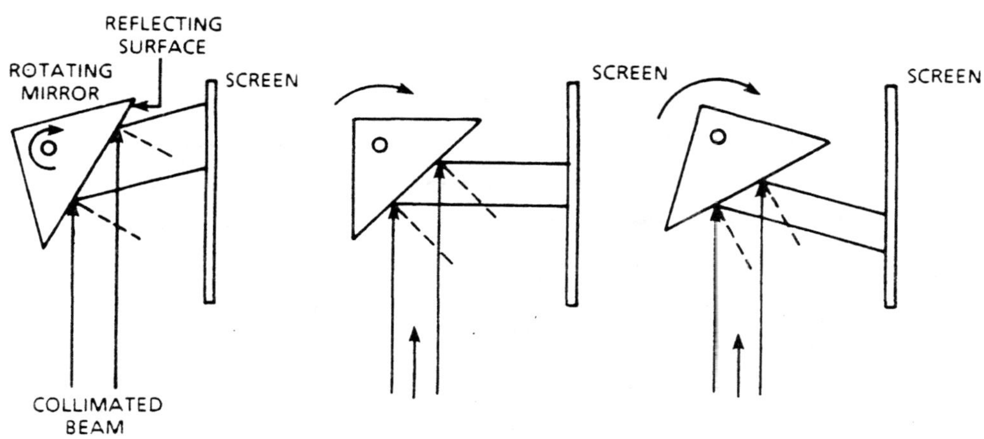

Fig. 18 A rotating mirror used to scan a beam across a surface.

The last specific application of a mirror we will mention makes use of the law of reflection to precisely point a surface in a given direction. For this purpose a mirror target such as shown in Figure 19 can be used. A collimated beam of light is aimed at the mirror target which is attached firmly to the surface being aligned. By rotation of the base surface so that the reflected light exactly returns to the source, the surface is caused to point in the direction of the source. Devices of this type are used in survey work.

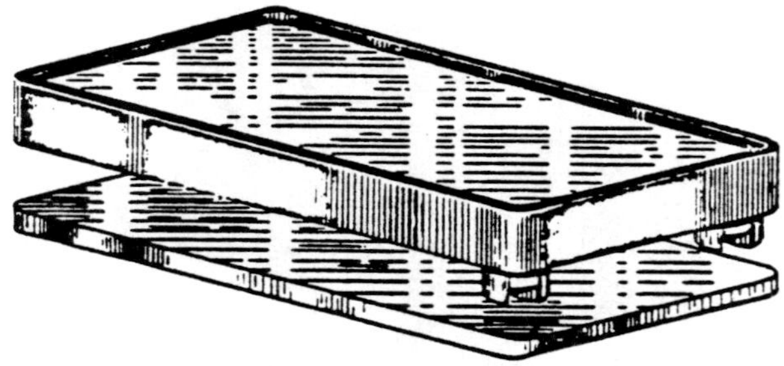

Fig. 19 Magnet-backed mirror for alignment.

Specifying a Mirror

The most important question to ask when selecting a mirror to serve a certain purpose is, "Which characteristic of the mirror do you want to use?" When you have decided this, you can write the mirror specifications around this requirement. For example, one of the most important characteristics to specify is the wavelength at which the mirror will be used. If you need a mirror to reflect in the visible but not in the infrared, a mirror described by Figure 20 would be specified.

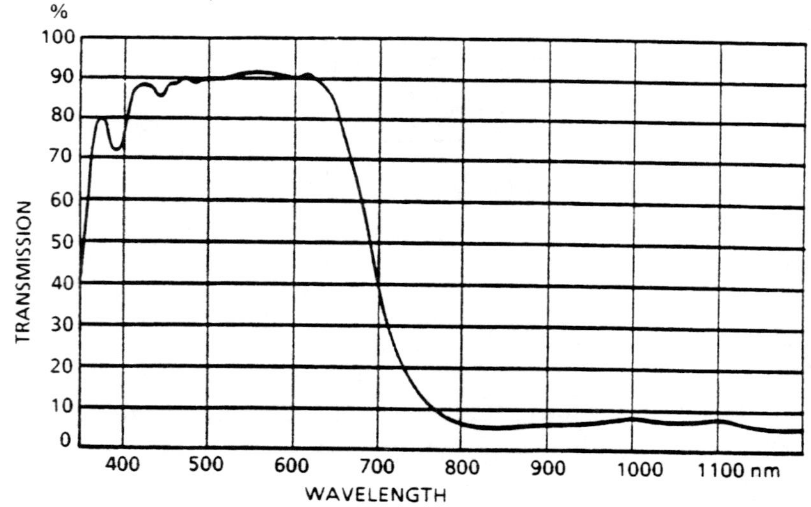

Fig. 20 A cold light mirror.

But if the need were the opposite of that stated above, you would use a mirror with properties similar to that described by Figure 21.

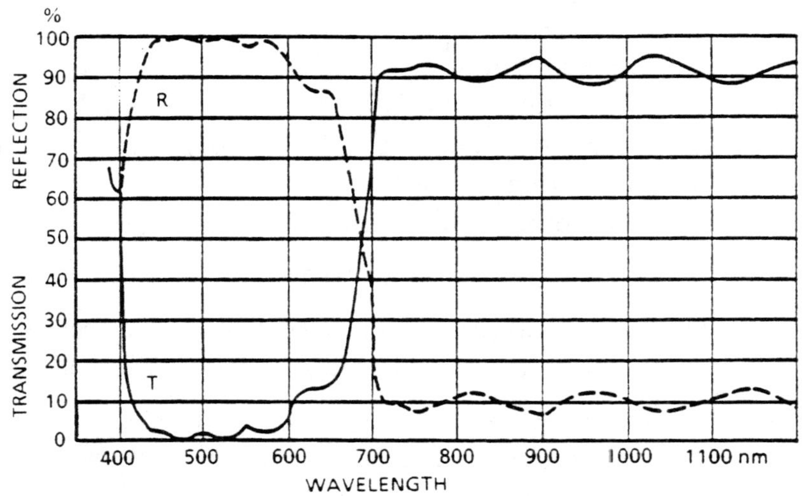

Fig. 21 A heat-reflecting mirror.

If the mirror is to be used in an imaging system, you must specify the focal length and quality of the surface. If the mirror is to be subjected to large temperature variations, you should specify a pyrex or quartz base instead of glass. If the mirror is to be subjected to a lot of handling, specify protective coatings offered by the manufacturer. If the angle of incidence is large for the intended usage, this too should be specified to ensure selection of the right mirror.

In addition to the items mentioned above, any special requirements posed on the mirror by your particular application should be described in the specifications.

EXERCISES

1. Discuss the advantages and disadvantages of front- and rear-surface mirrors.
2. Explain and give one application for each of the following:
   - Dielectric-coated metal mirror
   - Hot mirror
   - Cold mirror
   - Etalon
   - Autoreflection mirror

3. Outline a comparison of the quality and price of laser end reflectors with the given specifications.
   - Diameter 1/2 inch
   - Thickness 0.375 inch
   - For use with argon and krypton laser
   - High-reflectance mirror must be at least 99.7% reflective.
   - Output coupler must have r_1 flat but must have r, with 3 meters focal length and 5% transmission.
   - The cost must remain under $500 for the pair of reflectors.

   Compare at least three manufacturers while assembling your outline.

4. From which manufacturer would you recommend purchasing the end reflectors specified in Exercise 3? Using your outline, give the reasons for your choice.

5. Using outside references if necessary, list and explain at least two applications for retroreflectors other than those stated in the text.

6. From a *single* manufacturer obtain the price of three aluminum-coated mirrors of a standard size, say 2.5-cm diameter. The only difference in the mirrors should be surface flatness. Specify perhaps $^1/_4\ \lambda$, $^1/_{10}\ \lambda$ and $^1/_{20}\ \lambda$. Make a plot of cost per mirror versus flatness. Give a typical application of each mirror.

7. Using Figure 7, what would be the best metallic coating to use on a flat mirror intended to reflect N_2 laser light near normal incidence? What would be the approximate reflectance of the surface (without over-coating)?

8. Consider a plane, aluminum-coated pyrex mirror that's to be overcoated with MgF and used to reflect a ruby laser pulse at near-normal incidence. What is the approximate percent reflectance of the mirror?

9. A high-quality corner cube prism (similar to the one in Figure 22) is illuminated with a HeNe laser from a range of 100 meters. The beam strikes the corner cube in such a way that the exit and entrance spots on the cube's face are 1.00 cm apart. The manufacturer of the corner cube states that the angular deviation of the cube is 2.00 arc seconds. By how many centimeters will the center of the retroreflected spot miss the output aperture of the laser?

MATERIALS

HeNe laser
3 Front-surface aluminum mirrors
Scissors jack
3 Mirror mounts
Adjustable iris
Laser power meter
Second-surface mirror
Flat glass plate (optional)
Partially reflective dielectric mirror
Partially reflective aluminum coated mirror
HeNe laser end reflector.

PROCEDURES

Retroreflector

1. Mount two aluminum plane mirrors in mirror holders.
2. Lay the third aluminum plane mirror on the scissors jack and arrange the three as in Figure 22.

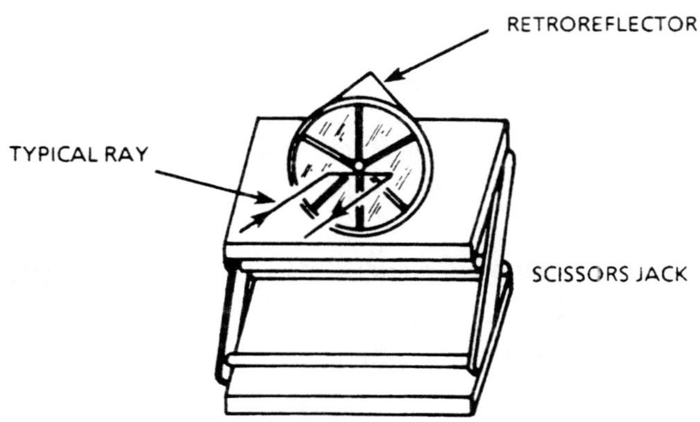

Fig. 22 Mirrors as retroreflectors.

3. Adjust the three mirrors to be at right angles to each other. When this is done properly you can see your own eye no matter at which angle you look into the corner reflector.
4. Set the retroreflector arrangement at one end of an optical table and a HeNe laser with an iris in front of it at the other end.

5. Direct the laser beam toward one of the mirror surfaces, observing all standard safety precautions.

6. Close the iris down around the incident beam so that the returning beam can be observed to lie on the iris very close to the incident beam. Mark the location of the reflected beam on the iris.

7. Move the laser and iris to several positions around the table and repeat. The result should be the same, no matter what the entrance angle.

Ghost Images

1. Direct a HeNe laser beam at a second-surface mirror. The incident angle must be large enough so that two or more reflected beams are distinguished. See Figure 23.

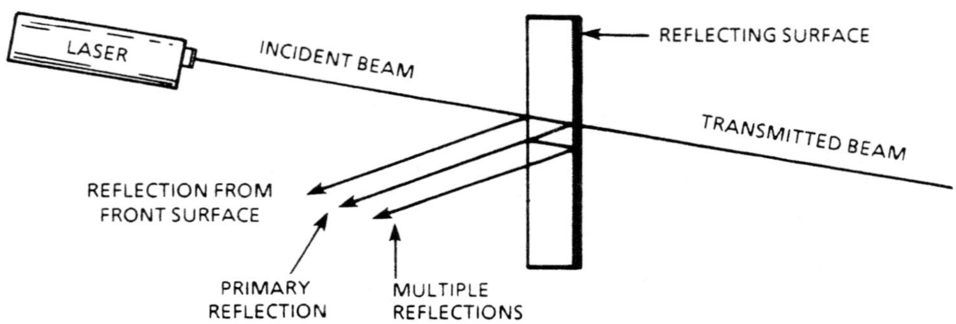

Fig. 23 Reflection from second-surface mirror.

2. Measure the intensity of the incident beam with a sensitive power meter. Record your results. For this procedure and those to follow, it will be necessary for you to prepare a data table in which to record your results.

3. Without changing the experimental arrangement, move the power meter detector to a position to measure the intensity of the reflection from the front surface. Record your results.

4. Move the detector to measure the intensity of the reflection from the primary reflecting surface. Then move it to measure the intensity of any multiple reflections and any transmitted beam.

5. Record all intensity measurements in your data table.

6. Calculate the percentage of the incident beam intensity that's reflected from the first surface, the primary reflecting surface and any multiple reflections.
7. (Optional) Perform the same procedure with an uncoated, thick, flat glass plate. In this case, the transmitted beam intensity will be of major concern.

Front-surface Mirrors

1. Replace the second-surface mirror in the above procedure with a partially reflective dielectric mirror. Use a slight angle of incidence (about 10°) and keep it constant throughout the experiment. Again measure the intensity of the incident beam, the transmitted beam and the reflected beam. Calculate the intensity of the beam lost due to absorption and scattering. Assume the difference between the incident beam intensity and the sum of the reflected beam and transmitted beam intensities is due to absorption and scattering.
2. Calculate the percentage of the incident beam that's transmitted, reflected, absorbed and scattered.
3. Compare your results above with the results you obtain using a partially reflecting aluminum-coated mirror.
4. Use a HeNe laser end reflector, and again compare results.

REFERENCES

Kingslake, Rudolf. "Applied Optics and Optical Engineering." Optical Components, Volume 3. New York: Academic Press, 1965.

Smith, W. J. Modern Optical Engineering. McGraw-Hill, 1966.

LASER

ELECTRO/OPTICS

MODULE 6-6

FILTERS AND BEAM SPLITTERS

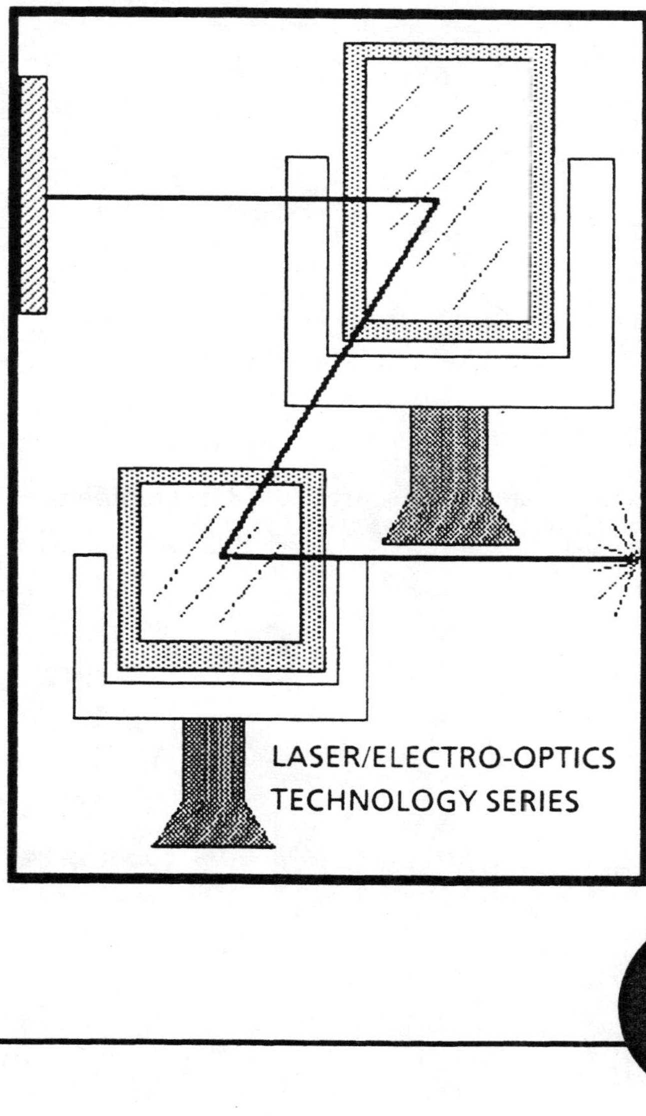

CORD

The Center for Occupational Research and Development
601 C Lake Air Drive
Waco, Texas 76710

Printed in the U.S.A.

ISBN 1-55502-024-0

FILTERS AND BEAM SPLITTERS

MODULE 6-6

A filter is a device that separates a substance trying to flow through it by allowing part of the substance to be transmitted while selectively inhibiting the transmission of the rest. Filters are used with liquids (such as gasoline and water) to remove solid impurities. Electrical filters restrict the frequency spectrum of current flowing in a circuit. Optical filters are used to transmit specific frequencies, uniformly reduce the intensity of a light beam or control the beam's polarization.

Common examples of colored filters are the filters on colored spotlights for lighting fountains, buildings and theatrical stages. Examples of attenuation filters are sunglasses and tinted windows. (These examples also may include polarization and color isolation.) The envelope of a heat lamp is used as a filter to pass infrared rays and reject the transmission of ultraviolet and most visible rays.

Optical filters are used in many laser/electro-optical applications where accurate numerical measurements are needed. This requires that filter characteristics be accurately controlled and specified. You must know accurately the degree of attenuation reflection or polarization induced by the filters as a function of frequency.

This module will familiarize you with the functional classification of optical filters, the various types of filters within each classification, the advantages and disadvantages of each type and how to specify and use them.

Before you study filters and beam splitters, you must know basic laser eye safety. You should have studied elements and operation of a laser and laser safety.

OBJECTIVES

When you complete this module, you should be able to do the following.

1. Define the following terms as they are related to filters:
   - Optical density
   - Neutral region
   - Half bandwidth
   - Spectrophotometric curve
   - Central wavelength
   - Peak transmittance
   - Reflection ratio
2. Describe the following types of filters and tell when they are normally applied:
   - Neutral-density
   - Gelatin
   - Optical interference
   - Glass absorbing
   - Short-wave pass (high pass)
   - Long-wave pass (low pass)
   - Bandpass
   - Narrow pass (narrow band)
3. Predict the effect of several neutral-density filter combinations on the output power of a helium-neon laser. Measure the effects in the laboratory and compare the results to the predictions. Results should be within ± 5% of each other.
4. Determine the transmittance of a narrow-pass filter at several angles of incidence as well as for normal incidence.

DISCUSSION

Several of the applications of scientific optical filters are in photography, optical communication, radar systems, holography, and general laser and electro-optical laboratory experiments.

Filters can be classified into three categories, according to the way they're used:

- Attenuation filters
- Wavelength-selective filters (both transmissive and reflective)
- Polarization filters

The following discussion will be organized to describe filters and their uses according to these classifications.

Attenuation Filters

Attenuation filters are used to reduce the intensity of a light beam. High-quality attenuation filters are said to have a "flat response." This means that they attenuate all wavelengths of light over their usable spectral range by the same amount.

Attenuation filters are used over a photosensitive surface when the light signal received is too intense. This would prevent overdriving a photodetector or overexposing photosensitive film. You can use calibrated attenuation filters to determine if the photosensitive surface is reacting linearly to the exposure.

An example of this application is a light signal being measured by a photodetector. If the photodetector is responding linearly, the insertion of a 50% attenuation filter in the light beam should cause a 50% reduction in the output electrical signal.

Attenuation filters can be divided into two groups according to the mechanism used to reduce the beam power.

Geometrical Filters

In the first group are those filters that physically block a fraction of an aperture through which the light beam passes. If you assume that the beam is uniform in intensity across the aperture, then the percentage of the aperture area that's blocked is the percent reduction in beam power.

One example of a geometrical filter is the iris diaphragm, shown in Figure 1. The iris or "stop" is used to reduce the aperture of a light beam. Consequently it reduces the transmitted beam power.

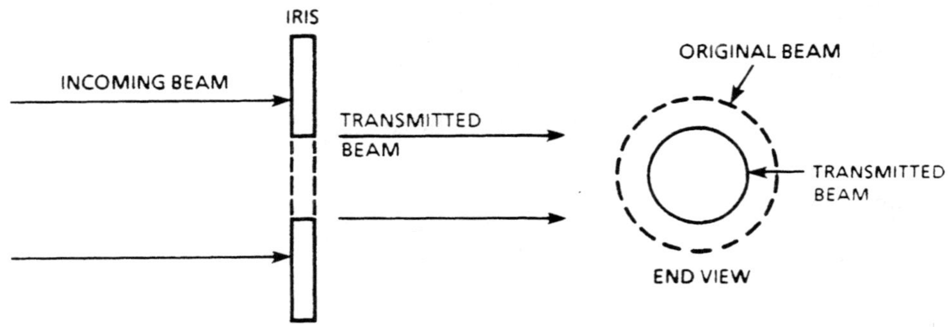

a. Iris diaphragm

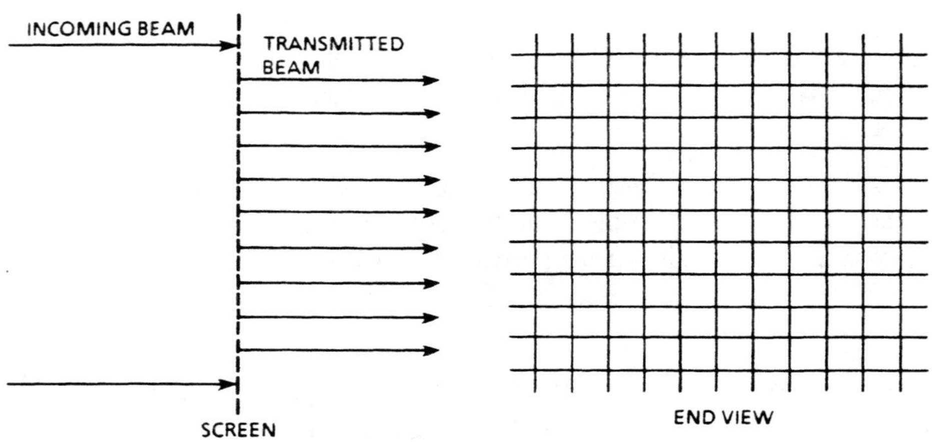

b. Screen or mesh

Fig. 1 Geometrical filters.

Another example is the screen or mesh filter, shown in Figure 1. This type of geometrical filter actually breaks the relatively large beam into many smaller ones, resulting in a nonuniform distribution across the large aperture in the near field immediately preceding the screen. However, as the beam travels farther away from the screen, it becomes more uniform. Because of the possible development of undesired interference patterns caused by this type of filter it's wise to use it only near the detector. When used properly, however, such filter screens can reduce the light level in front of imaging systems without significantly reducing the resolution of the system.

Neutral-density Filters

Neutral-density filters are uniform, "grey" filters that absorb and/or reflect a fraction of the energy incident upon them. The term "neutral" is designated because the absorption and/or reflection characteristics of the filter are constant over a wide wavelength range.

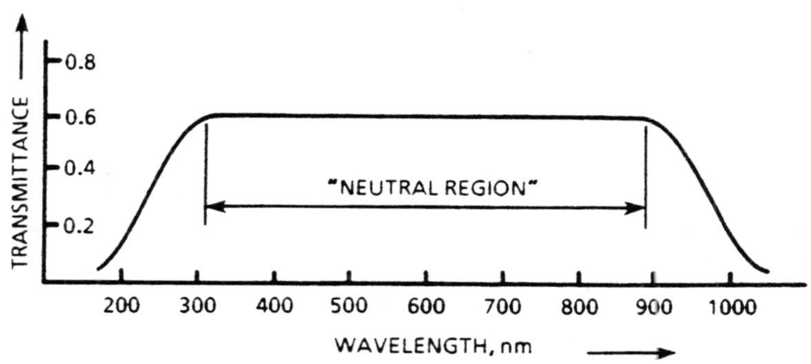

Fig. 2 Idealized spectrophotometric curve for a neutral-density filter.

It's useful here to define "optical density" as it applies to filters. Optical density is the degree of opacity of a translucent medium. It is given by the equation OD = $-\log_{10}T$ where T is the transmittance of the filter.

Table 1 lists the values of optical density for transmission filters. Note, for example, that if the transmission through the filter is 10%, the transmittance is 0.1 and the corresponding optical density is

$$OD = -\log_{10}\left(\frac{1}{10}\right) = -\left(-1\right) = 1.$$

A filter with 100% transmittance quite properly has an optical density of 0.

When light irradiance travels through a filter a part of the beam is reflected at the two air-glass interfaces (plastic-air interface of gelatin filters) and another part of the beam is absorbed. Assuming that scattering and fluorescence are negligible, the remainder of the irradiance is transmitted through the filter. Accordingly, the following equation is applicable:

$$I_O = I_R + I_A + I_T \qquad \text{Equation 1}$$

where: I_O = Incident light irradiance

I_R = Reflected light

I_A = Absorbed light

I_T = Transmitted light

Table 1. Optical Densities of Transmission Filters

% Transmission	Transmittance	Optical Density
0.01	0.0001	4
0.1	0.001	3
1.0	0.01	2
10	0.1	1
20	0.2	0.7
40	0.4	0.4
70	0.7	0.15
90	0.9	0.05
100	1.0	0

Neutral-density (ND) filters usually are rated in optical density (OD) numbers. If two or more ND filters are used together, the density of the composite of the filters is the sum of their individual filters. As an example, if three filters of ND 0.1, 0.3, and 0.4 are stacked together they comprise an effective 0.8 ND filter. Several types of ND filters are available.

Plastic or Gelatin Neutral-density Filters

The most inexpensive (and commonly used) type of ND filter is the plastic or gelatin filter. These often are called No. 96 Wratten filters by the Eastman Kodak Company. These filters are made of colloidal carbon dispersed in gelatin. They can be purchased in two-, three-, and four-inch square sheets, 0.1 ± 0.01 mm thick. The calibration on neutrality of these filters is valid for the visible wavelengths only. M-type carbon, gelatin filters are also available. They can be used in the near ultraviolet, visible and near infrared. However, the M-type filters use larger carbon particles (~ 1 μ) and tend to slightly scatter a light beam. If you must use ND filters on an image-forming beam, it is recommended that you use the No. 96 type filters.

You can get gelatin filters easily in densities from 0.1 – 4.0. You can specially order them with densities as high as 6.0. Although the gelatin film filter is protected by a thin lacquer coating, you should handle it only by the edges and extreme corners. Keep it flat and stored in a dark, dry place. Continued stress can deform it permanently, and moisture tends to cloud it. If you must cut the filter, place it between two pieces of clean, fairly stiff paper and cut with a sharp pair of scissors.

To maintain the stability or calibration of the filter we recommend that you not subject it to temperatures above 120°F. The output of high-power lasers (pulsed lasers exceeding roughly 0.1 joule and CW lasers exceeding 0.1 watt, depending on beam diameter) should not be used directly on gelatin film ND filters. Such laser outputs will permanently damage the filter.

You also can get gelatin film filters cemented between glass plates. This keeps the surface from being easily scratched. Cemented filters are lacquered on the edges to resist the entry of water, which will cause the gelatin to swell and separate from the protective cover.

Absorbing Glass, Neutral-density Filters

Neutral-density filters made of absorbing glass are also available. A typical example of this type is the Schott "NG series" glass. The neutrality of these filters is valid to a wavelength of ~ 2.8 µm. These filters are more rugged than the gelatin type. They are less susceptible to scratching and offer more resistance to damage from high-power laser radiation. These filters are also much less susceptible to damage from high temperature than the gelatin film filters. They can be cleaned similarly to other optical glass.

The most likely possibility of a defect in these filters is inhomogeneity of the glass. You can detect inhomogeneities by looking through the filter toward intense light and inspecting for streaks and bubbles in the glass. Corning Glass Company manufactures similar types of colored and neutral-density glass filters.

Metallic Film on Glass

Metallic-film ND filters have the flattest response over a wide range of all the classes of filters. These filters are composed of a vacuum-deposited metal film on a glass or quartz substrate. (Quartz is used if we want to operate the filter near UV wavelengths.) Most manufacturers use the alloy inconel (nickel-chromium-aluminum) for the deposited films.

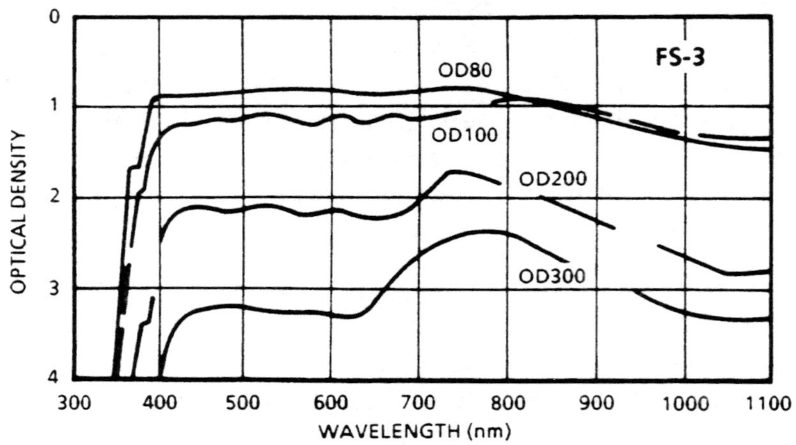

Fig. 3 Transmission curves for actual absorbing glass, neutral-density filters. Note that the "flat" region becomes more restricted for the high OD filters. These filters are constructed from Schott colored glass. Courtesy of Newport Corporation, Fountain Valley, CA.

Attenuation is by both reflection and absorption. You must consider the reflected beam energy in any application if there's any chance of it being reflected in an undesirable direction.

These filters generally are not susceptible to changes brought about by temperature or aging. They may be used for extended periods at temperatures from –50°C to 120°C. Damage will occur if the temperature reaches approximately 160°C. The filters operate over a wavelength range from 0.275 µm to 2.5 µm.

Filter Wheel

For applications such as holography, it's often convenient to have a way to continuously vary the intensity of a light beam. One way to do this is by use of a "filter wheel" as shown in Figure 4.

This particular variable ND filter is produced with a variable-density evaporated aluminum coating deposited on 270° of the disc. The remaining 90° is clear. The aluminized surface of the wheel is protected with an SiO overcoating. The opposite side is AR coated (reflectance less than 0.75%, 450 nm to 700 nm). Because much of the light is reflected rather than absorbed by the metallic coating, these wheels can be used as variable beam splitters as well as filters. The substrate material is either BK7A crown glass or OPTOSIL-1 quartz.

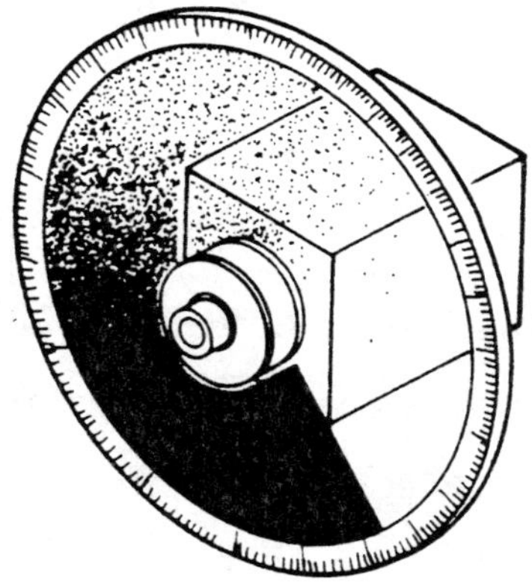

Fig. 4 Variable ND filter/beam splitter. This 5-inch-diameter wheel varies from OD 0.05 to 1.0 (±0.15 at 633 nm) over 270° range. The coating is of the variable-density evaporated aluminum type.
Courtesy of Newport Corporation, Fountain Valley, CA.

Wavelength-selective Filters (Color Filters)

The most complete description of the optical behavior of a filter is given by its spectrophotometric curve (also called transmittance): that is, a plot of wavelength versus transmittance. An example of a family of rather idealized spectrophotometric curves for a specific absorption filter is shown in Figure 5.

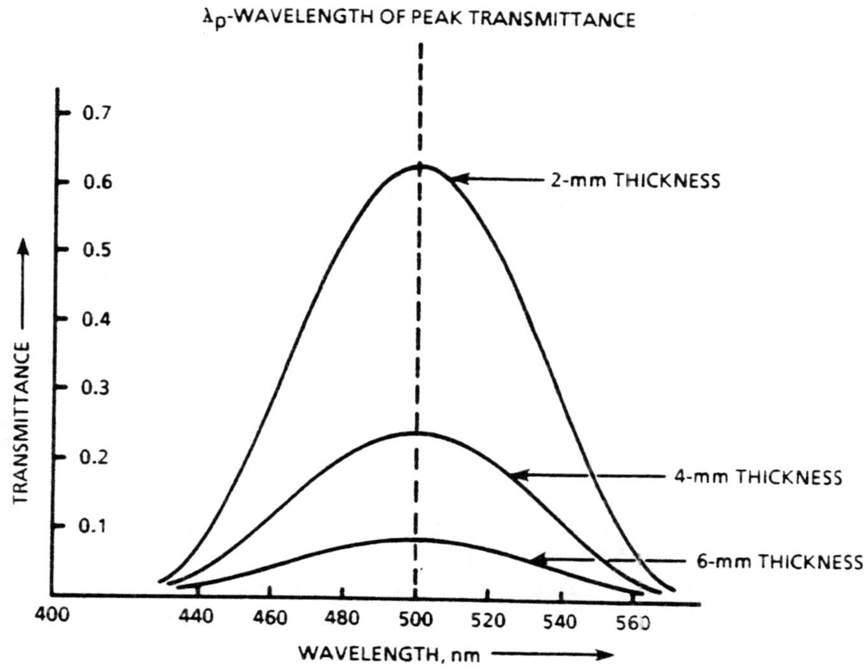

Fig. 5 Spectrophotometric curve for a family of various thickness absorption filters.

For absorbing filters, the transmittance is an exponential function of thickness,

$$T = t_1 t_2 e^{-\alpha_\lambda x} \qquad \text{Equation 2}$$

where:

t_1 = Transmission of front surface of filter

t_2 = Transmission of back surface of filter

α_λ = Filter absorption coefficient (cm^{-1}) at wavelength λ

x = Filter thickness (cm)

The product $t_1 t_2$ sometimes is called the "filter correction factor." For light incident normally to the filter surface, it's given by,

$$t_1 t_2 = 1 - 2\left(\frac{n-1}{n+1}\right)^2 + \left(\frac{n-1}{n+1}\right)^4 \qquad \text{Equation 3}$$

where: n = Index of refraction of the filter glass (relative to air).

The top curve in Figure 5 is for a filter two millimeters thick. The second curve is for a filter four millimeters thick. It has 63.2% less transmittance. The curve in Figure 6 is typical for a wavelength-selective filter. This curve would transmit light at 500 nm very well. But is would transmit very little light at wavelength less than 440 nm or greater than 560 nm.

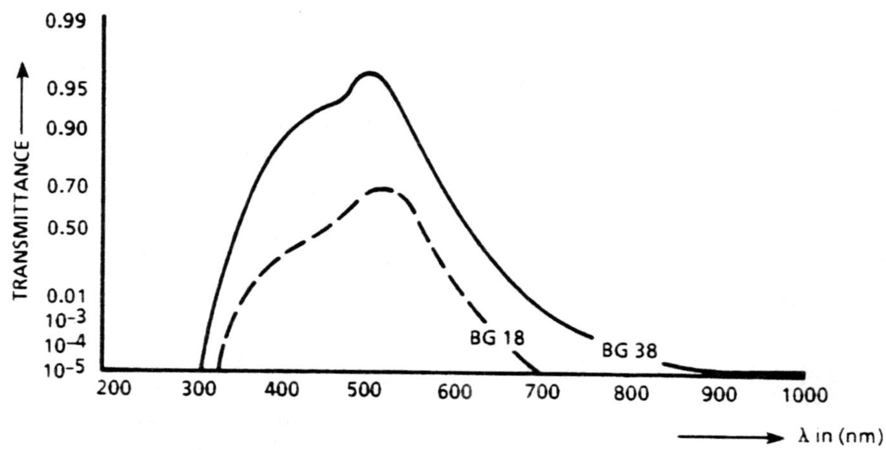

Fig. 6 Actual transmission curves for two blue-green transmitting colored glass filters. Filter dimensions are 50 × 50 × 3.0 (±0.25) mm. Designations BG 18 and BG 38 refer to standard types of Schott colored filter glass. These curves show internal transmission of the filter only. They don't include loss due to surface reflections. Courtesy of Melles Griot, Irvine, CA.

This type of filter is much different in its transmission characteristics from that of the neutral-density filter shown in Figures 2 and 3. The idealized ND filter in Figure 2 has a "flat" response from 300 nm to 900 nm. This flat region is called the neutral region.

Wavelength-selective filters are used to produce or select specific color or band of color from a white light source, to isolate a specific wavelength, or to reject a specific wavelength or band of wavelengths. We will study three general classes of wavelength-selective filters according to their usefulness:

1. **Cut-off filters**—The first class consists of cut-off filters that have an abrupt division between the regions of high and low transmission. If a filter transmits the shorter wavelengths and rejects the longer wavelengths, it's called a short-wave-pass filter. If it transmits the longer wavelengths and rejects the shorter wavelengths, it's called a long-wave-pass filter. Note that some books use annotation derived from electronics where frequency rather than wavelength is signified. In this notation a long-wave-pass filter is called a "low-pass filter," and a short-wave-pass filter becomes a "high-pass filter." Examples of short-wave-pass and long-wave-pass filters are shown in Figures 7 and 8.

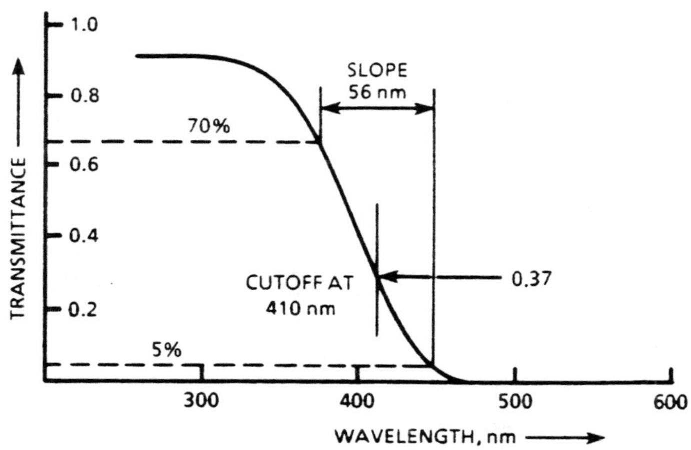

Fig. 7 Idealized short-wave-pass filter (high-pass filter).

The most important characteristic of a cut-off filter is the wavelength describing the position of the sharp cut-off. Different manufacturers define this position to be the wavelength at 3, 5, 30 or 37% transmission (0.37 transmittance). This definition is quite arbitrary, but we recommend that you use the 37% point.

It's also important to describe the sharpness of the cut-off or the steepness of the curve at the transition between low and high

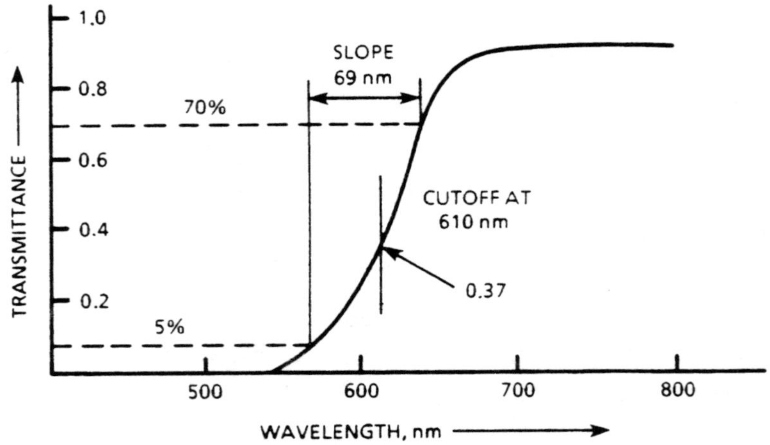

Fig. 8 Idealized long-wave-pass filter (low-pass filter).

transmission. This has been set arbitrarily as the wavelength difference between 5 and 70% transmission points. It is called the slope of cut-off.

2. **Bandpass filters**—The second class of filters are those called "bandpass." An example of a bandpass filter is shown in Figure 9.

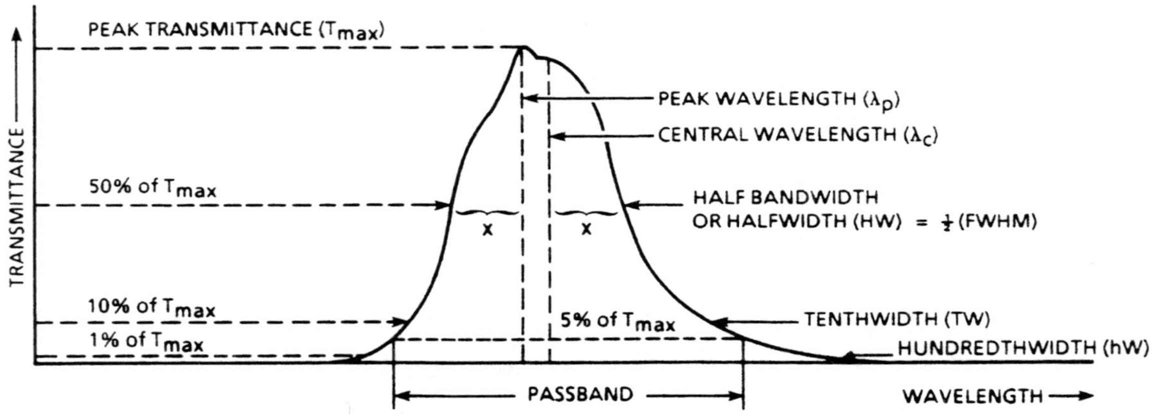

Fig. 9 Description of a bandpass filter.

These filters are characterized by the following parameters:

- The *peak transmittance* (T_{max}) and its corresponding wavelength (λ_p).
- The *half bandwidth* or halfwidth (HW), the wavelength interval between the two points on the transmittance curve where the transmittance is 50% of T_{max}. This sometimes is called the full

width at half maximum (FWHM), so that is is not confused with half width at half of the maximum transmittance. In like manner, we can define the tenthwidth (TW) or hundredwidth (hW).

- The *central wavelength* (λ_c), which is midway between the halfwidth points on the curve.
- The *passband*, which normally is defined as the wavelength interval between the values on the curve where the transmittance is 5% of the peak transmittance. However, for narrowband filters, the passband is defined arbitrarily as three times the half bandwidth.
- The *rejection ratio* for narrowband filters is the transmittance outside the passband divided by the peak transmittance. Rejection ratios can be as low as 10^{-5}.

Bandpass filters can be produced that transmit only a very narrow wavelength range. Such filters are called by many equivalent names. "Narrowpass filters," "spike filters," "notch filters," are some of the more common. One important application of such filters in electro-optics is the isolation of individual laser lines. Interference filters are available "off the shelf" with a bandwidth of 1 ± 0.2 nm (FWHM) for use with common visible laser lines. The narrow bandwidth is achieved by multi-layer dielectric coatings. The negative result of the multi-layer coatings is high cost and relatively poor peak transmission (usually less than 50%).

3. **Compensating filters**—In this class are filters that have gradually sloping spectral curves. Compensating filters have many widely varying applications, but there are no distinct quantitative terms to describe them. However, one group that we can classify is the color-temperature correcting filters used in color film photography. They are used to compensate for incorrect lighting such as when indoor film is used outdoors or when outdoor film is used indoors (under tungsten lamps). These filters are of the amber type that lower color temperature, and the blue type that raise color temperature.

4. **Other special-purpose filters**—Many other filters are used in the optics and electro-optics industries for specialized purposes. A common example is the so-called "hot mirror" that reflects strongly in the wavelength range of approximately 750 to 1200 nm while transmitting in the visible and near UV. Such a mirror can be used as a filter to block IR radiation from, for example, reaching the film plane in a high-intensity projector. Conversely "cold mirrors" are really filters that transmit in the near IR but block radiation in the near UV and visible

range. Both of these filters or "mirrors" are really examples of pass filters.

Rather than reflecting certain wavelengths, some specialized filters absorb particular wavelength ranges. As an example, heat-controlling filters often are made out of common Schott glasses, such as glass types KG1, KG2, KG3, and KG4. Such filters can get very hot if placed close to a bright source of IR radiation. Design of an optical system using heat-absorbing filters must take into account the heat-dissipation problem.

In pulsed solid lasers, such as Nd:YAG, it's now common practice to surround the flashlamp with a flow tube that is coated to block UV light. Such a flow tube is really a specialized example of a pass filter. UV light from a xenon or krypton filled flashlamp has at least three deleterious effects on a solid laser: (1) it often causes "solarization" of the rod material, (2) it can increase thermal loading of the rod, and (3) it can accelerate deterioration of the interior of the pump chamber by interacting with contaminants in the cooling system. Inclusion of the UV reflecting coating on the flow tube decreases all of the deleterious effects while actually increasing the efficiency of the laser at high pump levels. The increase in efficiency is in great part due to the reflection of the UV pump light back into the lamp where it is reemitted at longer wavelengths and more efficiently absorbed by the laser rod. For example, a small Nd:YAG laser with a 20-joule input may increase its output energy/pulse by 15 – 20% when a UV-blocking flow tube is included in the laser head design.

We have introduced wavelength-selective filters according to the characteristics of their spectrophotometric (or transmittance) curves. These filters can be physically described by grouping them according to the way they alter the light, as follows:

- Absorption
- Reflection
- Combination of Absorption and Reflection

Absorption Filters

Others types of wavelength-selective absorption filters besides those previously discussed in detail are available. One such filter—the **liquid-dye filter**—is made by dissolving a dye into a liquid solvent and filling a glass cell with the solution. Liquid filters are inexpensive and easy to prepare and change. Density of the filter can be altered by changing the strength of the solution. For example, a liquid filter made up of 240 gm/liter of N_aSo_4 and 45 gm/liter of $CoSo_4$ transmits over the broad band from 220.0 to 340.0 nm.

Disadvantages of liquid filters are that they have rather broad spectral transmittance curves, they are often chemically (and optically) unstable with age, and they can be altered by intense light causing them to change chemically.

You can find more complete discussion of liquid filters and spectral descriptions of several chemical solutions in Wood's Physical Optics. (See References.)

Reflection Filters

Reflection filters for wavelength selection (sometimes called "optical interference filters") are used when you need sharp cut-off or very narrow bandpass. They often are called "spike filters" or "narrow-pass" filters. They generally are made by depositing alternating layers (thin coatings) of dielectric materials on a glass or quartz window. Selection of materials and thickness of the coatings are chosen to provide reflection or transmission at the desired wavelengths. When the number of layers is increased, the cut-off or the passband becomes sharper, but the peak transmittance also decreases. An interference filter may have as many as 30 layers of coatings. These multilayer coating techniques are the same as are used to produce high-reflective laser mirrors.

Absorption and scattering from these surfaces is less than 1%. Consequently, when these coatings are used as filters, practically all of the light is either transmitted or reflected. The reflective properties constitute the filter. Since the transmitted beam shows negligible distortion from the interference coating, these filters can be used in imaging systems.

Compared to metallic films, many interference coatings are quite resistant to damage from high-power lasers. However, they are vulnerable to mechanical and chemical damage. Because they are easily scratched, they should not be wiped with lens tissue. You can remove dust by blowing clean, dry air across them or by gently wiping with a soft, camel's hair brush. Do not wash them in water but rather clean them by pouring clean, dry ethyl alcohol, (preferably "nano grade," one part in 10^9 contamination) over the surface and allowing it to drain and evaporate. It's best to store them in a dry container. Poor-quality coatings will develop cracks with age. Recent improvements have been realized in developing "hard" coatings that won't scratch or wipe off as easily. A good, hard coating should not be damaged when a piece of masking tape is applied to the surface and removed (a standard "MIL spec" test).

Normally, an interference filter is designed to be used at 0° angle of incidence. The spectrophotometric curve of these filters can be shifted to longer or shorter wavelengths by varying the temperature and/or the angle of incidence at which they are used. To a lesser extent, the shape of the curve also is altered by these variations.

The curve of a multilayer interference filter shifts toward shorter wavelengths as the angle of incidence increases. The magnitude of this shift depends on the type of filter (bandpass, long-wave-pass, short-wave-pass, etc.), the filter design, and the refractive indices of the coating materials used. The relationship between "percent shift to shorter wavelength" and "angle of incidence" is shown graphically in Figure 10. The crosshatched area represents the variation in this relationship. Figures 11, 12 and 13 show the measured incident angle effects on three typical infrared filters. For incident

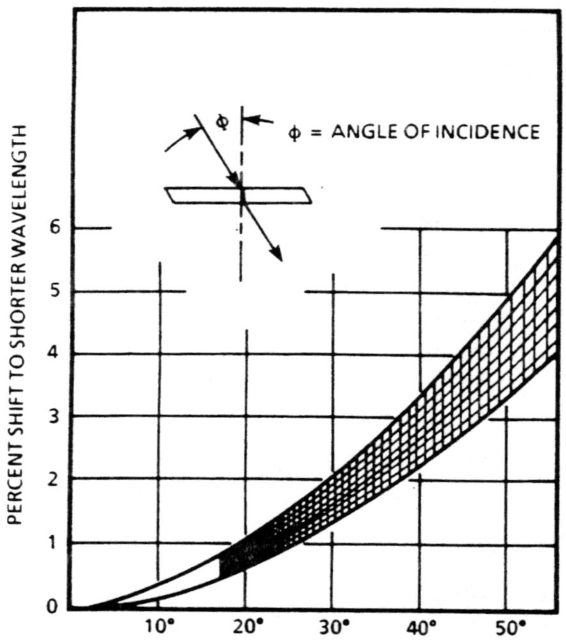

Fig. 10 Wavelength shift as a function of angle of incidence.

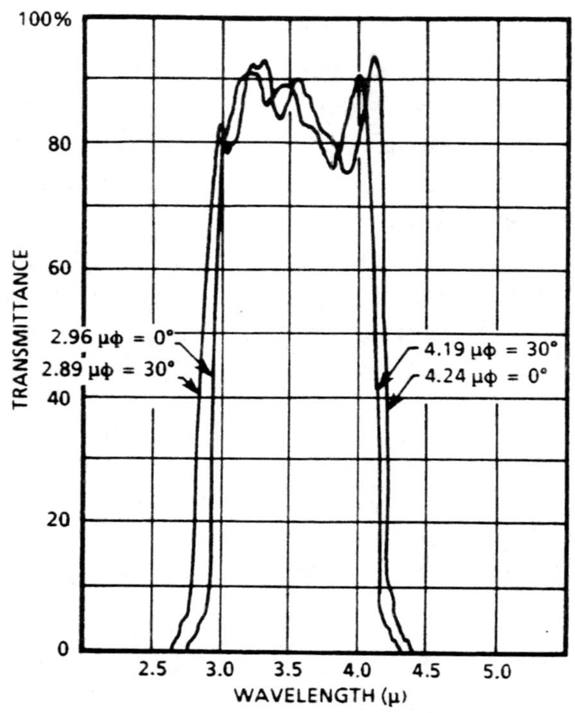

Fig. 11 Wavelength shift of passband filter curve as a function of angle of incidence.

angles up to about 40°, the shape of the curve is not appreciably distorted except for a slight decrease in transmittance. (See Figures 11 and 12.) However, when the incidence of angle becomes greater than 40°, changes begin to occur due to polarization effects and increased mismatch in the optical thickness of adjacent layers. (See Figure 13.)

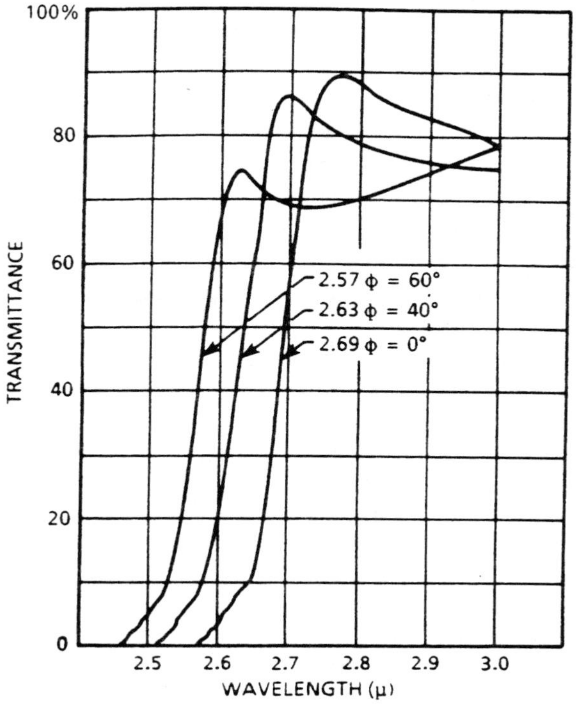

Fig. 12 Long-wave-pass filter slope shift as a function of incidence angle.

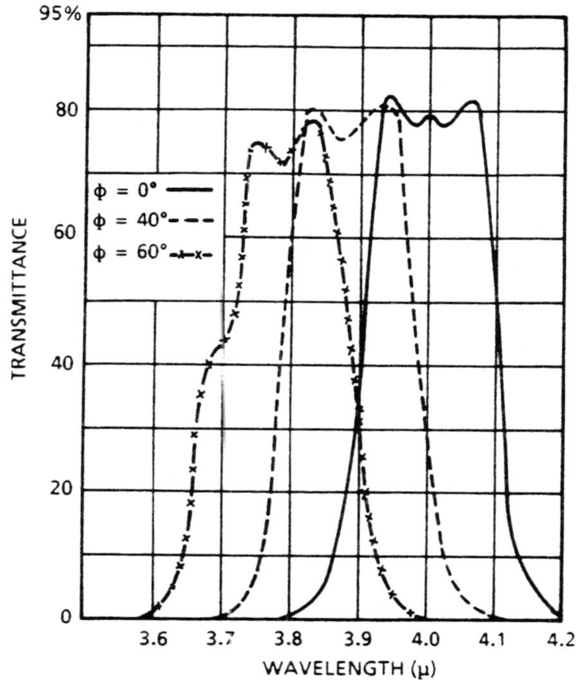

Fig. 13 Passband filter curve distortion as a function of large changes in incidence angle.

The wavelength shift of a multilayer interference filter transmittance curve is dependent on film temperature inasmuch as the materials that form the various layers have indices of refraction that change with temperature. Generally interference filter curves will shift to longer wavelengths when the film temperature is increased, and to shorter wavelengths when the film temperature is decreased. An interference filter is designed for its transmittance curve to be valid at one temperature, usually 20°C. The relationship between temperature variations above and below 20° and the shift of the transmittance curve are shown in Figure 14. The crosshatched area shows the variation in this relationship for a number of tested filters.

Generally, interference filter curves shift but don't change their transmittance and shape characteristics over a wide range of temperatures from −190°C to 40°C. (See Figures 15 and 16.) However, significant transmittance losses may occur at filter temperature ranges exceeding 50°C. (See Figure 17.) Typical examples of applications where filter temperatures

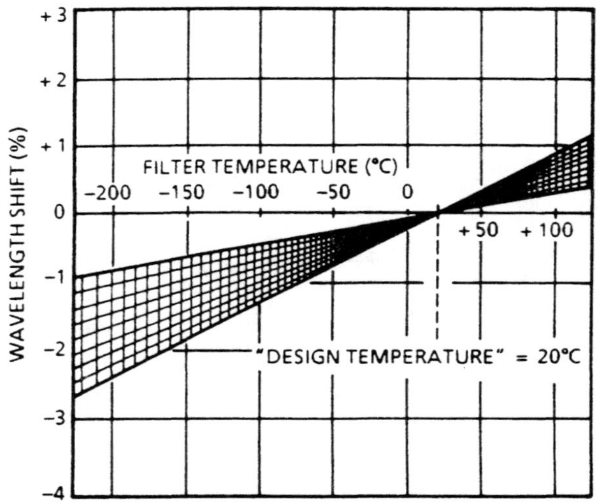

Fig. 14 Wavelength shift of an interference filter as a function of temperature.

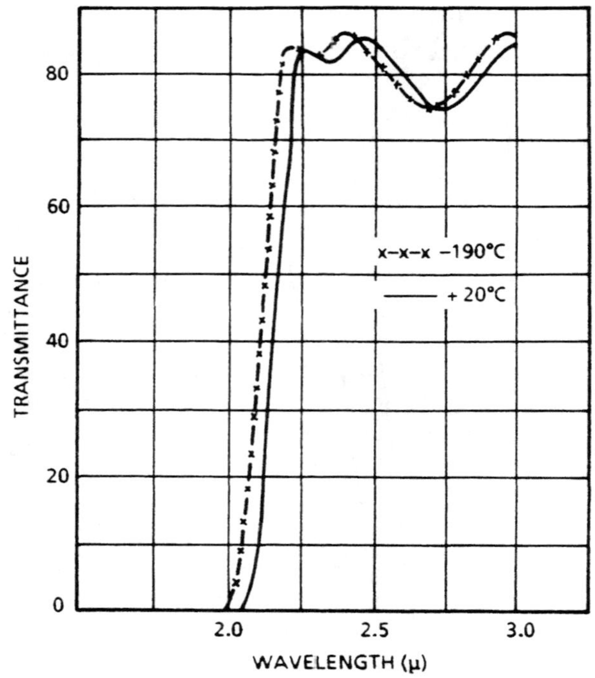

Fig. 15 Effect of temperature on a long-wave-pass IR filter.

may be extreme are cryogenically cooled infrared systems (cold), and spaceborne optical systems that are exposed to the sun (hot).

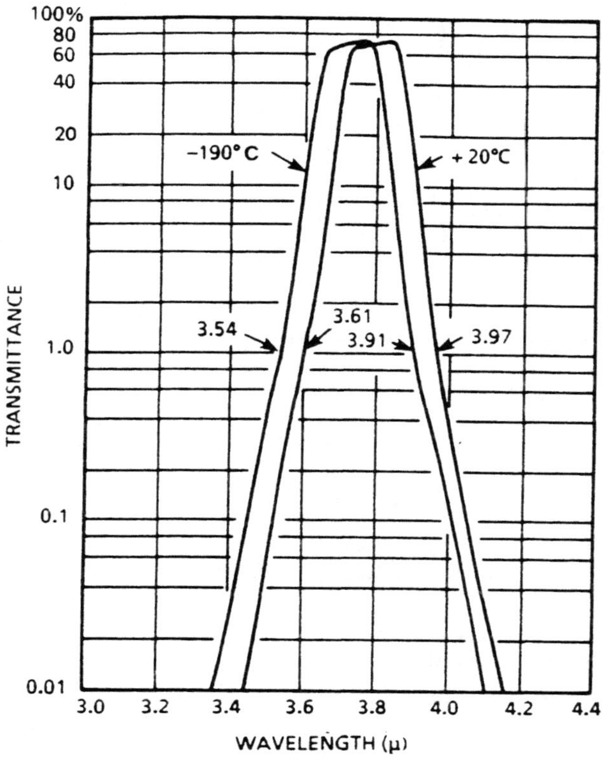

Fig. 16 Effect of temperature on a band-pass filter.

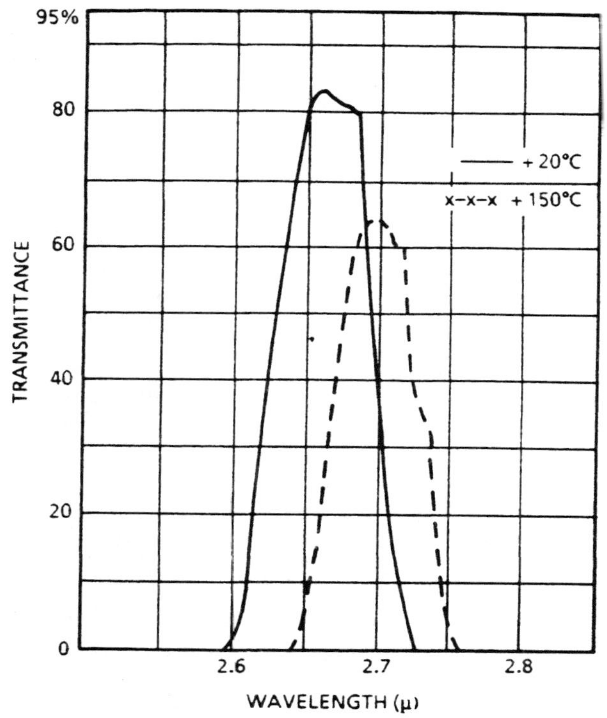

Fig. 17 Distortion of transmittance curve due to temperature rise greater than 50°C above design temperature.

Combination Absorption/Reflection Filters

In the following typical application, it's desirable to use a very narrow passband filter with a low rejection ratio outside the passband.

- To isolate the 488-nm line of an argon laser output while rejecting the 514.5-nm line.
- To filter out as much background light as possible from a neodymium laser radar receiver photodetector. This will provide a relatively high receiver signal to noise ratio.

Very narrow transmittance passbands can be achieved in interference filters by increasing the number of layers of dielectric coatings. An example of such a filter is shown in Figure 18. It is used for isolating the 488-nm argon laser line from the 514.5-nm line. Note that there are four other transmission lines also in this filter that don't affect this line isolation. However, if you

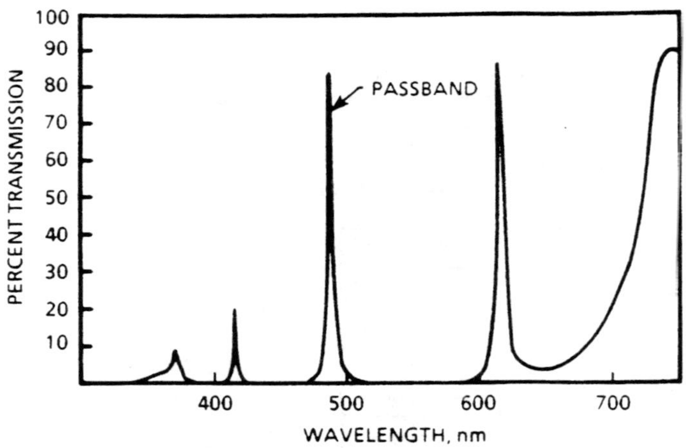

Fig. 18 Interference filter deposited on a glass substrate, with no blocking filters. The passband of interest is at 489.4 nm.

want to isolate this line from a broad, continuous light source such as the sun, then it's desirable to:

- Eliminate the other four transmittance lines.
- Reduce the passband of the 488-nm transmittance line, thus lowering the rejection ratio.

You can accomplish these two task by combining a glass absorption filter with the interference filter. The glass filter has a broader transmittance line about 488 nm but reduces its transmittance everywhere outside this region to very low values. The passband of the glass filter and a portion of the interference filter are shown in Figure 19. When these filters are combined, a composite filter curve is achieved similar to that shown in Figure 20.

The absorbing filter is called a "blocking filter" because it blocks out the other four transmittance bands and it reduces the width of the passband. Blocking filters can reduce the reflection ratio of the filter to 10^{-5}.

Narrow-passband filters often are called "spike" filters because of the shape of their transmittance curve. As previously discussed, these filters are available with a half width of 0.1% or less of λ_c. As an example, a spike filter with a central transmittance wavelength of 1.06 μ (output wavelength of Nd:YAG laser) could have a 1.0-nm HW (halfwidth) or less. In these very narrow spike filters, the peak transmission is reduced to approximately 40 to 50% instead of the usual 70 to 80% for those filters with halfwidths that are 1.0% of the central wavelength.

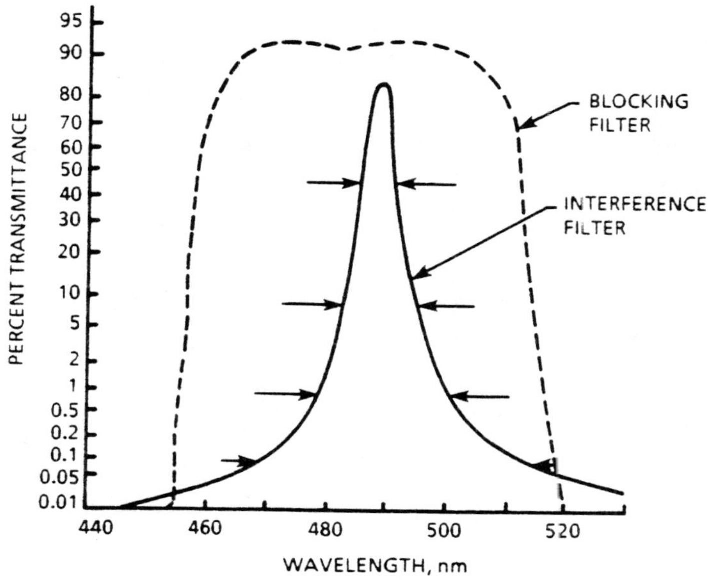

Fig. 19 Transmittance of interference filter (unblocked) and absorption filter used as a narrow-passband filter.

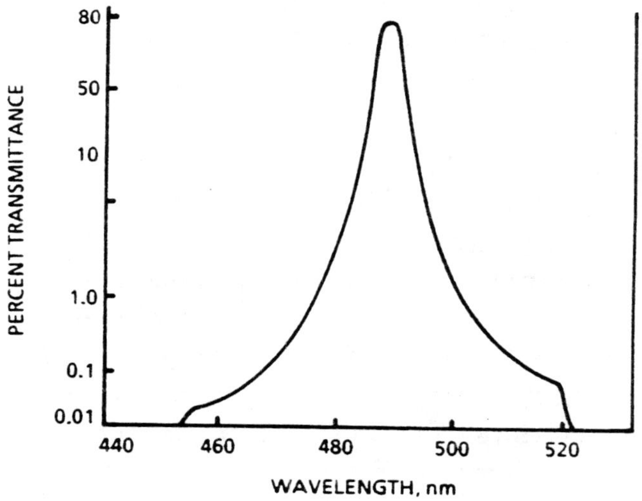

Fig. 20 Transmittance of blocked, narrow-passband filter composed from the two filters shown in Figure 19.

Polarization Filters

We can make the following summary statements about the phenomenon of polarization and the use of optical elements as polarizers:

- Polarizing prisms (such as Wollaston, Glan and Nicol) act as polarizing filters by separating the two polarization components of a light beam into two beams leaving the prism at different angles in space.
- Polarizing films on glass substrates or glass stacks act as polarizing filters by reflecting one polarization component and transmitting the other.
- Polarizing film is a thin, plastic sheet that acts as a polarizing filter because it absorbs one polarization while transmitting the other.

Beam Splitters

A beam splitter is a device that's used to divide an optical beam into two or more components according to given criteria. Beam splitters can be grouped into the following four classifications. These classes include those beam splitters that divide the beam according to beam intensity, wavelength, polarization or physical size.

The most common beam splitter is a partially reflective mirror. This mirror allows a predetermined percentage of the light to be transmitted while reflecting the remaining portion. The actual percentage of the beam reflected and transmitted depends on the coatings applied during the manufacturing process. Coatings commonly available include inconel and achromatic dielectric. Beam splitters with these coatings can be made to give a transmitted to reflected beam ratio of 30% - 70%, 40% - 60% and 50% - 50%. Variable-density beam splitters are also available where reflectivity of the coating is varied across the surface of the beam splitter. This allows you to change the beam ratio by using different locations on the beam splitter.

Partially reflective beam splitters have serious limitations. They provide a high loss factor, multiple reflections and distortions. To understand these limitations see Figure 21. Multiple reflections cause both the reflected and transmitted output beams to look similar to cross-sectional views shown in Figure 22. Areas where the beams overlap produce interference patterns when the incident light is monochromatic, coherent laser light. If the light is not monochromatic, coherent light, ghost images can be produced.

Distortion also can be introduced by a beam splitter that's not made of the highest-quality material or by one that is not perfectly flat.

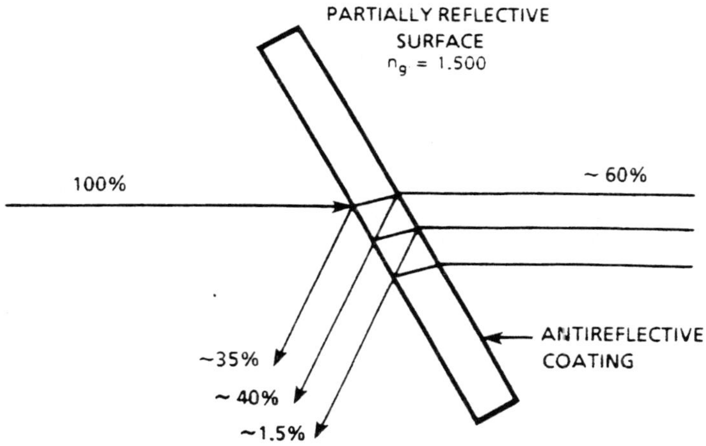

Fig. 21 Multiple reflections produced by partially reflective beam splitters.

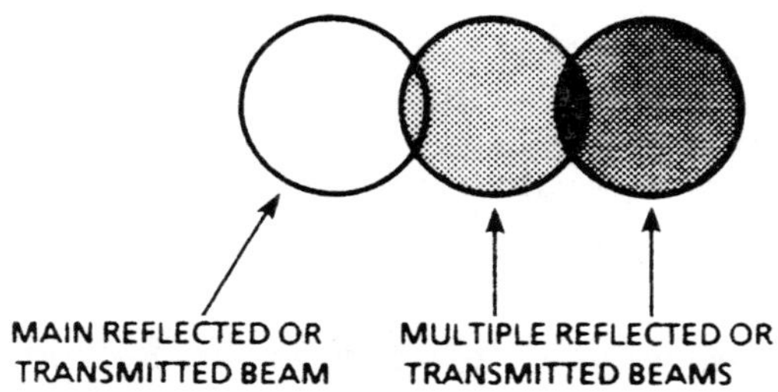

Fig. 22 Cross-sectional views of output beams from a partially reflective beam splitter.

A pellicle is a precise optical component that works similarly to the partially reflective glass beam splitter. It is made of a plastic membrane stretched over an optically flat metal frame. The membrane has a thickness of approximately 7 microns. Because of their thinness and flatness, they demonstrate several advantages over glass beam splitters. For example, they produce almost no change in the optical path length of a light ray that occurs when thick glass beam splitters are used. The thin plastic membrane provides a very low absorption loss factor, and multiple reflections are no longer a problem. A pellicle is, however, more easily damaged and more costly than a glass beam splitter.

Holography and interferometry are very important applications for partially reflective beam splitters. Figure 23 shows a typical arrangement of optical components for exposing a hologram. Note the position and function of the beam splitter.

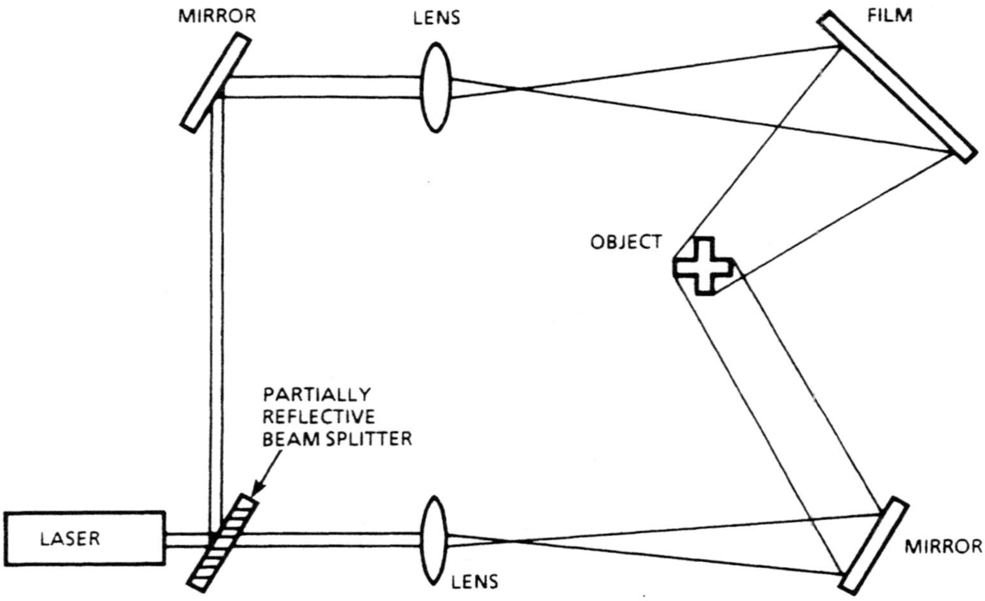

Fig. 23 Beam splitter used in a typical holographic arrangement.

Certain types of dielectric filters can be considered to be beam splitters because they divide a beam according to its wavelength. The dielectric coating permits wavelengths longer than a specified value to be transmitted. It reflects wavelengths shorter than that value. Figure 24 illustrates this type of beam splitter. Notice that the graphs in the figure show the relationship of the intensity of the respective beams as a function of the wavelength of that beam.

Certain optical solids known as birefringent crystals separate a randomly polarized light beam into two plane polarized components. The crystal presents a different index of refraction to each polarized component, causing the components to be refracted at different angles. As a result, the polarized components exit the crystal in separate directions. One of these two polarized components is called the "extraordinary ray." It is plane polarized perpendicular to the other ray, which is called the "ordinary ray." Figure 25 illustrates a polarizing beam splitter.

The final beam splitter to be discussed in this module is one that divides a beam according to its position or geometry. For instance when a mirror is placed in a portion of a light beam it obviously will reflect the part of the beam that strikes it while leaving the remaining part of the beam undiverted. Figure 26 illustrates this beam splitter more fully.

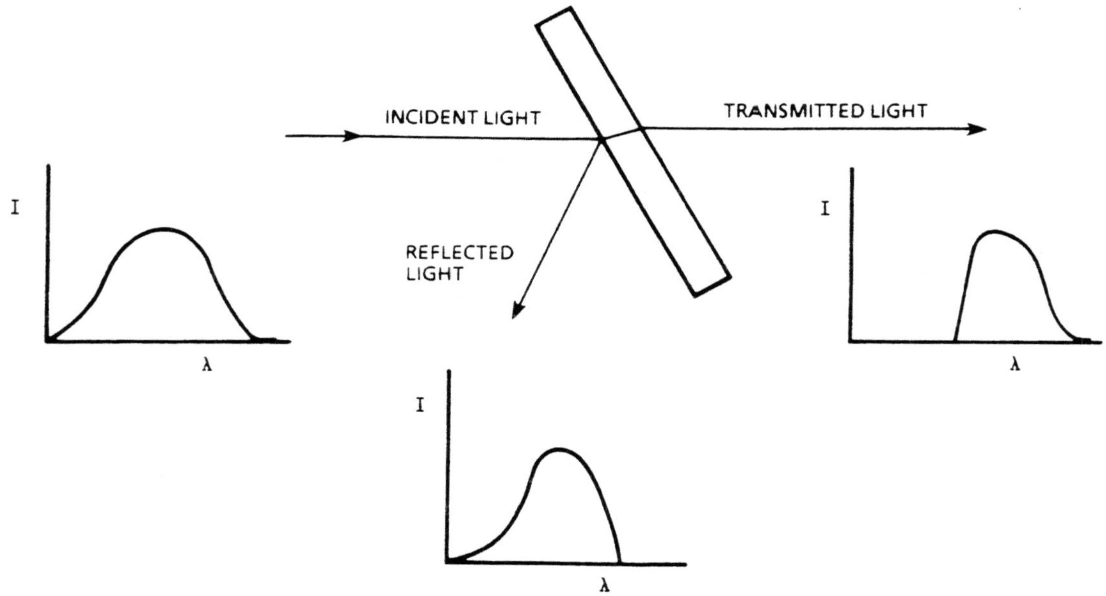

Fig. 24 Intensity versus wavelength of the incident, reflected and transmitted beams for the dielectric filter beam splitter.

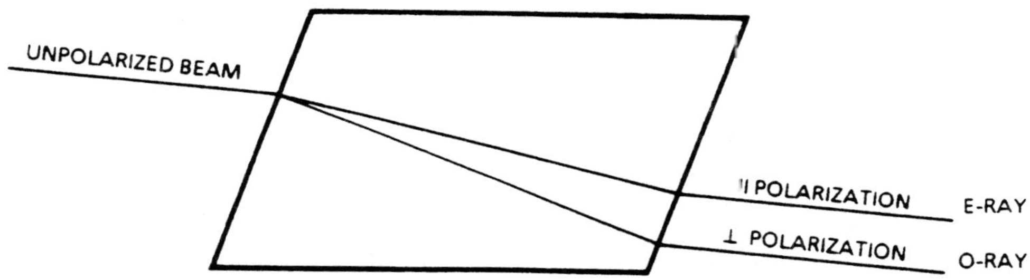

Fig. 25 Beam splitting by double-refracting crystals.

The beam splitting mirror arrangement shown in Figure 27 sometimes is used as the output coupler on a CO_2 laser.

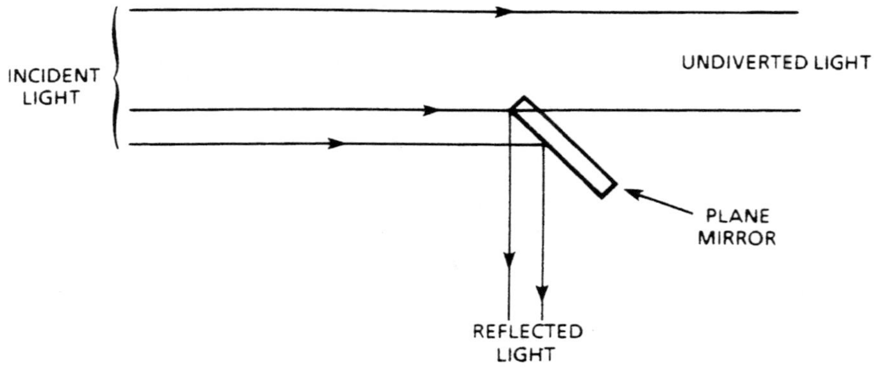

Fig. 26 Beam splitting by position of a mirror.

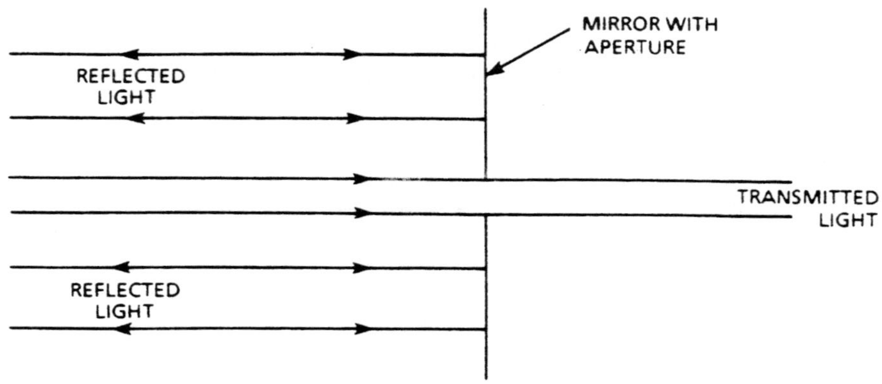

Fig. 27 Beam splitting by geometry of a mirror.

EXERCISES

1. Define the following terms as they are related to filters so that the meaning is the same as that conveyed in the text.
   a. Optical density
   b. Neutral region
   c. Half bandwidth
   d. Spectrophotometric curve
   e. Central wavelength
   f. Peak transmittance
   g. Rejection ratio

2. Describe the following types of filters and tell when they are normally applied.
   a. Neutral density
   b. Gelatin
   c. Optical interference
   d. Glass absorbing
   e. Short-wave-pass
   f. Long-wave-pass
   g. Bandpass

3. Using the following ND filters (0.1, 0.2, 0.3, 0.4, and 0.5):
   a. Calculate the transmittance of each filter separately.
   b. Calculate the transmittance of a combination of the (0.1, 0.3 and 0.4) filters.
   c. Calculate the transmittance of a combination of all five filters.
   d. Show mathematically the validity of adding the OD values to find an equivalent value for the combination (easy!).

4. What conclusions can you draw from your experimental results for the transmittance of a spike filter at various angles of incidence?

5. Compare the price and quality of filters from two different suppliers. For this comparison, use a one-inch-diameter laser line filter for 6328 Å with a passband of 3 nm (FWHM) or better. From the specifications given in the supplier's catalog, which of the two would you recommend for purchase?

6. A colored filter with a thickness of 5.00 mm has an absorption coefficient of 1.25 cm^{-1} at 6328 Å. For a red HeNe laser beam, what is the equivalent optical density number of this filter? Ignore reflections from the surface of the filter.

7. A glass filter with an absorption coefficient of 1.55 cm^{-1} at 488 nm is required to have the same effect as a neutral-density filter of OD = 2.50 when placed in the argon laser beam. Neglecting surface reflections from the filter glass, how thick should the filter be?

8. Consider the absorption curve for Schott type-BG38 glass given in Figure 6. Assume that the filter we required is to be used for attenuating light from the 5145-Å argon laser line. At that wavelength, the index of refraction of the glass is approximately 1.53. Taking into account reflections at the surface of the glass filter (used at normal incidence), how thick should the filter be to allow 50.0% transmission?

9. What is the relative index of refraction (index of the glass divided by that of the surrounding medium) of a glass filter that has a correction factor (t_1t_2) of 0.900?

10. Obtain quotations from three vendors for a Pyrex flow tube that will surround a 4.0"-long xenon linear flashlamp with an outer lamp diameter of 9.0 mm. Use your judgment regarding the dimensions of the flow tube. The laser head is designed for an Nd:YAG rod and will be pumped with up to 200 joules of energy. The flow tube must be coated to minimize "solarization" and other deleterious effects. Which vendor would your recommend?

MATERIALS

1 helium-neon laser, 1 mW
1 power meter, range 1μW to 1 mW
1 set of Kodak Wratten 96 neutral-density filters—ND 3, 2, 1, 0.9, 0.8, 0.7, 0.6, 0.5, 0.4, 0.3, 0.2, 0.1
1 filter holder and stand
1 helium-neon spike filter, $\lambda_c = 633$ nm
1 rotary mount for spike filter, with scale calibrated in degrees in plane of rotation
1 lab jack

PROCEDURES

1. Determine the transmittances of three combinations of neutral-density filters, and compare the experimental and calculated values of each.

   a. Set up the filter and helium-neon laser, about one foot apart, so that the beam is incident perpendicular to the filter.

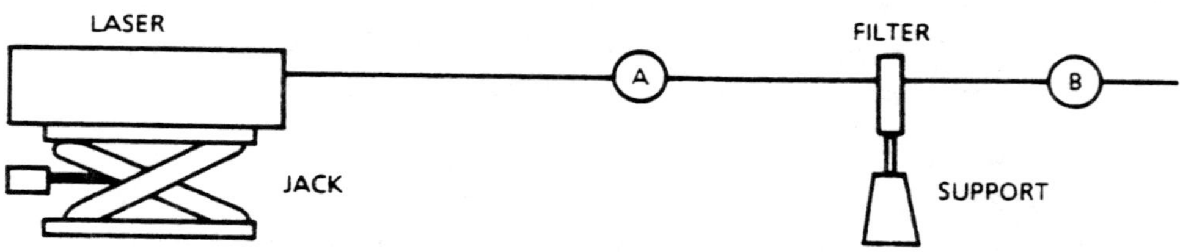

Fig. 28 Experimental arrangement.

   Adjust the spot reflected from the filter's first surface onto the laser's exit aperture, to obtain good alignment.

b. Reduce room illumination or cancel ambient light with the power meter suppression control. Measure the beam power at a point (A) before the filter. Record power (P_{in}), making sure the entire cross section of the beam is collected by the detector.

c. Measure power transmitted through the filter with the detector at (B). (If you can't reduce or suppress ambient light, use a cylindrical shield of black paper mounted on the detector to provide extra shielding.) Record the transmitted power (P_{out}) in Data Table 1.

d. Repeat Steps a through c for each filter combination.

e. Calculate the transmittances of each set by $T_{meas.} = P_{out}/P_{in}$, and compare (find the percentage of error) with the values obtained by:

$$(OD_1 + OD_2 + OD_3 \ldots) = \log_{10} T,$$

which can be rearranged to:

$$T = \frac{1}{10^{(OD_1 + OD_2 + OD_3 \ldots)}}$$

where the combined optical density is the sum of the individual optical densities ($OD_1 + OD_2 + OD_3 \ldots$).

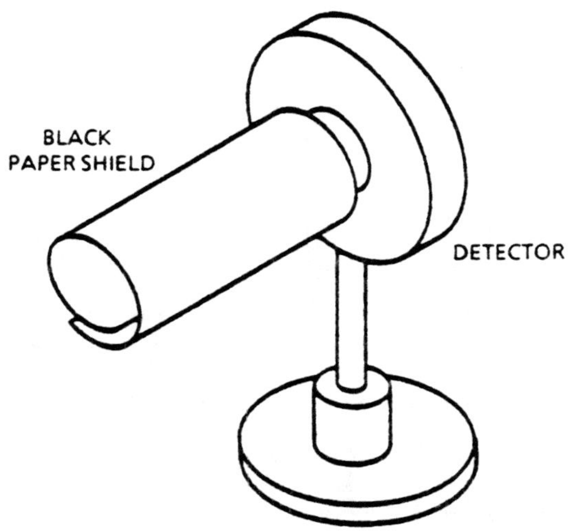

Fig. 29 Experimental arrangement.

2. Determine the transmittance of the spike at normal incidence.

a. Mount the filter on the calibrated rotary holder. Orient it so the laser beam is normally incident on the first surface of the filter, with the degree scale at zero.

b. Taking precautions to eliminate effects of ambient light, measure P_{in} and P_{out} as in Part 1. Record data in Data Table 2. If you know the transmittance of the filter, calculate and record the percentage error between the measured transmittance and the known value of T.

3. Determine the effect of the angle of incidence on the transmittance of a spike filter.

a. Using the same setup as in Part 2, measure and record P_{in}.

b. Measure P_{out} at 89.5°, 89°, 88.5°, 88°, and successive 0.5° intervals of angle of incident, down to 80°. (If the degree scale is too small, use 1° intervals.) Record each value in the Data Table 3.

c. Ensure that the detector reading is maximized for each angle of incidence, since a slight beam displacement may occur as you rotate the filter.

d. Repeat Steps b and c for an equal number of angles of incidence on the other side of the normal. Record P_{out} in Data Table 3.

e. Make a plot of the transmittance versus angle of incidence for this filter.

DATA TABLES

Data Table 1. Neutral-density Filters

	Filters	P_{in}	P_{out}	$T_{measured}$	$T_{calculated}$	% error
Combination #1						
Combination #2						
Combination #3						

Data Table 2. Transmittance at Normal Incidence

FILTER	______________
P_{in}	______________
P_{out}	______________
$T_{measured}$	______________
T_{known}	______________
% error	______________

Data Table 3. Effect of the Angle of Incidence on the Transmittance

Filter ______

P_{in} ______

Angle of Incidence	P_{out} Left of Normal	P_{out} Right of Normal	T Left	T Right

REFERENCES

Corning Glass Filter Catalog, Corning, NY.

CVI Laser Corporation Laser Products Catalog. CVI Laser Corp., 200 Dorado Place, SE, Albuquerque, NM 87192, (505) 296-9541.

"Ealing Optics Catalog," Current Year, Cambridge, MA.

Kingslake, Rudolf. Applied Optics and Optical Engineering, Vol. I. NY: Academic Press, 1965.

The Newport Corporation Catalog No. 100. Newport Corporation, 18235 Mt. Baldy Circle, Fountain Valley, CA 92728-8020, (714) 963-9811.

Optical Coating Laboratory, technical report entitled "Effects of the Variation of Angle of Incidence and Temperature on Infrared Filters," Santa Rosa, CA, November 1962.

"Optical Filters Designers Guide and Catalog," Waltham, MA: Infrared Industries, 1973.

Optical Filter Catalog. Bedford, MA: Baird Atomic, 1972.

Optics Guide 3. Melles Griot, 1770 Kettering Street, Irvine, CA 92714, (714) 261-5600

Oriel Corporation Catalog Vol. III. Oriel Corporation, 250 Long Beach Blvd., Stratford, CT 06497-0872, (203) 377-8282.

Schott Glass Catalog, West Germany.

Wood, R. W. Physical Optics. 3rd Edition, NY: Dover Publications, 1961.

LASER
ELECTRO/OPTICS

MODULE 6-7
Prisms

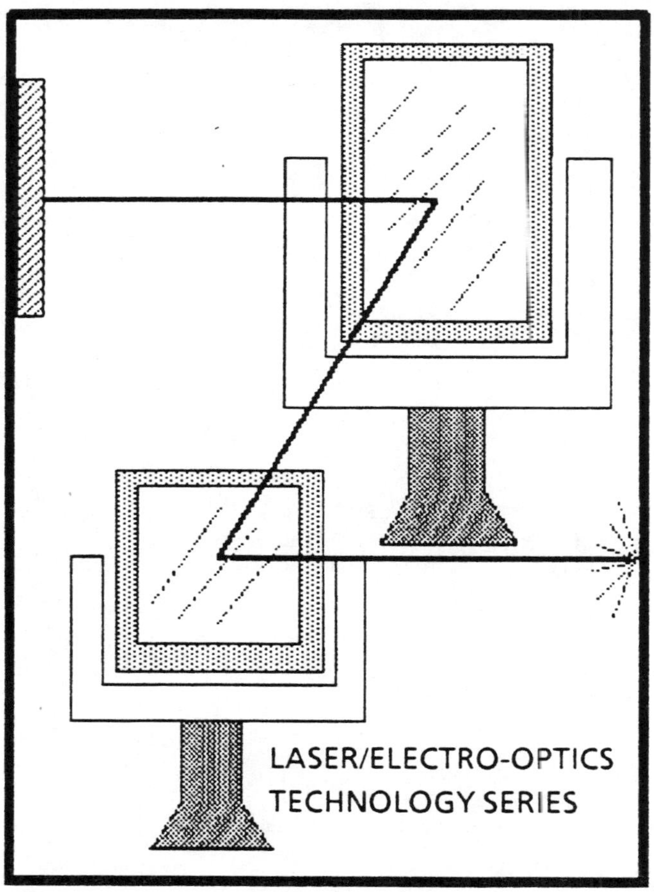

The Center for Occupational Research and Development
601 C Lake Air Drive
Waco, Texas 76710

Printed in the U.S.A.

ISBN 1-55502-024-0

PRISMS

MODULE 6-7

A prism is a solid, transparent object that has polished, plane surfaces. The surfaces generally are tilted with respect to one another. It is used to refract and/or reflect light.

Many forms and shapes of prisms are used singly or in combination in optical instruments. They may be completely transparent, or may have one or more surfaces silvered for high reflection to prevent loss of light through transmission. They may be used to displace the path of light around obstacles or to invert, reverse or rotate images. Prisms also may be used to separate light into its component frequencies or polarizations. In other words, they may be used for reflection, refraction, chromatic dispersion, polarization or some combination of these.

This module will acquaint you with functional classes of prisms. It also describes specific configurations and applications of several commonly used prisms.

OBJECTIVES

1. From drawings, identify the following prisms:
   a. Wedge
   b. Equilateral
   c. Right-angle
   d. Penta
   e. Porro
   f. Cube corner-reflector or triple-mirror
   g. Amici
   h. Dove

2. Write an explanation of each of the following terms or concepts. Convey the same meaning as described in the text:
    a. Total internal reflection
    b. Chromatic dispersion
    c. Inverted image
    d. Reversed image
    e. Rotated image
    f. Lateral displacement of light beams
    g. Ordinary ray
    h. Extraordinary ray

3. Calculate the minimum deviation angle and the index of refraction of a wedge prism, using an experimental arrangement described in the text.

4. Draw ray diagrams, tracing the light rays from a two-dimensional object through the following prisms:
    a. Right-angle
    b. Dove
    c. Amici or roof

    The diagram should show when image inversion, reversal or rotation occurs.

5. By comparing the catalogs of optical equipment suppliers, determine the supplier from whom you would recommend purchasing a Porro prism, a cube corner-reflector or a penta prism. Consider such things as quality, size, cost, delivery time, etc.

DISCUSSION

A prism is used to change the path of part or all of a beam of light that is transmitted through it. Prisms can be classified into the following categories according to their use.

a. Prisms that are used to deviate a light path by refraction.
b. Prisms that are used to disperse (separate) light into its component frequencies.
c. Prisms that are used to deviate light by reflection.
d. Prisms that are used to separate polarization components of a light beam.
e. Prisms that are used to rotate an image.
f. Prisms that can accomplish two or more of the uses mentioned above.

The following discussion will be divided into these classifications of prisms. We will give examples of instruments in which the prisms are used.

Prisms to Deviate Light by Refraction

Figure 1 is a simple sketch of a prism used to deviate light by refraction. Refraction is a change in the direction of light rays as they pass from one material to another, across a well-defined surface or interface.

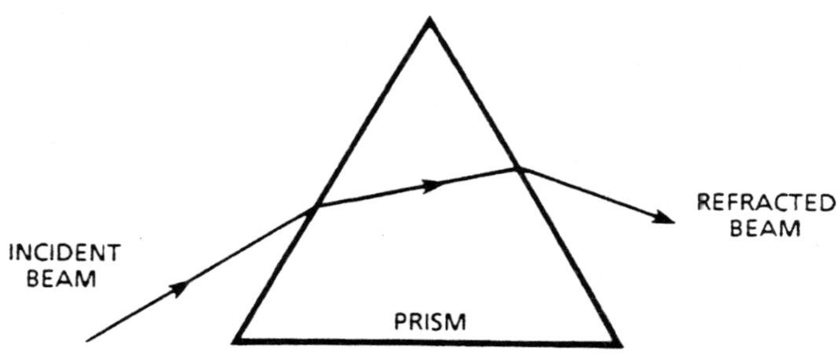

Fig. 1 Prism used to deviate light by refraction.

When you studied refraction, you learned that there's a relationship between the angle of incidence, angle of refraction, and the indices of refraction of the materials. This relation—known as Snell's law—is shown in Figure 2. The angle of incidence θ_1, and the angle of refraction θ_2 are measured between the light rays and the normal to the surface. The relationship between the angle of incidence and the angle of refraction is given by Snell's law as

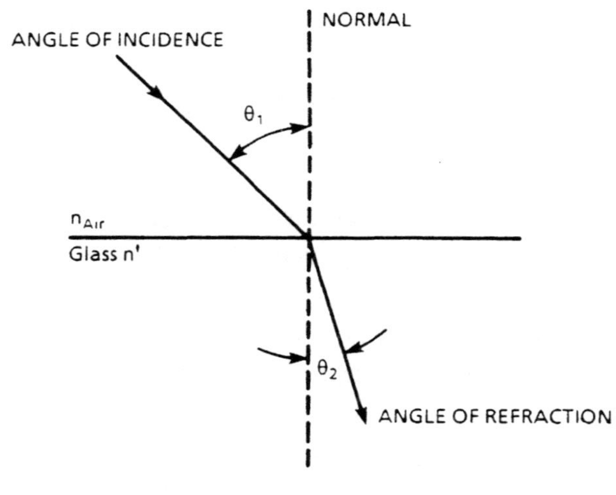

Fig. 2 Model for refraction of monochromatic light.

$$n_1 \sin\theta_1 = n_2 \sin\theta_2 \qquad \text{Equation 1}$$

where: n_1 = Index of refraction of air.

n_2 = Index of refraction of glass.

Refraction by a prism usually involves a light ray passing from air into glass and from glass into air. In a prism, the entrance and exit plane surfaces are inclined at some angle (called the apex angle) so that the deviation produced by the first is not canceled by the second but is further increased. The path of a light ray refracted through a prism according to Snell's law is traced in Figure 3.

Refraction of the ray at the two surfaces obeys Snell's law, so that in terms of the angle shown

$$\frac{\sin\theta_1}{\sin\theta_2} = \frac{n_2}{n_1} = \frac{\sin\theta_4}{\sin\theta_3} \qquad \text{Equation 2}$$

The incident ray is deviated an amount β by the first surface and an amount γ by the second surface. The total angle of deviation δ between the incident and emergent rays is the sum of the deviation angles at each surface,

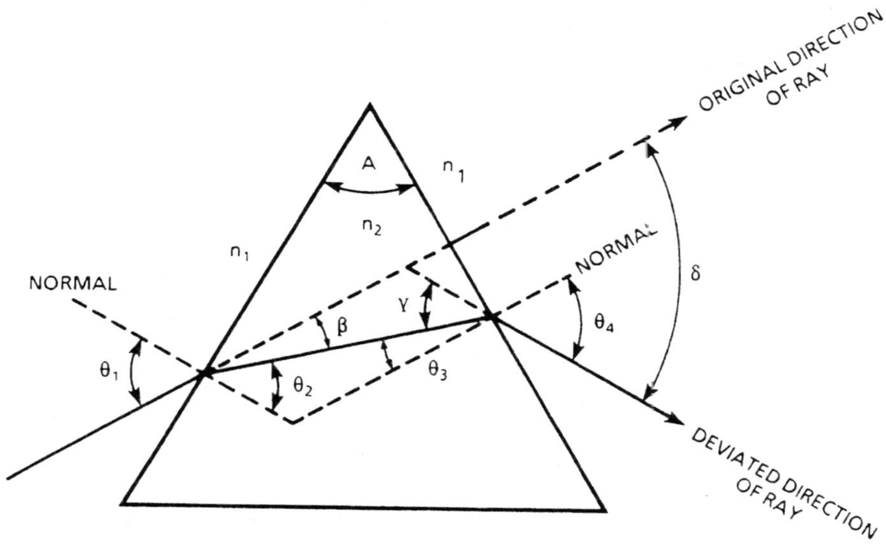

Fig. 3 Geometry associated with refraction by a prism.

$$\delta = \beta + \gamma \; . \qquad \text{Equation 3}$$

With the help of trigonometry, we also can express the deviation angle of a prism, δ, in terms of the incident angle, θ_1, the emergent angle, θ_4, and the apex angle of the prism, A.

$$\delta = \theta_1 + \theta_4 - A \qquad \text{Equation 4}$$

The index of refraction (n_2) that occurs in Snell's law is not the same for all wavelengths of light. For most transparent materials, the index increases slightly as the wavelength decreases, that is, it's slightly greater for violet light than for red light. As a result of this difference in refractive index, rays of violet light incident on a prism at a given angle are bent more sharply than are rays of red light incident at the same angle. Table 1 shows the index of refraction of borosilicate crown glass (a typical prism material) for selected wavelengths in the visible spectrum.

Table 1. Index of Refraction for Borosilicate Crown Glass as a Function of Wavelength

Color	Wavelength, nm	Index of Refraction, n
Red	640	1.50917
Yellow	589	1.51124
Green	509	1.51534
Blue	486	1.51690
Violet	434	1.52136

You can see in Figure 4 that the total angle of deviation for red light (δ_{red}) is less than the total angle of deviation for violet light (δ_{violet}). The difference in these deviation angles ($\delta_{violet} - \delta_{red}$) is a measure of how well the prism disperses white light.

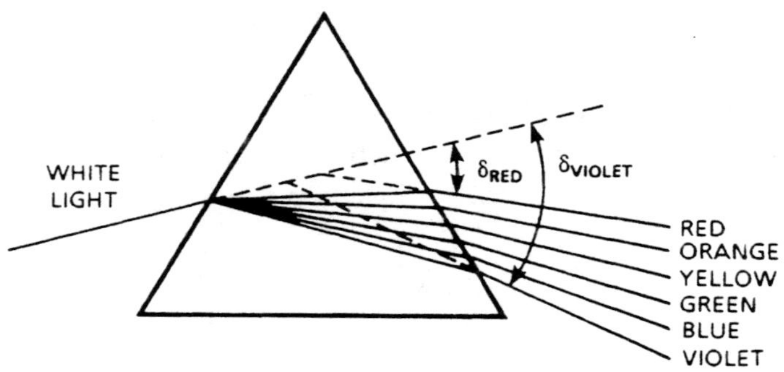

Fig. 4 Prism used to disperse white light into its component frequencies or wavelengths.

Prisms Used to Disperse Light into Its Component Frequencies

This phenomenon—known as dispersion—can be either an advantage or a disadvantage of prisms depending upon their application. For example, an equilateral prism such as in Figure 4 can be effectively used for chromatic dispersion in a spectrograph. A spectrograph uses collimated light transmitted through a prism arrangement as shown in Figure 5. The composite light beam then is focused to a line image having the red components on one end of the line and the violet components on the other end. The dispersion is

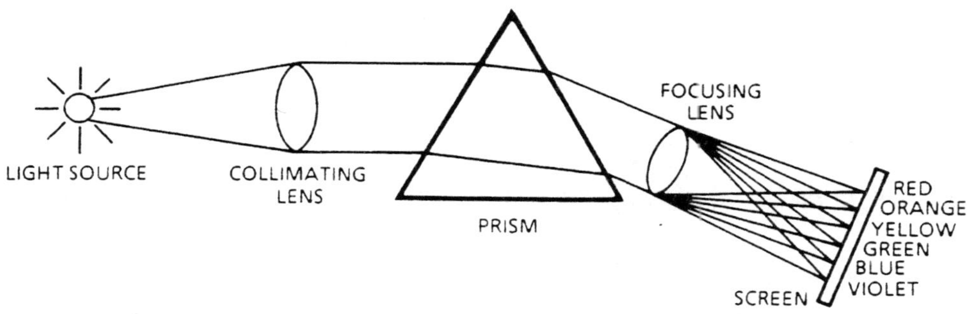

Fig. 5 Experimental arrangement for dispersion of light with a prism (i.e., a spectrograph).

too great in the equilateral prism, however, for that prism to be used when forming an image.

Minimum Deviation Angle of a Prism

For a symmetrical prism (one in which the apex angle lies at the top of an isosceles triangle), the total deviation angle δ of the light is minimized when the light ray inside the prism travels parallel to the prism's base. (See Figure 6.) Note that $\theta_1 = \theta_4$ and the ray inside the prism is parallel to the base.

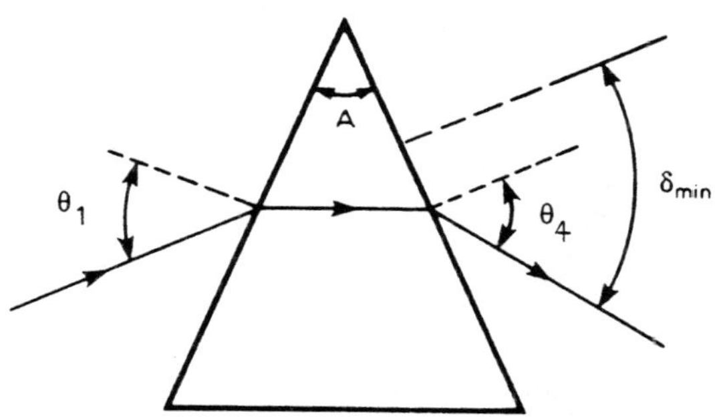

Fig. 6 Condition for the deviation of a symmetric prism to be minimized.

More advanced books on optics show that you can calculate the relative index of refraction of the prism if you measure the minimum deviation angle.

Equation 5 gives us an accurate way to determine the index of refraction of an unknown prism material if the minimum deviation angle is accurately measured.

$$n_r = \frac{\sin\left[\frac{1}{2}(A + \delta_{min})\right]}{\sin\left(\frac{1}{2}A\right)} \qquad \text{Equation 5}$$

where: $n_r = n_{glass} / n_{surroundings}$

The Dispersive Power of a Prism

The dispersive power of a prism is a measure of how well the device separates light into its component wavelengths. Dispersive power Δ is defined mathematically as the ratio of the angular dispersion of light exiting the prism to the total deviation angle.

$$\Delta = \frac{D}{\delta} = \frac{n_F - n_C}{n_D - 1} \qquad \text{Equation 6}$$

where: D = Angular separation between two reference wavelengths exiting the prism.

δ = Deviation of a third standard wavelength.

n_F, n_C, and n_D = Indices of refraction measured at the three reference wavelengths.

The relationship between D and δ is shown in Figure 7. Note that a prism with high deviation does not necessarily have good dispersion.

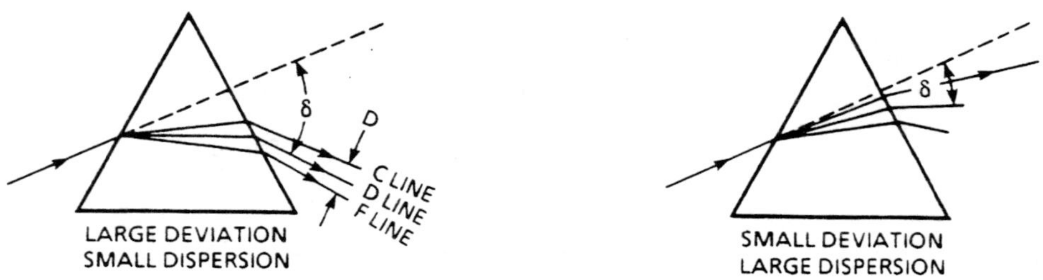

Fig. 7 Extreme cases showing the dispersion D for three wavelengths and the deviation δ for the intermediate wavelength.

Wavelengths that are traditionally used in the definition of dispersion of optical glass come from the so-called Fraunhofer lines. These three wavelengths, designated F, C and D were among the lines studied by J. von Fraunhofer in the solar spectrum. They are convenient because the lines

F and C lie at either end of the visible spectrum while D lines lie near the middle. The F and C lines originate from atomic hydrogen. The D line originates from atomic sodium. Table 2 gives the wavelength of the F, C and D lines as well as the index of refraction of crown and flint glass at those wavelengths.

Table 2. Index of Refraction of Crown and Flint Glass Measured at Three Standard Wavelengths

(nm)	Characterization	n	
		Crown glass	Flint glass
486.1	F, blue	1.5286	1.7328
589.2	D, yellow	1.5230	1.7205
656.3	C, red	1.5205	1.7076

A wedge prism, as shown in Figure 8, is a prism with a very small apex angle. Wedges are often truncated prisms. They are specified according to either the wedge angle (apex angle of prism) or the refracting angle for light entering perpendicular to one face. You can use a wedge prism to form an

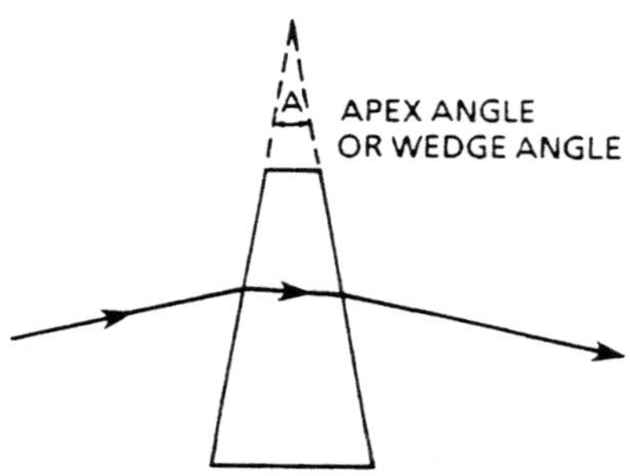

Fig. 8 Diagram of a wedge prism.

image if the image being formed is not greatly magnified. A small amount of dispersion often can be corrected by using a compound wedge prism made up of two wedges with different indices of refraction, placed together with opposite orientations as shown in Figure 9. Wedges can be rotated about an axis perpendicular to the entrance face to scan a transmitted light ray in a conical locus as in search radar.

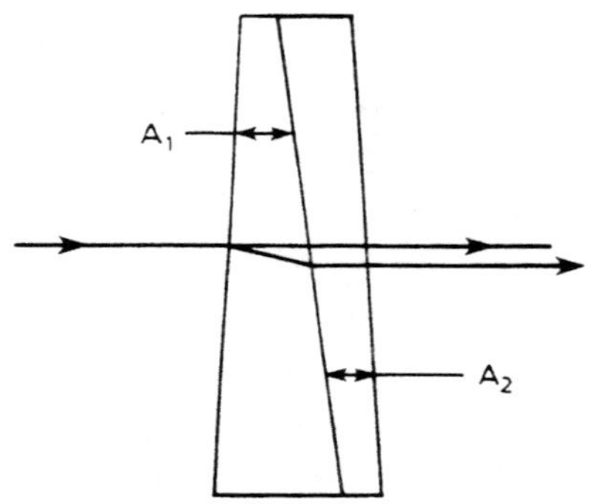

Fig. 9 Compound wedge prism.

Prisms Used to Deviate a Light Path by Reflection

Figure 10 shows a prism used to deviate a light path by reflection. In this type of prism, one or more of the plane surfaces acts as a mirror(s). Reflection from these surfaces can be achieved in either of two ways:

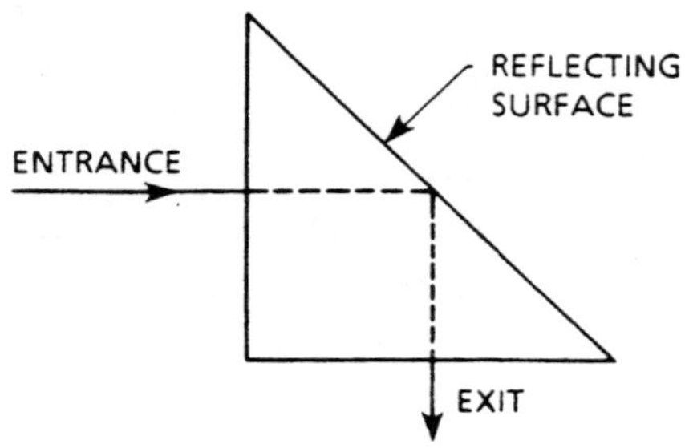

Fig. 10 Prism used to deviate light by reflection.

1. Coating the reflecting surface of the prism with a highly reflective, metallic film.
2. Designing and using the prism (without reflective coatings) in such a way that the light beam incident upon the reflecting surface exceeds the critical angle.

Right-angle prisms as shown in Figure 11 are basic prisms that can be used in several different applications.

The Porro prism is a right-angle prism in which the light enters through the face opposite the right angle. The light then is reflected twice by total internal reflection from the opposite faces. Then it exits from the same face through which it entered. It's used often as a rotating reflector, Q-switch in lasers. See a Porro prism in Figure 12.

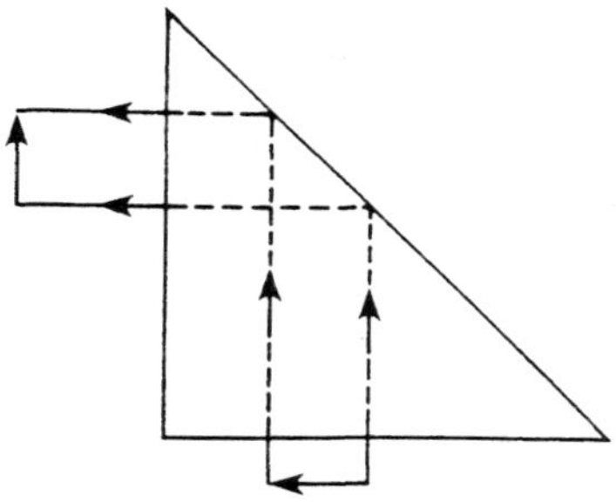

Fig. 11 Right-angle prism reflects the light path and rotates the image through 90°.

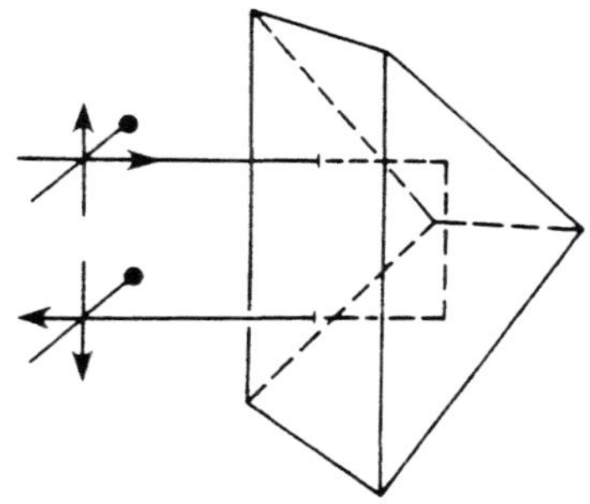

Fig. 12 Porro prism.

Notice that a Porro prism reverses an image in the plane in which the reflection takes place. When two Porro prisms are placed at right angles to each other and in such a way that the light exiting the first prism will enter the second, the image is rotated and reversed as you can see in Figure 13.

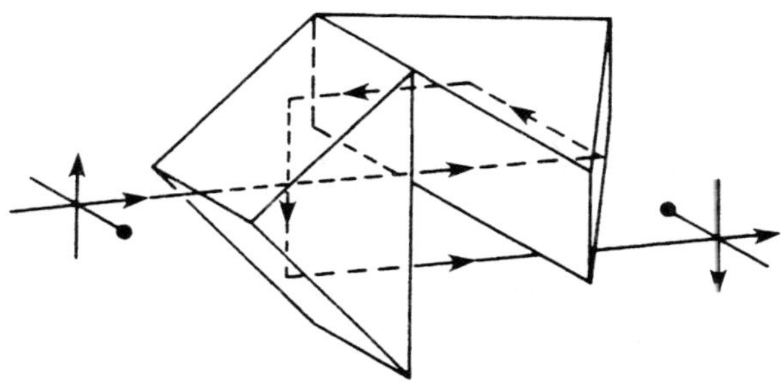

Fig. 13 Porro prism system.

This particular combination of Porro prisms commonly is used in binoculars. The image formed by the objective lens is reversed. But after the light passes through the Porro prisms, it will be reverted and will have the same orientation as the original object seen by the observer.

The Dove prism, as shown in Figure 14, is a special case of a 45° – 90° – 45° prism with the triangle truncated. The Dove prism interchanges the two rays. If the prism is rotated about the direction of the light, the rays rotate around each other twice the angular velocity of the prism. This is shown in Figure 15. This rotation feature makes a Dove prism usable in the viewfinder of a camera. Any slight change in the camera's horizontal position will be exaggerated by the prism and allow the operator to correct for it.

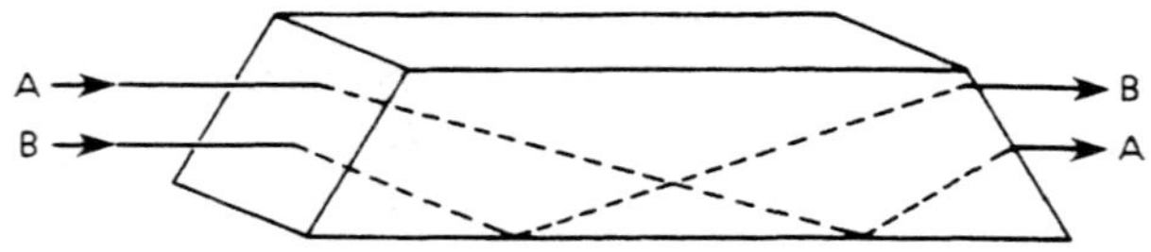

Fig. 14 Dove prism—light ray is both refracted and reflected.

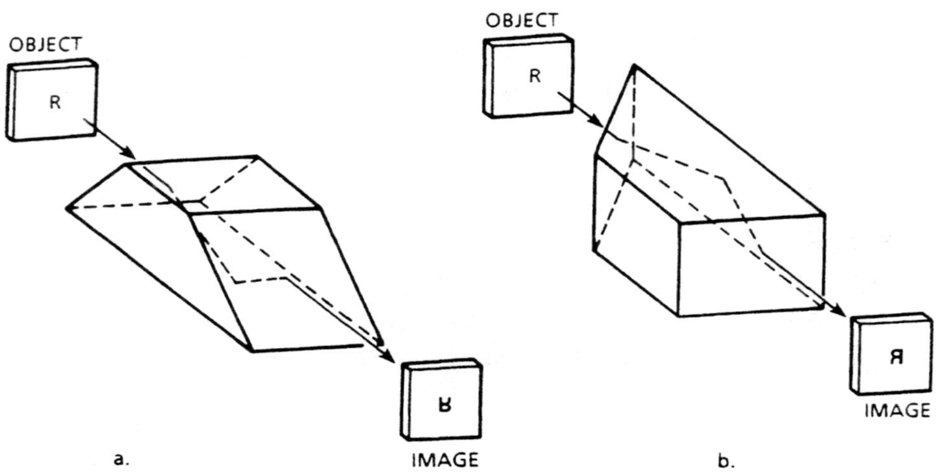

Fig. 15 Dove prism used to rotate an image.

The Amici prism is a right-angle prism that usually has a 90° roof on the face opposite the right angle. Figure 16 shows such a prism. The fact that these prisms often have inverting roofs results in their being called roof prisms. However, many other prisms can be fitted with roofs. The purpose of the roof is to invert the image left to right. If the two reflecting surfaces of the roof don't intersect within a few seconds of 90°, the beams reflected from them will not match and a double image will be formed.

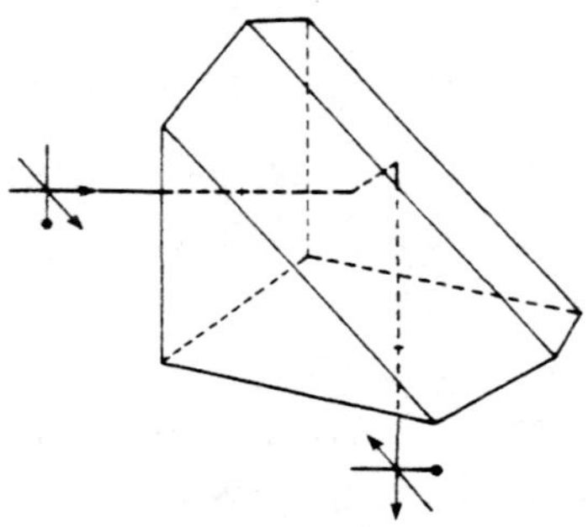

Fig. 16 Amici prism or roof prism.

Amici prisms often are used in simple direct-vision spectrographs and in sighting equipment for the military.

The cube corner reflector uses three mutually orthogonal, reflective surfaces. It redirects a beam through 180° regardless of incident angle to the prism. The cube corner reflector is used as a passive cooperative target for laser radars, range measurement, and angle-tracking systems. (See Figure 17.)

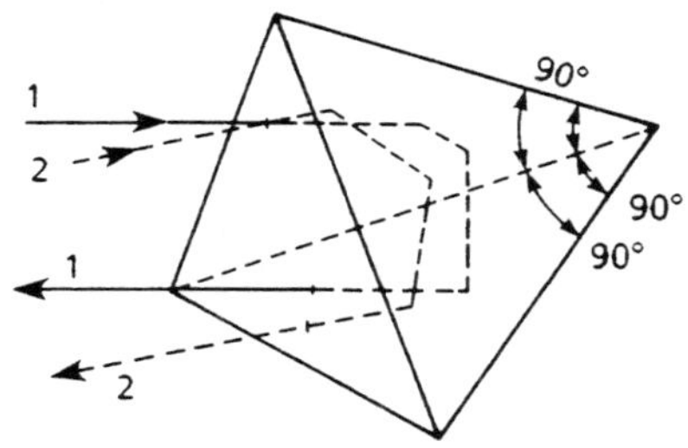

Fig. 17 Cube corner reflector.

A Penta prism, as shown in Figure 18, has five sides with one angle equal to 90°.

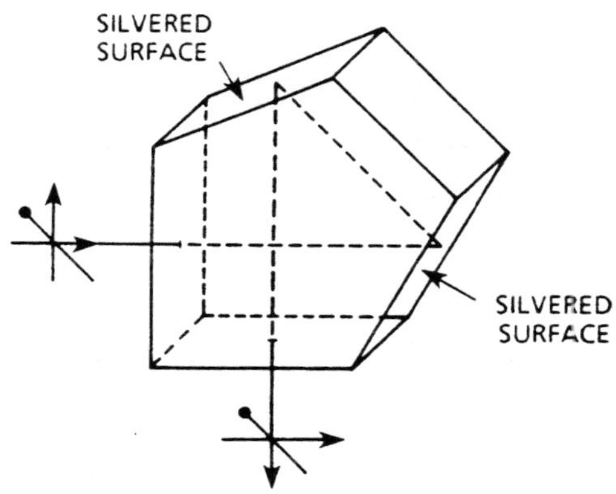

Fig. 18 Penta prism.

This prism deviates the beams exactly 90° in the principal plane even if the beam doesn't strike the end faces exactly normal. Both reflecting surfaces must be silvered because the angle of incidence is not large enough to permit total internal reflection.

High-precision rangefinders use Penta prisms. The angles measured by these rangefinders are so small that, if such prisms were not used, the normal bending of the tube of the instrument would produce large errors. Penta prisms minimize the bending effect.

When you merely want to displace the axis of a beam without introducing deviation and without reverting or inverting the image, you can use a rhomboidal prism. (See Figure 19.) They often are applied in eyepieces where we need to provide a way to adjust to various observers.

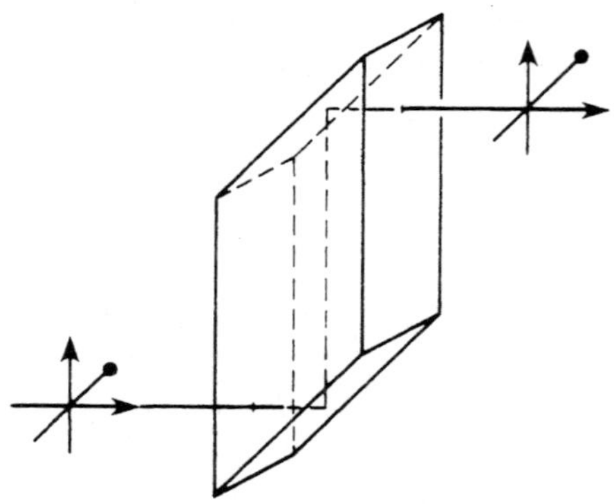

Fig. 19 Rhomboidal prism.

Prisms Used to Separate Polarization Components of a Light Beam

A beam of ordinary (unpolarized) light is composed of random, transverse vibrations of the electric field. Each vibration is caused by one of many different atoms in the light source.

The amplitudes of each of the many random transverse vibrations can be divided into components along two orthogonal axes, each perpendicular to the direction of propagation. This is called "dividing" the light into two polarized components.

It's enough for this module for you to know that certain prisms can be used to separate the two perpendicular polarization components of a light beam. These prisms are made of an optical material that has a characteristic called "birefringence" or "double refraction." Birefringence in an optical material means that the material behaves as if it has a different index of refraction for one polarized component of a light beam than for the other polarized component. If the material is made into a prism, it will refract or bend one polarized component of a light beam more than the other polarized component, since the angle of refraction, in Snell's law, depends on the indices of refraction. As a result of this, a single, ordinary beam will enter a polarizing prism and two polarized beams will emerge, either displaced and traveling parallel or traveling in slightly different directions.

The two rays that emerge are called the "O-ray" or ordinary ray, and the "E-ray" or extraordinary ray. Only the ordinary ray follows the path predicted by using Snell's law.

The optical materials in which the phenomenon of double refraction is observed are crystalline quartz and calcite. Calcite is used more frequently because the separation of the two polarization components is greater. Calcite (calcium carbonate) has a crystal structure that causes it to form easily or cleave easily into a rhombohedral shape (as shown in Figure 20). Each face of the crystal is a parallelogram whose angles are 71° and 109°. Figure 20 shows the double refraction of light in a calcite prism. Both the E and O rays are shown, with the E-ray emerging above the O-ray.

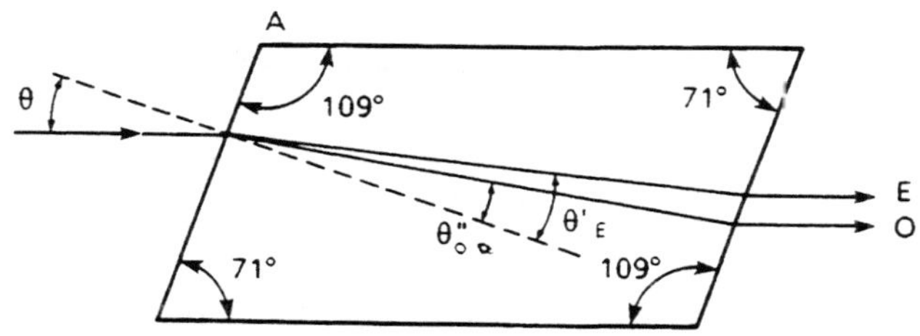

Fig. 20 Double refraction of light in a calcite crystal prism.

Figure 21 shows a Nicol prism, also made of calcite. In the Nicol prism, the calcite rhombohedron is precisely cut into two halves. Then it's polished and cemented back together with Canada balsam. Because of its geometry, and the relative indices of refraction of calcite and Canada balsam, the Nicol prism will remove one of the two refracted rays by total reflection. The Nicol prism is therefore used as a polarizer or analyzer.

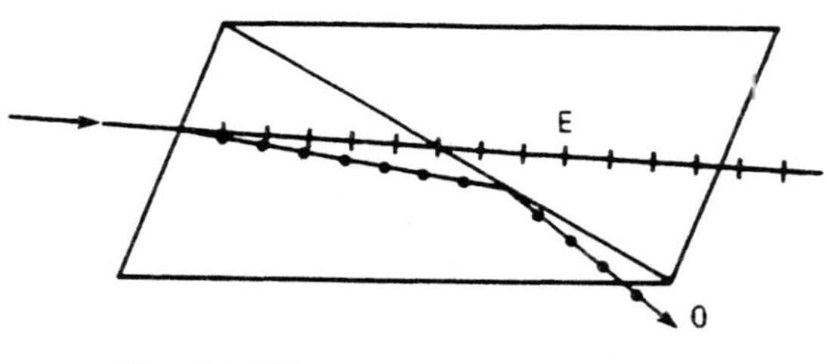

Fig. 21 Diagram of a Nicol prism.

We often want to split a light beam into two linearly polarized components that leave the prism in different directions. This is accomplished in a Wollaston prism, made from two pieces of crystalline quartz. The two pieces are cut from different crystal orientations and cemented together with glycerine or castor oil. See Figure 22. A Wollaston prism also can be used to combine two light beams of different polarization. The arrangement would be the same as that shown in Figure 21, but the arrows on the light rays now would be reversed.

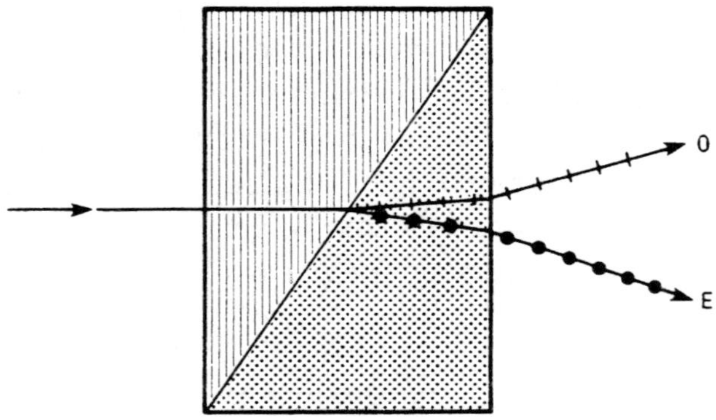

Fig. 22 Wollaston prism for angular separation of light beams of two different polarizations.

Prism polarizers are much more effective at polarizing light than are inexpensive sheets of Polaroid film. A common test of a polarizer's effectiveness is the so-called "extinction ratio." As defined in Equation 7, the ER for a pair of polarizers is the ratio of the maximum light transmitted by a pair of polarizers, divided by the minimum power transmitted.

$$ER = \frac{I_{max}}{I_{min}} \qquad \text{Equation 7}$$

where:

I_{max}, I_{min} = Maximum and minimum power transmitted by the polarizers when they are in the "parallel" and "crossed" orientation respectively.

Inexpensive polarizing sheets (such as Polaroid H, K or HR types) often have extinction ratios of 300 to 500. Prism polarizers can have ERs of 10^6 or more, but cost about one hundred times as much as the sheet-type polarizers. The inexpensive sheet polarizers work on selective absorption of light while the prism polarizers work on refraction and, often, reflection. So the prism polarizers are much less prone to damage from intense light, such as is present in a typical laser beam.

EXERCISES

Exercises 1 through 8 consist of identifying the eight prisms shown. In the space provided by each drawing, explain the important characteristics of that prism, as outlined in the text.

1. 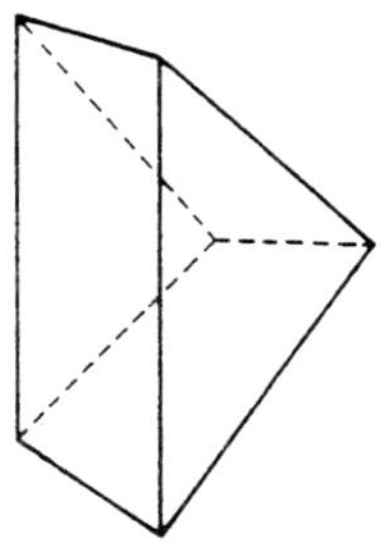

2. 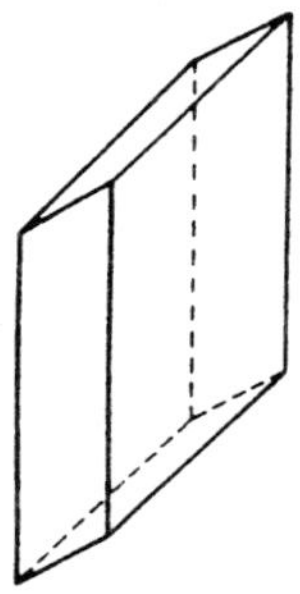

3. 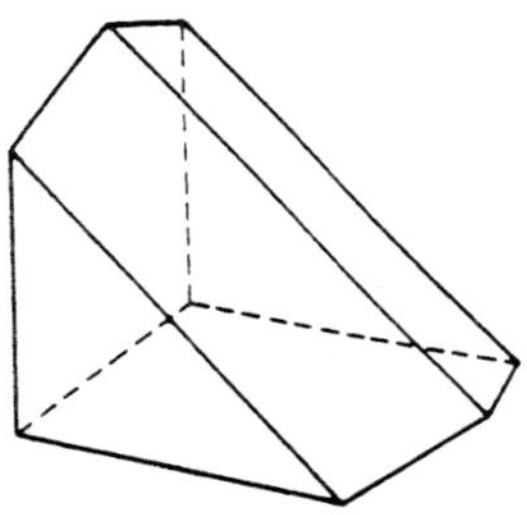

4. 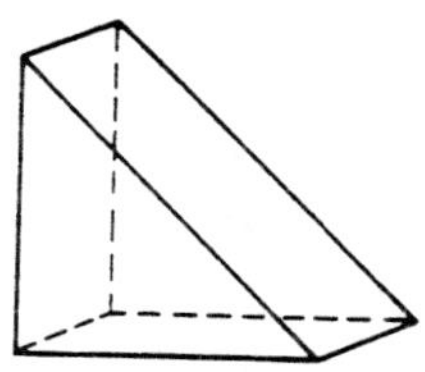

5. 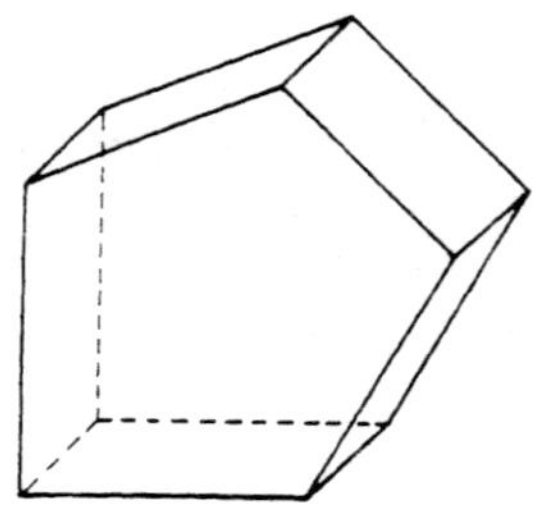

6. 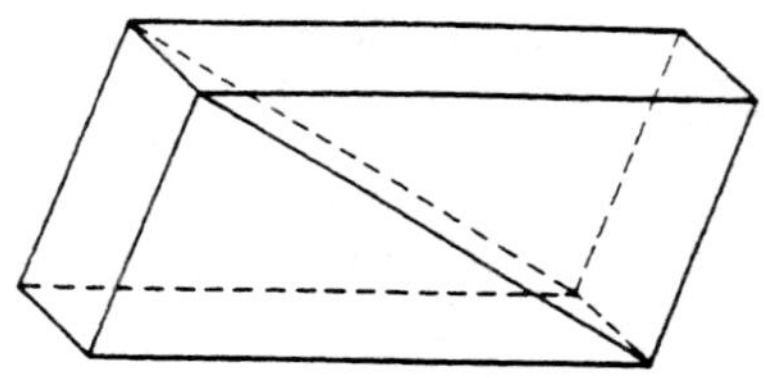

7. 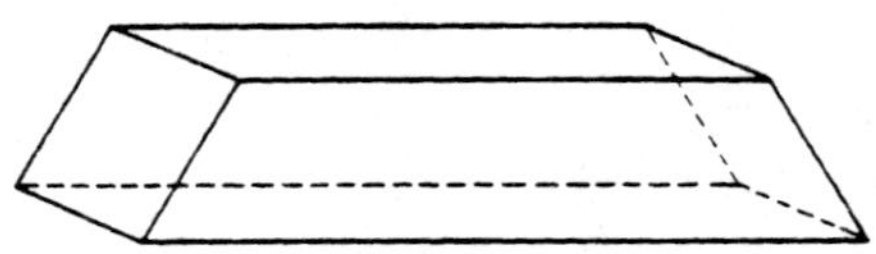

8. 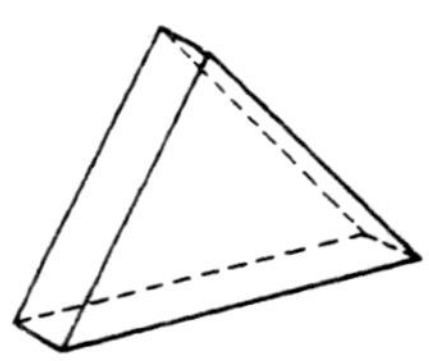

9. Describe the following terms or concepts:
   a. Total internal reflection
   b. Chromatic dispersion
   c. Inverted image
   d. Reversed image
   e. Rotated image
   f. Lateral displacement of light beam
   g. Ordinary ray
   h. Extraordinary ray

10. Calculate the minimum deviation angle and the index of refraction of a wedge prism. The experimental arrangement is the same as used in the Procedures. Distance from the prism to the screen (L) is 1.5 meters. Deviation of the beam by the prism (D) is 20 cm. Apex angle of the prism (A) is 15°.

11. Draw ray diagrams, tracing the light rays from a two-dimensional object through the following prisms:
    a. Right-angle prism
    b. Dove prism
    c. Amici or roof prism

    The diagrams should show when image inversion, reversal or rotation occurs.

12. Compare catalogs of three suppliers of optical equipment. Recommend a supplier from which to purchase a Porro prism, a cube corner-reflector and a Penta prism. Compare such things as quality, size, cost, delivery time, etc.

13. A prism with an apex angle of 60.0° is constructed of unknown glass. When the prism is illuminated with a narrow beam of collimated light at a wavelength of 4861 Å, the minimum deviation angle is measured to be 60.1°. What is the index of refraction of the prism (relative to air)? Of what type of glass is the prism probably made?

14. Using data given in Table 2, compute the dispersive power of crown and flint glass. Which material would be the best choice from which to fabricate a prism for separating visible light into its component wavelengths?

15. Get price and technical information from three vendors on a tuning prism to be used inside an argon laser cavity. The prism must be AR coated for the principal argon visible laser lines and should be equilateral and about 1 cm on a side. Determine as much useful information as possible from each vendor such as: damage threshold of coating, reflective loss at the prism surface, type of glass used, dispersive

power (calculate or obtain from vendors), surface quality, etc. From your information, which prism would you purchase?

MATERIALS

1Wedge prism
1 Right-angle prism
1 Amici prism
1 Dove prism
1 1-mW HeNe laser
2 One-meter optical benches
1 Incandescent light source
1 Prism holder with capability for rotating prism about horizontal axis
2 Lens holders
1 Ground glass screen
1 Screen holder
1 Rotation table component support
1 "F-shaped" source aperture
2 Lenses, positive, 6-inch focal length, 2-inch diameter
1 1-inch-diameter stop

PROCEDURES

1. Arrange a HeNe laser, a wedge prism on a rotary support, and a screen as shown in Figure 23.

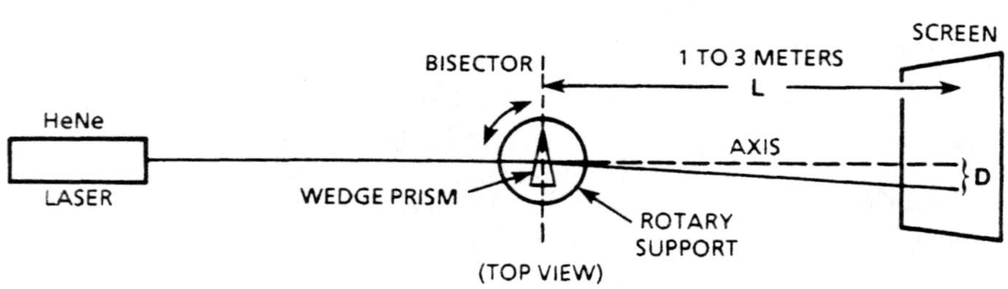

Fig. 23 Experimental arrangement for minimum deviation with a wedge prism.

   a. Adjust heights of laser, prism and screen so that the beam passes through the prism and falls on the screen. The beam should be approximately level, and distance L should be at least one meter.

b. Rotate the prism support so that the reflected beam from the first surface falls on a vertical line passing through the exit hole of the laser beam. **Observe safety precautions!** The reflected spot probably will be above or below the exit hole. If it falls on the hole, omit Step c.

c. Adjust the tilt of the laser so that the reflected spot enters the exit hole. (Shim the back end if the spot is high, the front end if the spot is low.) The beam is now perpendicular to the first face of the prism.

d. Rotate the prism support back and forth a few degrees while you watch the spot on the screen. As the prism bisector approaches 90° to the axis of the system, the distance D on the screen will decrease, reaching a minimum at 90°. (Try several rotations through this position, to get the most accurate setting.) The prism now is oriented to produce its minimum deviation of the refracted beam.

e. Carefully mark the point on the screen where the beam strikes.

f. Accurately measure the distance L, from the apex of the prism, perpendicularly, to the screen. Record this distance.

g. Remove the prism. Mark the point where the undeviated beam strikes the screen.

h. Measure, as accurately as you can, the distance D. Record.

i. Calculate the angle of minimum deviation of the prism, δ_{min}, using:

$$\tan \delta_{min} = \frac{D}{L}$$

j. Measure the apex angle, A, of the wedge prism. With the angle δ_{min} calculated in Step i, and the apex angle A, calculate the index of refraction of the prism from Equation 5.

$$n = \frac{\sin\left(\frac{\theta min + A}{2}\right)}{\sin \frac{A}{2}}$$

From this value of n, determine the type of glass of which the prism might be made.

2. a. Visually examine the effect on the "F"-source aperture of the right-angle, Amici, and Dove prisms. Use the viewing positions shown in Figure 24.

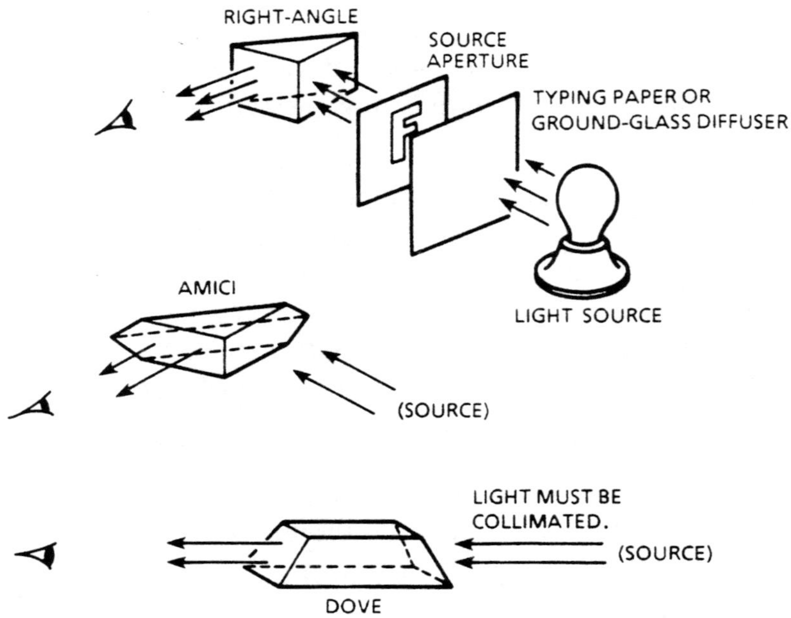

Fig. 24 Experimental arrangement.

Record the orientation of the "F" in terms of upright-inverted and normal-reversed.

b. Explain the results of Step 2a, tracing rays through each prism in appropriate sketches.

REFERENCES

Hecht, E. Optics, 2nd edition. Addison-Wesley, 1987.

Jacobs, Donald H. Fundamentals of Optical Engineering. New York: McGraw-Hill Book Company, Inc., 1943.

Jenkins, F. A., and H. E. White. Fundamentals of Optics, 4th edition. New York: McGraw-Hill Book Company, Inc., 1976.

Kingslake, Rudolf. Applied Optics and Optical Engineering, Vol. 3. New York: Academic Press, 1965.

Mauro, J. A. Optical Engineering Handbook. Syracuse, NY: General Electric Company, 1966.

Optics Guide 3. Melles Griot, 1770 Kettering Street, Irvine, CA 92714, (714) 261-5600.

Pedrotti, Frank L., and Leno S. Pedrotti. Introduction to Optics, Chapter 3. Englewood Cliffs, NJ: Prentice-Hall, Inc., 1987.

Shortley, G., and D. Williams. Elements of Physics. Englewood Cliffs, NJ: Prentice-Hall, 1955.

SPIE. Basic Optics and Optical Instruments, Vol. 1. Redondo Beach, CA: Northrop Corporation, 1965.

LASER/
ELECTRO-OPTICS

MODULE 6-8
LENSES

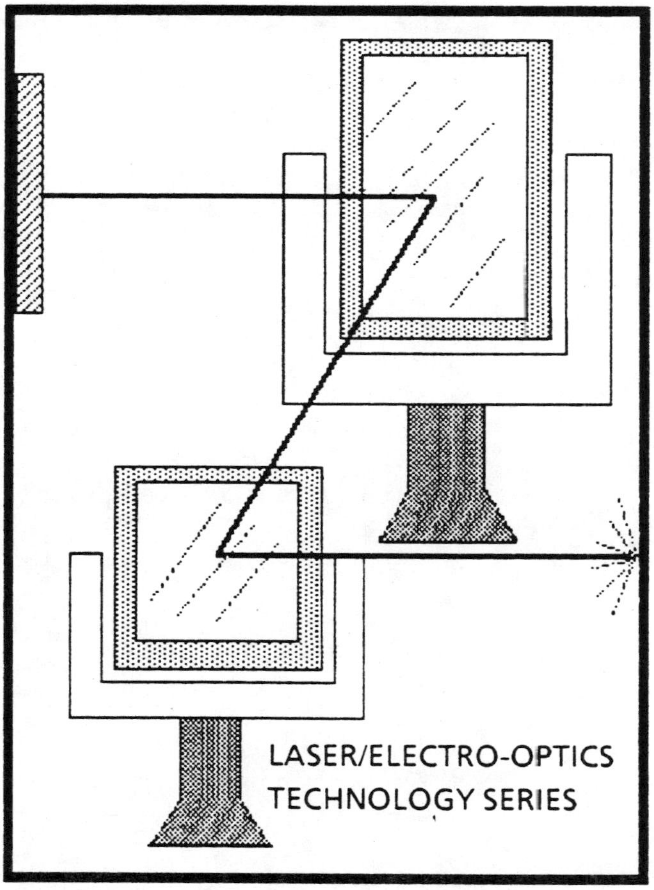

CORD

The Center for Occupational Research and Development
601 C Lake Air Drive
Waco, Texas 76710

Printed in the U.S.A.

ISBN 1-55502-024-0

LENSES

MODULE 6-8

People have understood and used simple lenses for hundreds of years. References to the use of a magnifying glass and telescopes occur in writings from the eleventh century. The Romans named the "*lens*" because they saw a resemblance between this optical component and the flat round seeds of the lentil plant.

An optical lens is a carefully ground or molded piece of glass, plastic, or other transparent material that has opposite surfaces. Either or both of these surfaces are curved. Lenses bend light rays so that they diverge or converge to form an image. Lenses are essential components for telescopes, microscopes, binoculars, eyeglasses, cameras, projectors and many other items of everyday use that employ optical components.

This module will acquaint you with functional classes of lenses and will describe specific configurations and applications of several commonly used lenses.

Before you begin this module, you should have studied the principles of refraction and ray tracing optical components.

OBJECTIVES

When you complete this module, you should be able to do the following.

1. Explain these terms, concepts or instruments. You should convey the same meaning as given in the text.
   a. Converging lens
   b. Diverging lens
   c. Power of a lens

d. Focal length of a lens
e. Lens systems
f. Collimation
g. Spherical aberration
h. Chromatic aberration
i. Coma
j. Astigmatism
k. Distortion
l. Field curvature
m. AR coatings
n. Damage threshold

2. From a cross-sectional diagram, identify the following lenses:
   a. Biconvex
   b. Plano-convex
   c. Convex meniscus
   d. Biconcave
   e. Plano-concave
   f. Aspherical
   g. Cylindrical
   h. Fresnel
3. Experimentally determine the focal length of a concave lens and a convex lens.
4. Construct a simple beam expander using two positive lenses.
5. Illustrate spherical aberration in lenses and a way to read it.
6. Write an explanation of the effects of spherical aberration on a camera.

DISCUSSION

Lenses are used to change the direction of rays of light. Figure 1 shows how lenses form images by refraction. The effect of a lens on light is embodied in Snell's law—the law of refraction. This law states that, in passing from a rarer medium into a denser one, light is refracted toward the normal. In passing from a denser to a rarer medium, light is refracted away from the normal. The degree of bending or refracting is in accordance with the equation:

Equation 1

$$n_1 \sin\theta = n_2 \sin\theta_2$$

where: n_1 and n_2 = Indices of the two media

θ_1 and θ_2 = Angles of incidence and refraction respectively

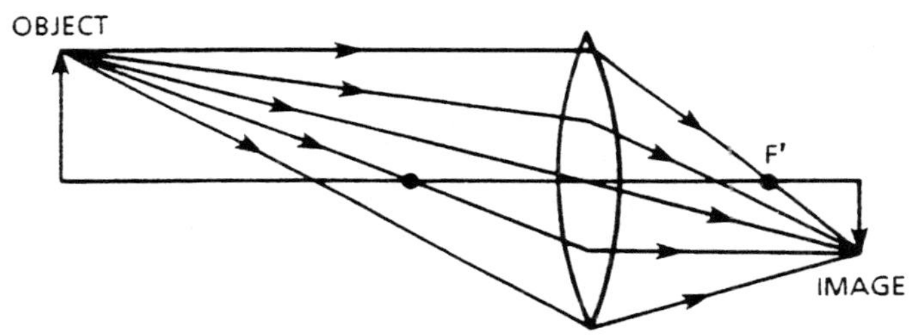

Fig. 1 Image formation by a simple lens.

Let's build a mental picture of a lens that has a number of straight-line prisms as shown in Figure 2a. When the number of prisms is very large, as shown in Figure 2b, the multiple straight-line segments begin to approach the curved surface of a lens. Figure 2 illustrates the law of refraction.

There are many types of lenses. But they all can be classified into two general types according to the effect they have on a parallel beam of light. The two types are *converging* and *diverging* lenses.

1. *Converging* lenses also are called "positive," "plus" or "convex" lenses. They are thicker in the middle than at the edges. As shown in Figure 3, they cause both parallel rays of light and diverging rays of light on the opposite side of the lens.

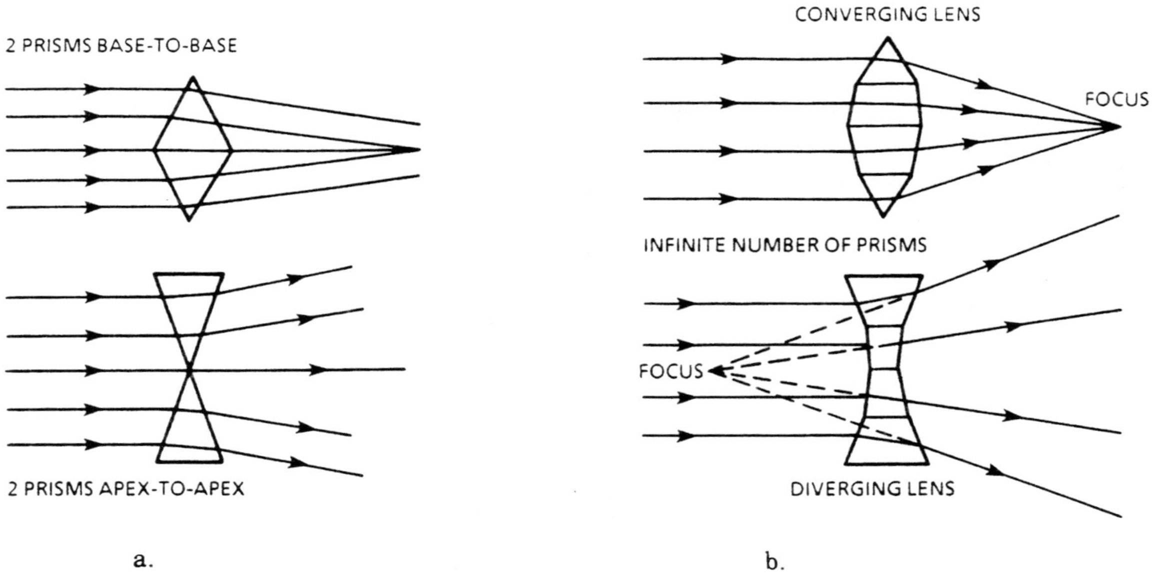

Fig. 2 Evolution of lenses from basic prism pairs.

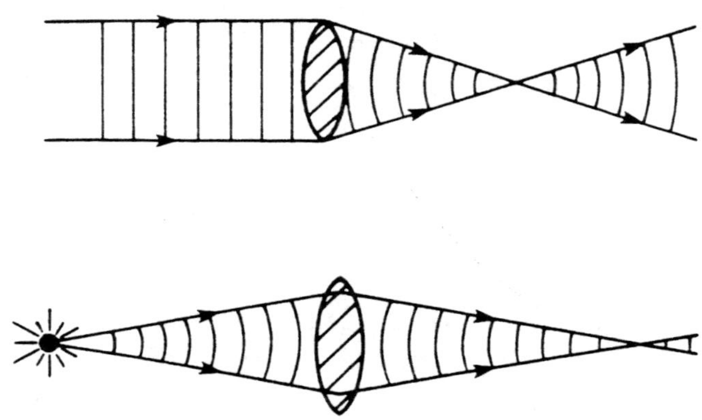

Fig. 3 Effect of positive (converging) lenses on plane and diverging spherical wave fronts.

2. *Diverging* lenses are known as "negative," "minus" or "concave" lenses. They are thinner in the middle than at the edges. Figure 4 shows us that a diverging lens causes parallel rays of light to diverge or spread in opposite directions on the other side of the lens. If rays initially are diverging toward such a lens, they will be made to diverge even more strongly after they pass through the lens. (See Figure 4b.)

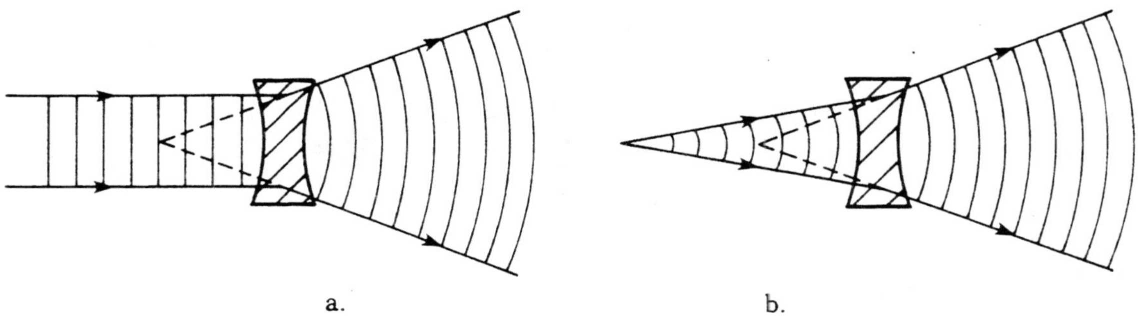

Fig. 4 Effect of negative (diverging) lenses on plane and diverging spherical wave fronts.

All lenses can be classified as converging or diverging. But further subdivisions of these two basic types can be made according to the curvature of the lens surfaces and the power of the lens. Spherical lenses are lenses whose surfaces are spherical in shape. Spherical lenses can be classified into the six sub-types shown in Figure 5. The biconvex lens—"i"—is used the most of these six.

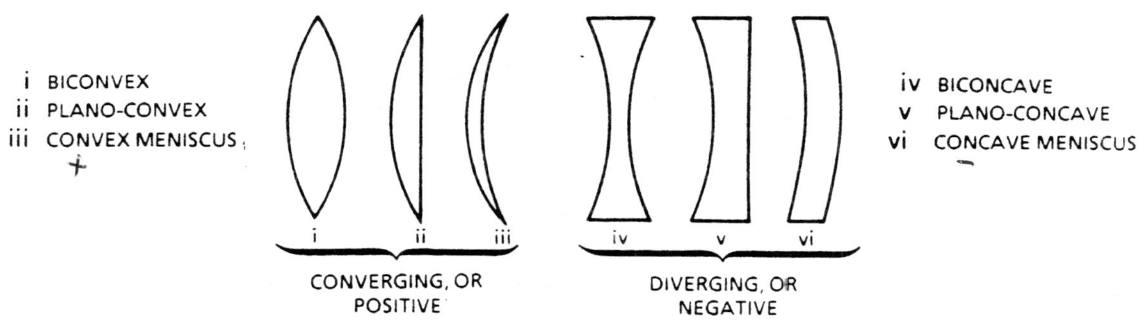

Fig. 5 Types of simple spherical lenses.

You should have studied the definition of focal points and focal lengths for simple lenses already. It would be a good idea for you to review that definition now. Here we'll briefly restate the essentials.

The *focal length* of a "thin lens" is the distance the focal point is removed from the vertical centerline of the lens. The focal point F' (Figure 6) of a positive lens is that point where parallel rays of light from the left are brought together. The focal point F on the left side of the positive lens is that point to which parallel rays, incident on the lens from right-to-left, would converge.

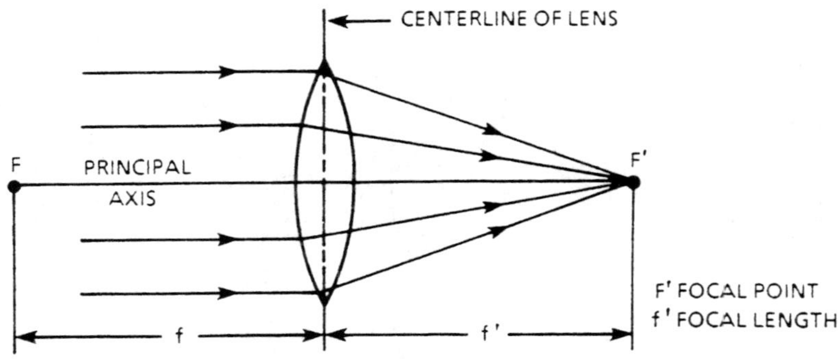

Fig. 6 Parallel rays of light brought to a focus by a positive thin lens.

The focal point F (Figure 7) of a negative lens is the point where parallel rays of light seem to originate. The focal point F' on the right side of the lens is the point from which light would seem to diverge, after passing through the lens, if it were incident on the lens from right-to-left.

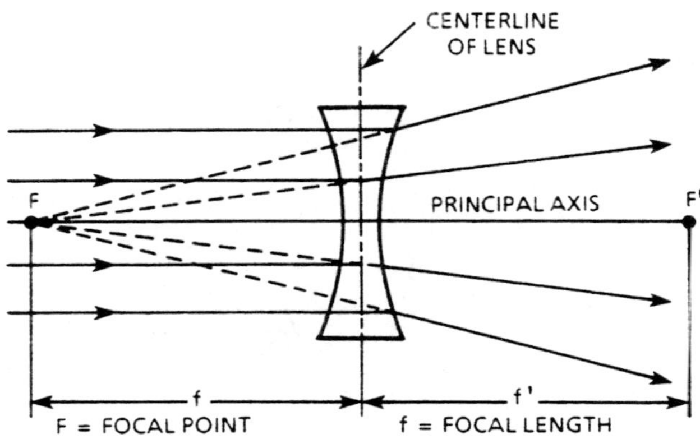

Fig. 7 Parallel rays of light incident on a negative thin lens diverge as if coming from a point F on the left.

Focal lengths are shown as f' and f in Figures 6 and 7. For a thin lens, f and f' are equal.

The *power* of a lens is the reciprocal of its focal length in meters. The power of a lens is a measure of its ability to converge or diverge light rays. For example, the higher the positive power, the more converging the lens. The unit of power is the "diopter." Diopter usually is designated as D. One diopter is the power of a lens with a focal length of one meter. So a converging lens with a focal length of 20 cm (0.2 m) has a power of $1/0.2\ \text{m} = 5\ \text{D}$. Note that a lens that causes light to converge has a positive power, and a lens that causes light to diverge has a negative power. As an example, a diverging lens with a focal length of – 25 cm (– 0.25 m) has a power of $1/-0.25\ \text{m} = -4$ D.

A second type of lens whose basic shape differs from the spherical lens is the cylindrical lens. Cylindrical lenses are so named because their surfaces are cylindrical in shape.

Figure 8 shows how a cylindrical lens focuses incident parallel light onto a focal line rather than onto a focal point as in the spherical lens. Cylindrical lenses are used in optical data processing and in other applications where we don't need the formation of an image in true dimensions. Cylindrical lenses

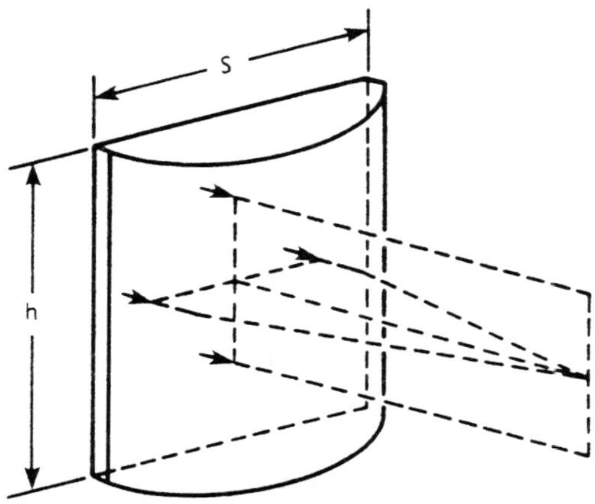

Fig. 8 Cylindrical lens.

sometimes are used also to image the rectangular mode shape of a pulsed nitrogen laser into a straight line. For example, they would be used in pumping an organic dye laser cell.

A third type of lens that's classified by its shape is the Fresnel lens. The Fresnel lens looks different from the spherical and cylindrical lenses. A Fresnel lens is a disk-shaped optical component with almost no thickness. It focuses and collimates light by means of a series of concentric (like a bull's-eye) grooves. Curvature of the grooves causes refraction of the incident beam, approximating that of a conventionally shaped lens as shown in Figure 9.

Figure 10 shows Fresnel lenses used as a collimator and as a collector. You can understand the effect of a Fresnel lens on light by again constructing the lens from many small prisms. The difference between conventional and Fresnel lenses is that the Fresnel lens has prisms laid out in a circular pattern in the same plane instead of being stacked together to form a regular lens.

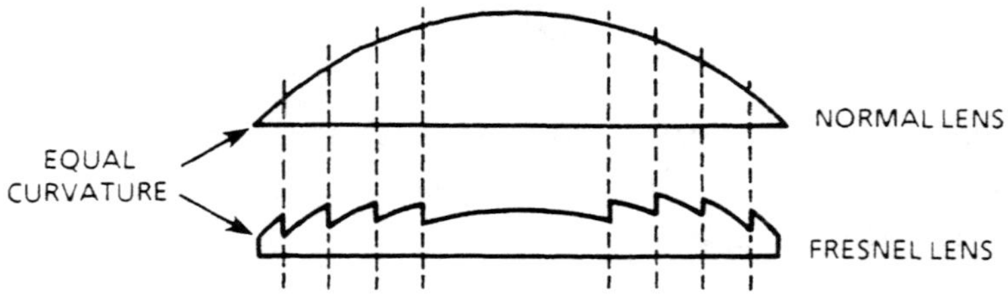

Fig. 9 Comparison of the shape of a normal lens to that of a Fresnel lens.

Fig. 10 Action of Fresnel lens.

Fresnel lenses often are constructed of rigid acrylic. They are available in positive, negative, and cylindrical configurations. Fresnel lenses are durable, lightweight ($^1/_{10}$ the weight of equivalent conventional elements) and comparatively inexpensive. Their optical quality is relatively poor. Fresnel lenses typically are used as field lenses, condensers and magnifiers, in viewers, microfilm readers, copiers, compact optical systems, and in wide-angle visual simulators. A most frequent application is in traffic lights.

Various types of lenses are used in different applications. The biconvex lens is the most common and most important lens. Since this is the type of lens in the eye for vision, we will take a brief look at its characteristics.

The lens of the eye is made of a transparent, flexible medium with attached muscle so that the lens can be made to change its shape (and consequently its focal length). The power of the lens thus is adjusted to focus objects located at different distances from the eye onto the back of the eye where the images eventually are interpreted by the brain. This changing of power is called accommodation of the eye. It's shown in Figure 11.

When the eye muscle is relaxed, the lens is shaped so that light coming from an object at infinity is focused onto the back of the eye as shown in Figure 11a. When an object closer to the eye is viewed, the eye must accommodate and thicken the lens so that the image point again is placed on the back of the

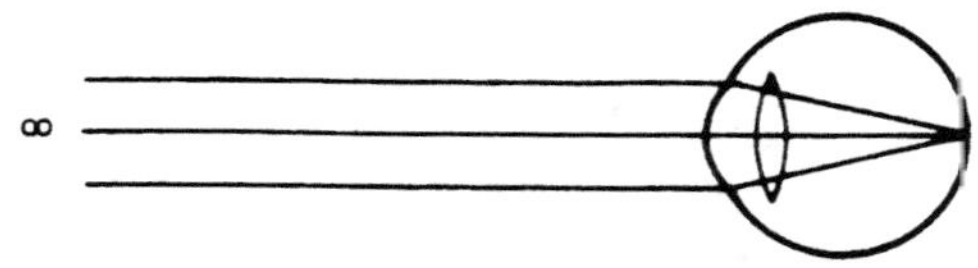

a. Lens focused on distant object.

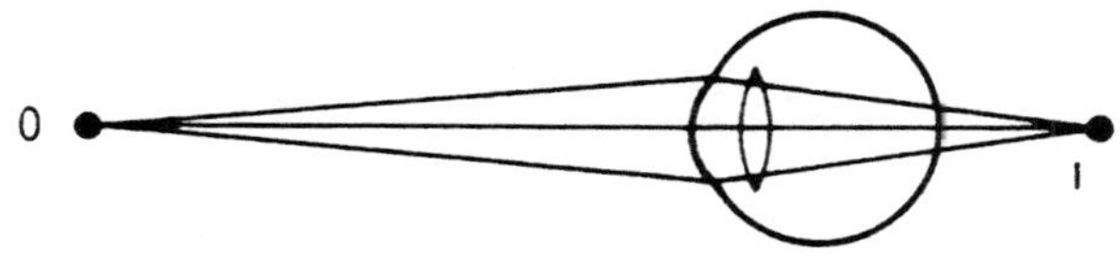

b. Unfocused lens, before accommodating to near object.

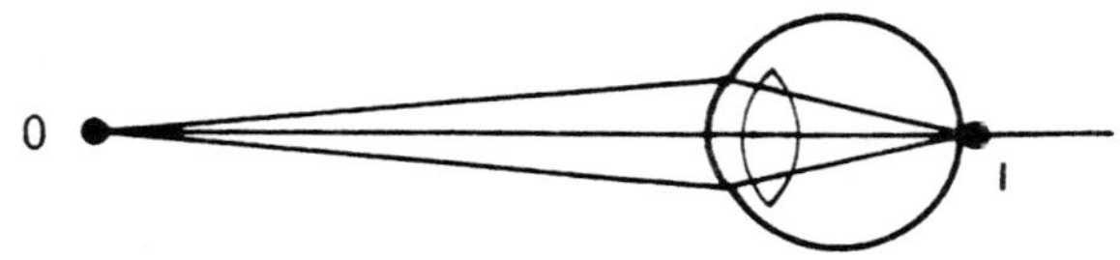

c. Lens thickened—accommodated to near object.

Fig. 11 Accommodation of the eye.

eye, as shown in Figures 11b and 11c. Accommodation in a normal eye lets us clearly see objects located anywhere from infinity to ten inches from the eye.

Next, we will examine several applications of the simple positive lens. A single lens forms the basis for a simple magnifier as well as a basic part of the microscope and telescope. So, first we will discuss the essential optics of the magnifier. Then we will examine the optics of a searchlight, essentially a scheme for "throwing" a collimated beam of light a long way. Following this we'll list other common uses of the positive lens.

The simple magnifier is a single lens that lets us see objects that are too small to be seen clearly by the unaided eye. To see a small object distinctly with the unaided eye, we have to place the object as near to the eye as possible. However, as we said in the last paragraph, the normal eye can focus clearly only on objects no closer than approximately 10 inches. The simple magnifier shown in Figure 12 is an "optical instrument" that, when focused on a nearby tiny object, produces a magnified, virtual image at a convenient distance for clear vision.

A searchlight is an optical device that collects light from an intense "point" source and transmits this light as a nearly parallel, well-collimated beam. In many cases this is accomplished with parabolic mirrors. But in principle it can be done with a large lens as shown in Figure 13.

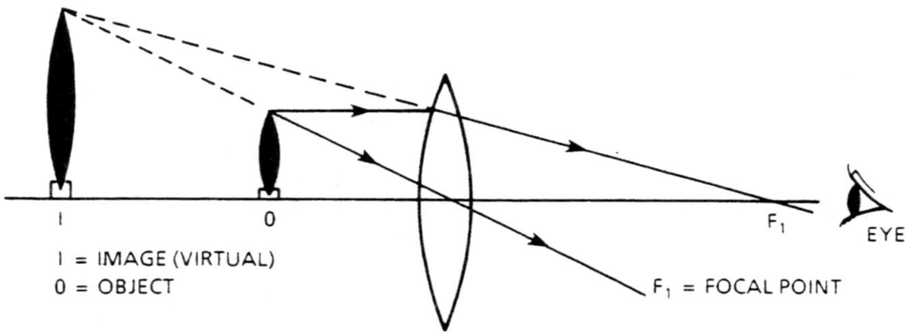

Fig. 12 Virtual image formed by positive lens acting as a magnifier.

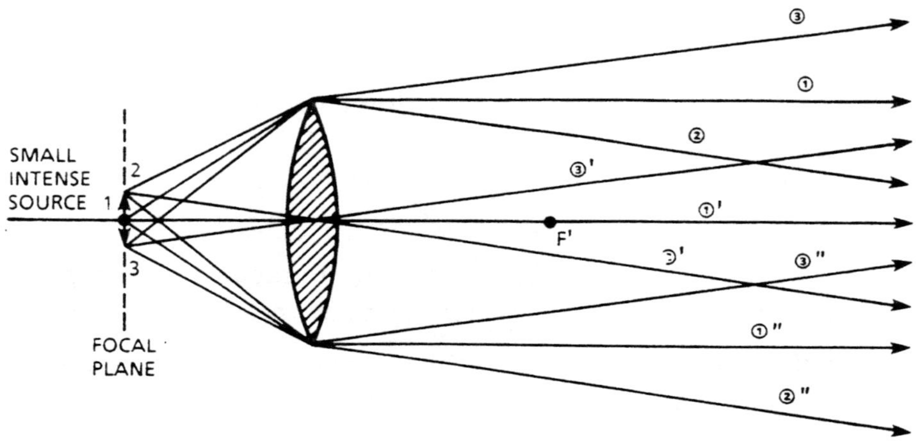

Fig. 13 Simple example of a searchlight.

The small source in Figure 13 is shown exaggerated in size for clarity. It's made up of points 1, 2, 3. Since the source lies in the focal plane of the lens, each point on the source leads to a beam of parallel light leaving the lens. For example, point 1 gives the parallel bundle ①, ①', ①". Point 2 gives the parallel bundle ②, ②', ②". Point 3 gives the parallel bundle ③, ③', ③". You can see that, if the source becomes very small and point 2 and 3 move toward point 1, nearly all of the rays will be parallel to rays ①, ①', and ①". The beam then is said to be well-collimated, progressing to the right with very little spreading or divergence.

Now consider the applications shown in Figure 14. Figure 14a shows the essential geometry of a motion-picture projector, where the object is located a distance more than one focal length (but less than two focal lengths) from the lens and is projected as a real, inverted, magnified image. If the object is moved to exactly two focal lengths from the lens, a real inverted image of the same size is formed two focal lengths on the other side of the lens. This scheme for one-to-one reproduction is used in a copy machine as shown in Figure 14b.

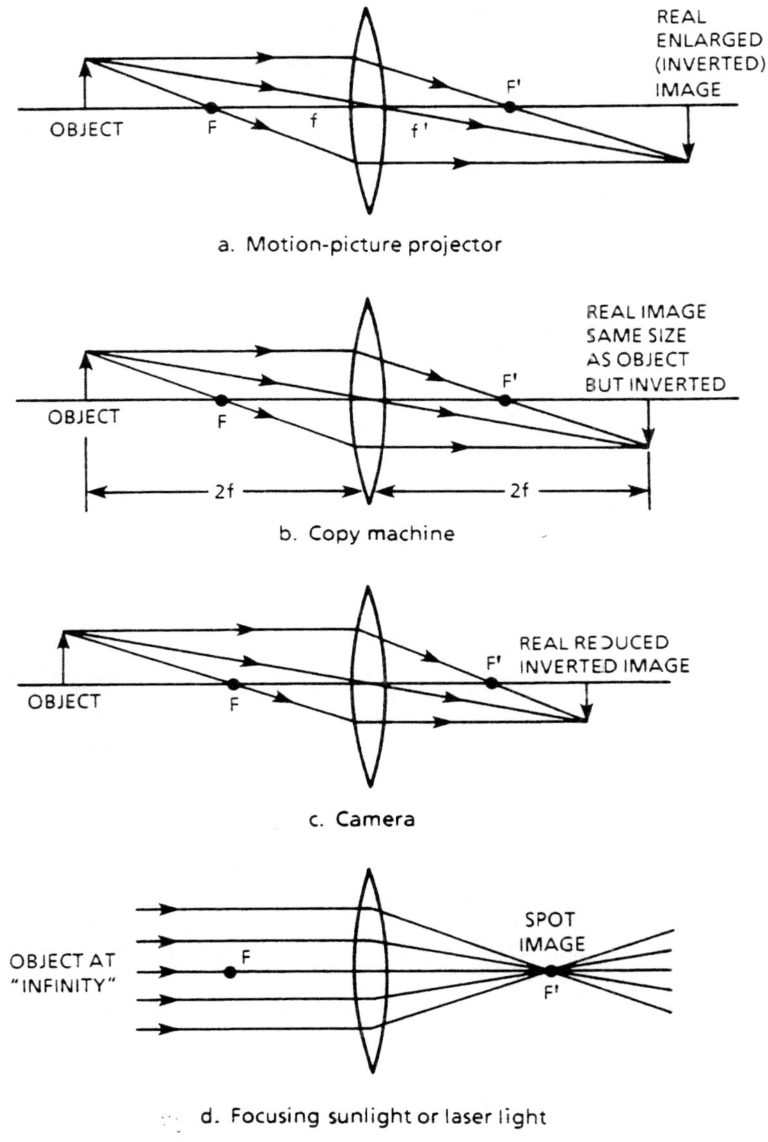

Fig. 14 Other applications of the positive lens.

Objects that lie some distance greater than twice the focal length from the lens are projected as real, inverted images of reduced size. This is often the case with optics in the ordinary camera, as shown in Figure 14c.

Finally, if the object is very far away from the lens, the rays of light that enter the lens are essentially parallel to one another and image in the focal plane of the lens, as a tiny spot. This is the example shown in Figure 14d, where a simple positive lens is used to focus the collimated light from the distant sun or the highly parallel light from a laser. You should see that the geometry in Figure 14d is the "opposite" of the geometry for the searchlight in Figure 13. Here parallel light is focused as a spot. In Figure 13, light diverging from a tiny spot is transmitted on a parallel beam.

Concave and convex meniscus lenses are used to correct myopia (nearsightedness) and hyperopia (farsightedness). The concave and convex meniscus lenses are used instead of regular concave or convex lenses to enable the eye to scan a large field of view and yet always be the same distance from the meniscus lenses. Figure 15 illustrates the operation of a normal eye, the problem with the nearsighted and farsighted eye, and the correction introduced with the appropriate meniscus lenses.

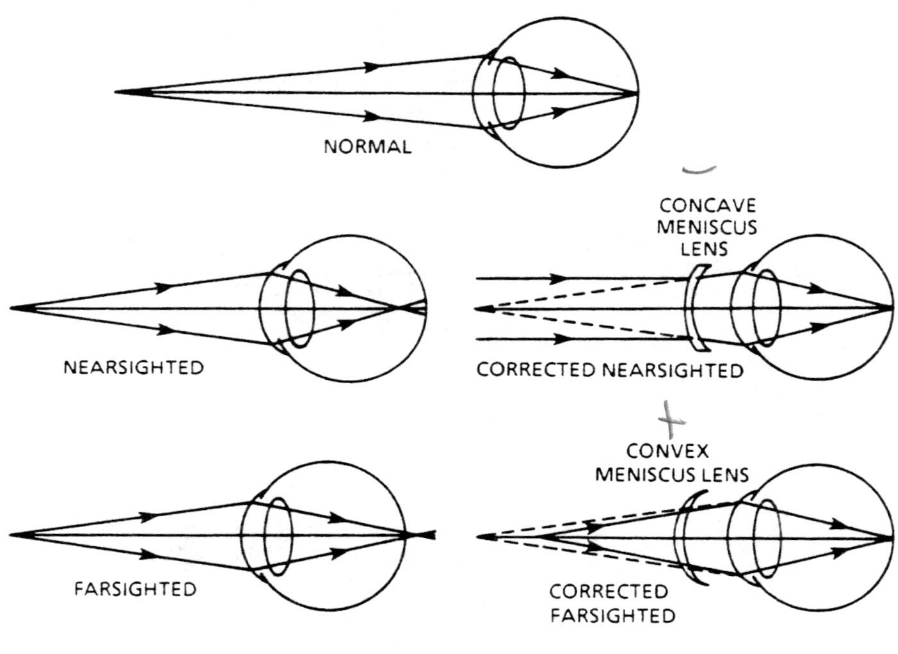

Fig. 15 Vision defects and corrections.

Uses of Lens Systems

Some uses for the single lens have been mentioned already. Complex optical tasks generally require the use of two or more lenses in an optical system. One such well-known lens system is the telescope, shown in Figures 16 and 17. The Galilean telescope, shown in Figure 16, gives an upright image of a distant object.

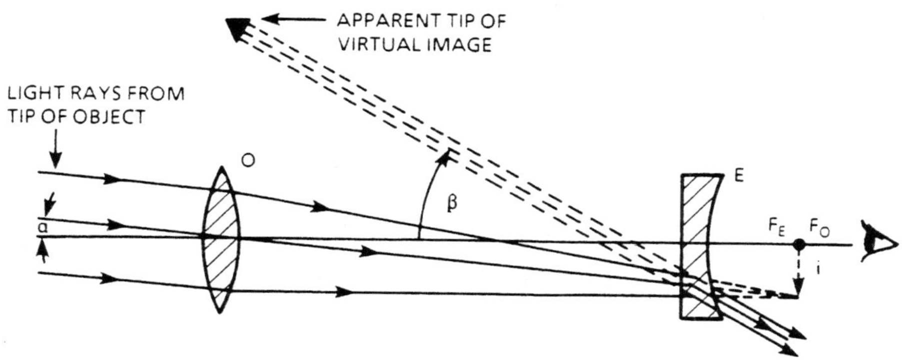

Fig. 16 Galilean telescope.

The Galilean telescope consists of a positive objective O and a negative lens eyepiece E. Their focal points F_E and F_O are in coincidence, as shown. If no eyepiece lens were present, rays from a distant object would, after refraction through the objective lens, meet to form a real, inverted image "i" in the focal plane. However, because these rays are intercepted at the eyepiece, in front of the focal plane, the rays are diverged and seem to the eye to be coming from an object of much larger dimensions. The dotted lines at angle β indicate the direction along which the tip of the virtual image is seen. With the Galilean telescope the image is erect and magnified. The amount of magnification is the ratio of angle β to angle α. This type of telescope is preferred for use as field or opera glasses because it is very short and compact and gives a bright, erect image.

The telescope in Figure 17 is called a celestial or astronomical telescope. This telescope uses two lens systems, the objective lens system and the eyepiece lens system. Rays from a distant object are shown entering a long-focal-length objective lens as a parallel beam. These rays are brought to a focus and form a real, reduced image in the focal plane of the objective, F_O'.

If we assume the distant object to be an arrow with the arrowhead upward, the image is real and inverted, as shown. The real image, viewed by the eye through the eyepiece lens system, is seen as a magnified, virtual image as shown by the dotted lines. The image, compared with the original object, is inverted. But this is not important when we look at stars, for example.

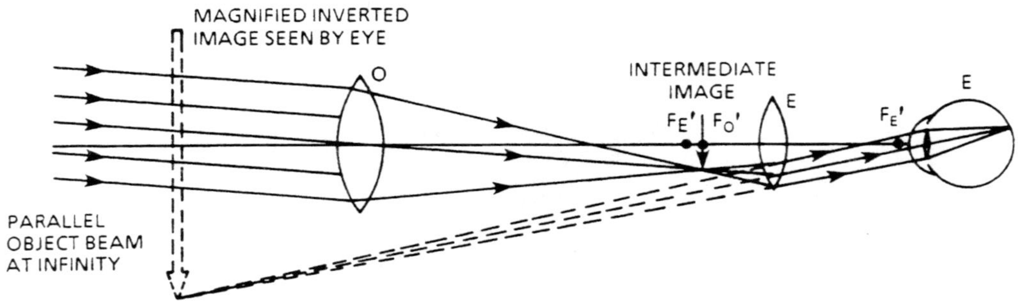

Fig. 17 Astronomical or celestial telescope.

The celestial telescope is the basis for many instruments used in optics. The transit, sight level, alignment telescope, and the collimator all use a telescope in one form or another.

One use of a telescope of this type in laser work is for the function of beam expansion. As shown in Figure 18, if a beam of light from the laser is used to illuminate the eyepiece and therefore is sent through the telescope "backward," it will emerge from the objective lens larger in diameter than when it entered and the rays will be somewhat more collimated or parallel.

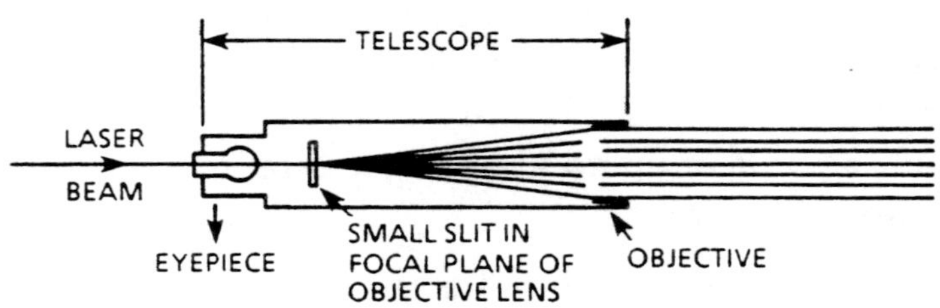

Fig. 18 Principle of beam collimation and expansion.

As shown in Figure 19, there are two different ways to produce beam expansion or beam reduction. The first method (Figure 19a) uses a convex lens followed by another convex lens. Note that the focal point F_1 of lens L_1 falls on the focal point F_2 of lens L_2. This beam expander/collimator often is called the Keplerian beam expander. It and the one shown in Figure 19b have names derived from alleged inventors of refracting telescopes that closely resemble the beam expanders.

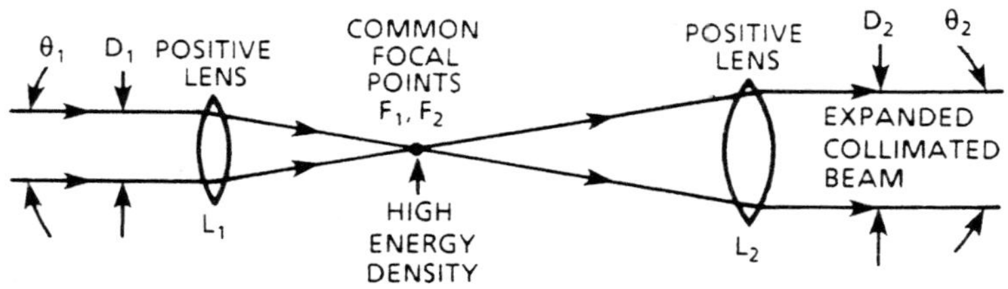

a. Often called "Keplerian."

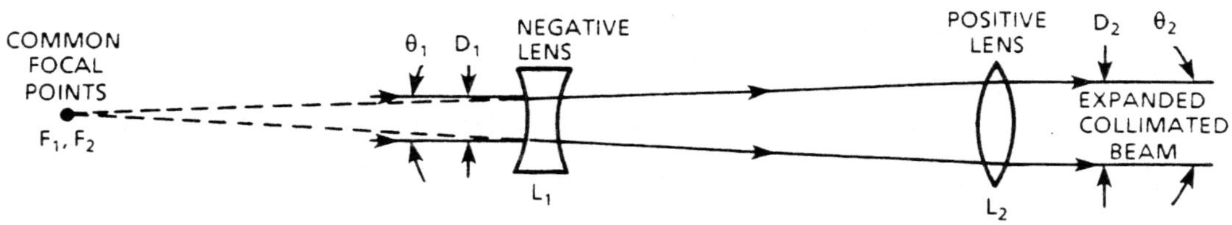

b. Often called "Galilean."

Fig. 19 Two common types of beam expanders/collimators.

The second method uses a concave lens followed by a convex. Note again that the focal point F_1 of lens L_1 falls on the focal point F_2 of lens L_2. This is called a Galilean beam expander.

In high-powered laser applications it's important to use the second method of beam expansion. This is because the small beam spot created at the common focal points in the first method results in high energy densities, which may cause the air to break down or ionize.

For either version of the two-lens beam expander, we can define the lateral magnification as

$$M = \frac{D_2}{D_1} \qquad \text{Equation 2}$$

Optics texts show that the lateral magnification of the beam expander is given simply by the ratio of the focal lengths,

$$M = -\frac{f_2}{f_1} \qquad \text{Equation 3}$$

We also can show that, to a good approximation, the divergence angle of the expanded beam is related to the divergence angle of the incident unexpanded beam by,

$$\theta_2 = -\frac{\theta_1}{M} \qquad \text{Equation 4}$$

Equations 2, 3, and 4 assume that the beam expander is focused properly. In other words, the distances between the center (approximately) of the lenses is equal to $f_2 - f_1$. Note that f_2 can be either positive or negative.

For laser applications it's important to realize that, as the beam expander increases the beam diameter, it also decreases the beam divergence angle by the same amount. It's usually necessary to first expand a beam before illuminating a distant target with a laser, if the beam size on the distant target is to be minimized.

Another use of lens systems is in the compound microscope. The microscope is very similar to the telescope except that the objective lens now has a much smaller focal length. So it has a definite focal point close to the front lens. In the compound microscope shown in Figure 20, the objective lens forms an enlarged image I_1 of the small object at O.

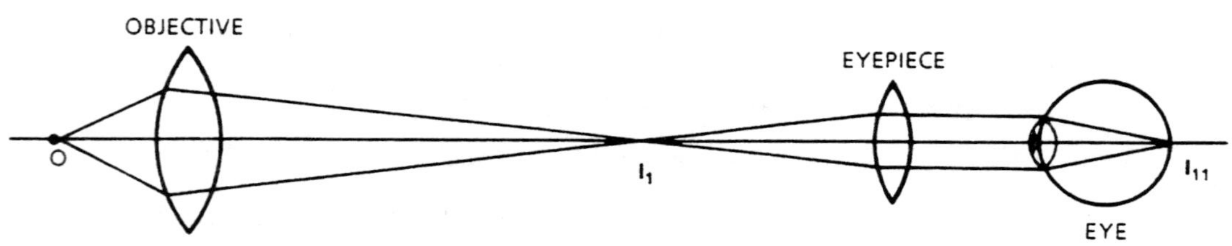

Fig. 20 Compound microscope.

The eye—viewing the image through the eyepiece—sees a much enlarged image according to the principles of the simple magnifier. Since the apparent size of the image is larger than the actual size of the object, the angle subtended at the eye is larger. So, there is magnification. It's interesting to note that placing a short-focal-length lens in the front of a telescope changes it to a microscope. This is shown in Figure 21.

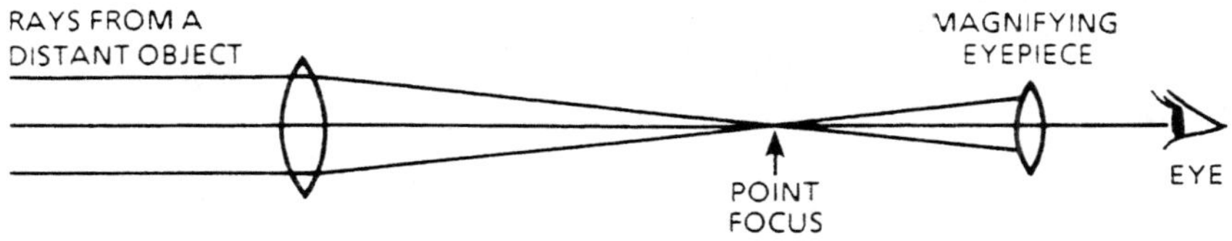

a. Celestial telescope

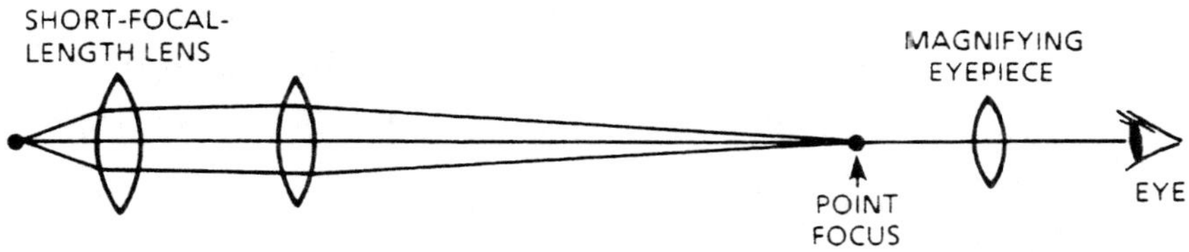

b. Compound microscope

Fig. 21 Converting a telescope into a microscope.

Problems in Lenses

Even the best lenses have certain amounts of aberrations or deviations from perfectly formed images. It is well-known that in practice, lenses normally don't perform exactly in accordance with a simple lens formula. This section will define some of these lens faults and discuss how to correct them.

Aberration can be defined as a faulty image formation caused by failure of all the incident light rays from a point on the source to focus as the corresponding point image in the image plane. The types of defects or aberrations in the image that we will discuss are classified as:

- Chromatic aberration } occurs in first order for light that is not monochromatic
- Spherical aberration
- Coma
- Astigmatism
- Distortion
- Field curvature

} third-order "primary" or "Seidel" aberrations (Spherical aberration through Field curvature)

Many beginning students of optics don't realize that they really are learning an approximate theory of ray tracing. The elementary formulation often is called "first-order theory." It gets its name from the expansion of the sin θ function,

$$\sin\theta = \theta - \frac{\theta^3}{3!} + \frac{\theta^5}{5!} - \frac{\theta^7}{7!} + \ldots$$

If the function sin θ is approximated by θ (measured in radians) then we are using a "first-order" theory. If we add the next term in the series, $-\theta^3/3!$, we're said to be using a "third-order" theory. Note that second, fourth, etc. theories don't exist because there are no odd power terms in the expansion.

Although often it's not shown clearly by instructors teaching beginning optics courses, most of the formulas you learn are only accurate to first order. Even the matrix theory that most laser students learn in optics is only accurate to first order. To this approximation, lenses and mirrors appear to image much better than they do in reality.

If the third-order terms are included in the mathematics of our theory, we start to predict the imaging "problems" that we mention in this module. In fact we predict, through the mathematics, the existence of the last five "problems" listed above. These often are called the five "primary" or "Seidel" aberrations (spherical aberrations, coma, astigmatism, distortion, and field curvature). These are named in honor of Ludwig von Seidel (1821–1896) who first studied the mathematics of these effects in detail (in the 1850s).

Although a mathematical description of these effects is not appropriate here, we will discuss briefly the five Seidel aberrations and the unrelated effect called "chromatic aberration." We also should note that modern lens-design software that runs on microcomputers is accurate to fifth order. It takes into account even higher-order aberrations than those studied by Seidel. Much of the task of modern optical engineers is minimizing these third-, fifth- and higher-order effects to produce exceptionally high-quality images, while keeping the size, weight and cost of the optical system within reason.

Chromatic Aberration

A prism deviates blue light more than red light. Simple lenses do the same thing with the result that blue light comes to a focus nearer to the lens than red light. This defect is called chromatic aberration of the lens.

Some photographic emulsions are most sensitive to light at the blue end of the spectrum. A lens with chromatic aberration may form an image that appears sharp to the eye, but it produces a blurred, out-of-focus image on the film negative. Figure 22 shows chromatic aberration. In the figure, white rays from a point object are split up inside the lens and come to separate foci that depend for their position on wavelength. Wavelengths range from about 4200 angstrom units at the blue end to 7500 angstrom units for the red end of the spectrum.

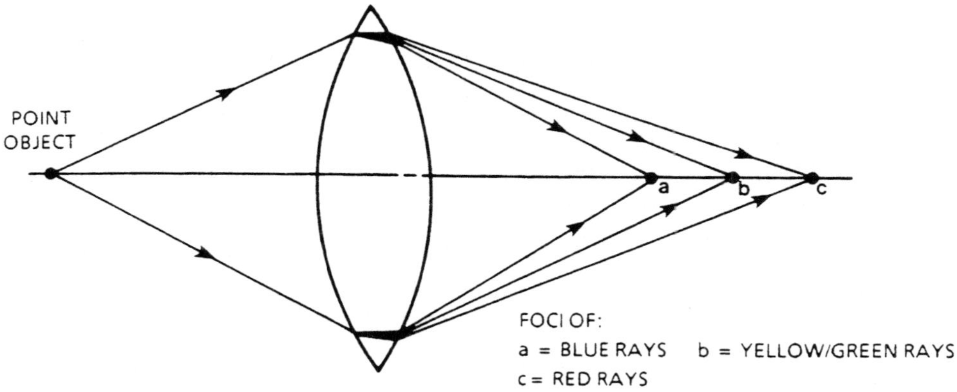

Fig. 22 Chromatic aberration.

The best correction for chromatic aberration is the use of a negative lens, made of a glass with a different index of refraction, in combination with a positive lens. The projection system shown in Figure 23 uses lenses, taken together, to correct for chromatic aberration.

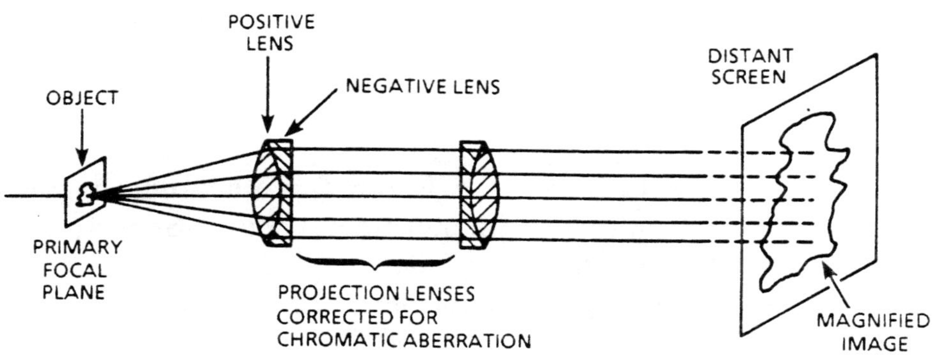

Fig. 23 Achromatic projection lens system.

You saw in Figure 14a that a single biconvex lens can be used as a projection lens. In practice, however, a projection lens consists of a pair of lenses that are corrected for chromatic aberration. Since the projected image is generally large compared to the object, aberrations (defects in perfect focus) will be quite noticeable. Figure 23 shows that a magnified image will be formed on a distant screen if a well-illuminated object or real image is placed in the focal plane of a projection lens.

Spherical Aberration

Rays that come from an object on the axis of a lens and pass through a portion of the lens near the center come to a focus at a certain point on the axis of the lens. Rays from this same object point that pass through the lens near the edges of the lens should come to a focus at the same point, if a good image is to be formed. In practice, because of spherical aberration, the marginal rays tend to come to a different point than do the "center" rays. The difference between these focal points is caused by spherical aberration of the lens. Longitudinal spherical aberration occurs when the marginal and paraxial rays passing through a positive lens fail to meet at a common focal point.

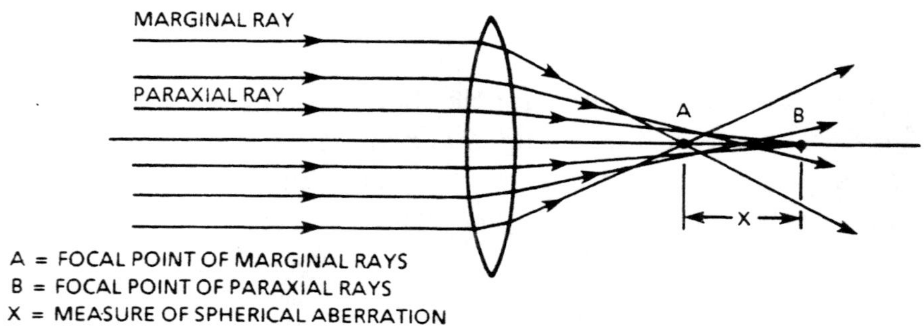

Fig. 24 Spherical aberration.

Spherical aberration is shown in Figure 24 for a bundle of parallel rays, all originating from an on-axis point of a distant object. Spherical aberration can be corrected with a diaphragm in front of or behind the lens that cuts out the marginal ray. Another possible correction occurs with a specially ground meniscus lens.

Coma

This type of aberration is restricted to "off-axis" image points. In some respects we can consider it as spherical aberration of an oblique bundle of rays. Since the lens has no symmetry about a line passing through the center of the lens and an off-axis point in the image, the effects of coma are complex and unsymmetrical. Coma is shown by Figure 25. (No hard rule for elimination of coma can be laid down.) It might help to cover part of the lens with a diaphragm. If this fails the only choice is to discard the faulty lens.

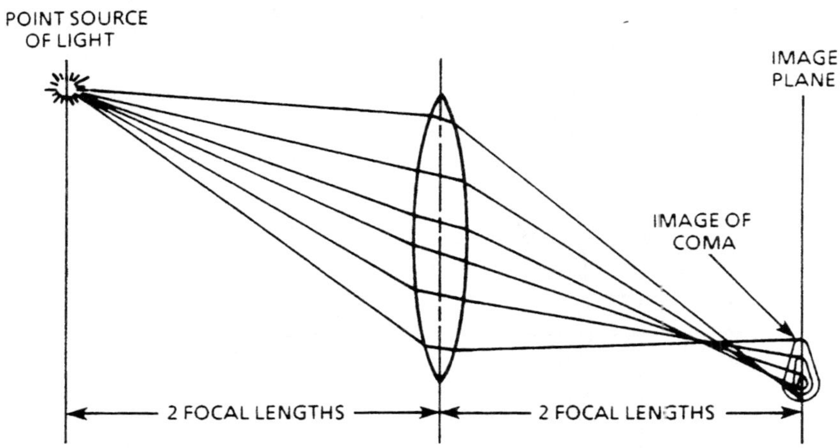

Fig. 25 Coma.

Astigmatism

Astigmatism is the inability of a lens to bring all the rays from a vertical plane to the same point as the rays from a horizontal plane. This problem is shown is Figure 26. Astigmatism can be improved with special lens shapes and by reducing the lens aperture, thus eliminating the offending marginal rays.

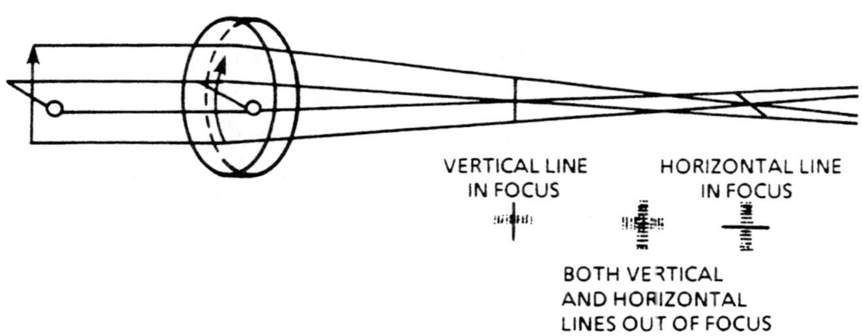

Fig. 26 Astigmatism.

Distortion

Distortion is the magnification of image points near the optical axis more (or less) than the magnification of image points at off-axis positions. Distortion causes parallel lines to bow and images to lose their proportion. When axial magnification is greater than off-axis magnification, the lens has positive or barrel distortion. When axial magnification is less, it has negative or pincushion distortion. The effects of these types of distortion are shown in Figure 27 for a uniform and parallel grid image.

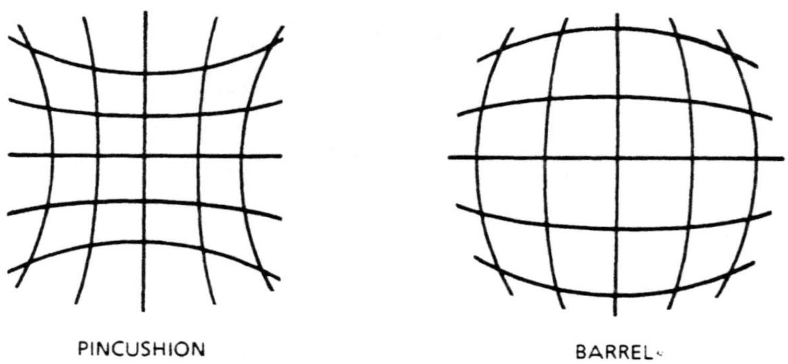

Fig. 27 Effects of distortion on squared patterns.

Distortion does not depend on lens aperture. It can be corrected only with specially ground lenses or filter optics.

Curvature of Field

Curvature of field often is called "Petzval field curvature" in honor of the Hungarian mathematician Josef Max Petzval (1807–1891) who first described it in mathematical detail. It is shown in Figure 28. Note that, unlike the assumptions made in an elementary first-order theory, objects lying in a plane don't generally image into a plane. The image "surface" is in general curved either concave or convex when viewed from the direction of the object.

To a good approximation, for spherical lenses, the image "surface" is paraboloidal in shape. This condition is unacceptable, for example, in the case of a 35-mm slide projector, where the object is a plane (the slide) and the image usually also is focused on a plane (the screen). If just a simple, spherical lens were used to project the image, the center and the edge of the screen could not be focused simultaneously.

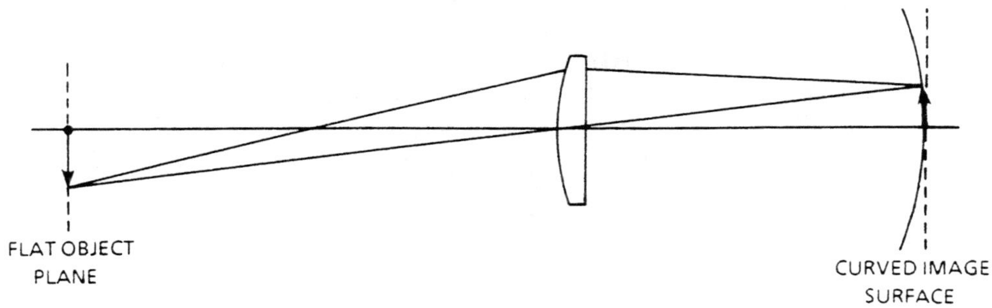

Fig. 28 Field curvature. Object in flat image plane focuses into curved image surface. Not drawn to scale.

To minimize this problem, combinations of positive and negative lenses usually are used. A typical high-quality 16-mm movie projector lens is shown in Figure 29. The negative elements of the lens combine with the positive focusing lenses to form a so-called "field flattener," that greatly minimizes the curvature of field.

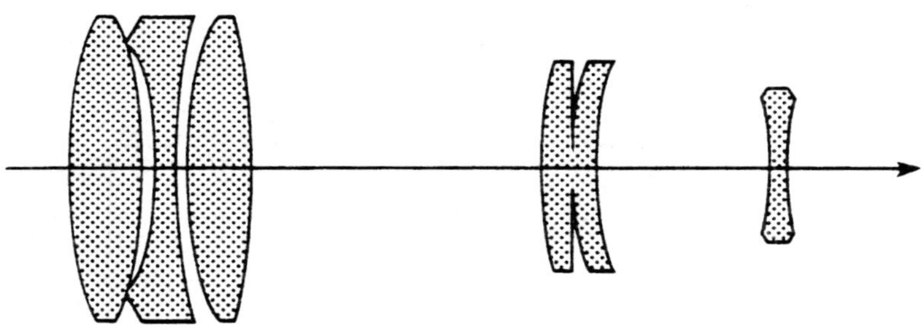

Fig. 29 High-quality projector compound lens containing field-flattening elements. (Hecht)

Reducing Aberrations

A complete description of ways to minimize lens aberrations is beyond the scope of this module. We can make a few closing statements about methods that can be used if aberrations become a particular problem in an optical system. Some of these aberration-reducing methods are relatively easy. Others require an intimate knowledge of the details of lens system design. In general, reduction of the Seidel aberrations (and chromatic aberration) requires one or more of the following four basic techniques:

- Careful choice of the curvature of the surfaces and the sequence in which the particular surfaces appear in the optical system.
- Using combinations of lenses.

- Reducing the system aperture with one or more stops.
- Producing lenses with aspheric surfaces.

Dramatic reduction in some aberrations can be achieved by simply turning the lens around (method #1). Reducing the aperture or "stopping down" a system usually will cause a significant reduction in aberrations (except curvature of field), but with a concurrent loss in image intensity.

Optical Coatings

An uncoated lens with an index of refraction of 1.50 will reflect about 4% of incident light, normal to each surface. In some applications, particularly where a large number of optical elements lie in the optical path, such loss may be unacceptable. To reduce reflective losses, optical elements generally are coated with one or more layers of transparent material. Figure 30 shows the effect of coating an optical substrate with an antireflection (AR) coating.

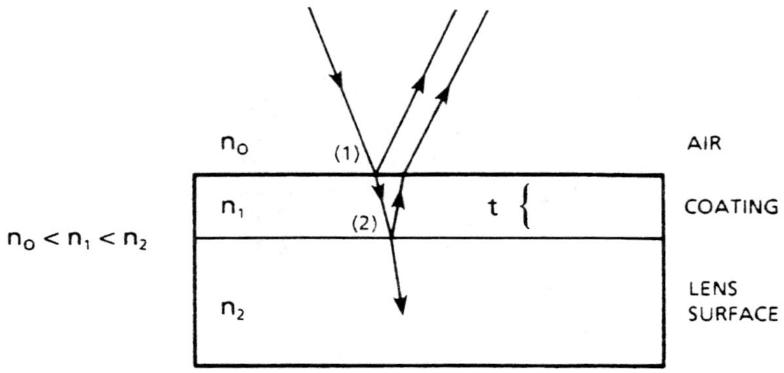

Fig. 30 Ray trace of a single layer AR coating. The thickness t of the coating is chosen such that the rays reflecting from interfaces (1) and (2) are 180° out of phase.

If the thickness and index of refraction of the coating are chosen properly, the rays that reflect from interfaces (1) and (2) will be 180° out of phase and cancel.

A simple analysis shows that, for light striking the surfaces at normal incidence, the minimum thickness t of the coating should be given by

$$t = \frac{\lambda}{4n_1} \qquad \text{Equation 5}$$

where: λ = Wavelength of light in a vacuum

n_1 = Index of refraction of the coating material

Maximum cancellation occurs if the two reflected waves are of the same amplitude. If we assume that little absorption occurs in the coating, the reflected rays will be equal in amplitude if,

$$n_1 = \sqrt{n_o n_2} \qquad \text{Equation 6}$$

For a typical case of a flint (n_2 = 1.72) glass lens surrounded by air, the ideal coating index should be

$$n_1 = \sqrt{(1.00)(1.72)} = 1.31 \qquad \text{Equation 7}$$

MgF_2 with an index of refraction of 1.38 is a close match to the ideal value for many types of common glass. Single-layer coatings usually are "tuned" for the middle of the visible spectrum, in applications such as camera and projector lenses. They don't work quite as efficiently at the red and blue ends of the visible spectrum. So the reflection from these coated optics often appears purple to the eye.

A single layer of MgF_2 can greatly reduce reflections from a surface, as shown in Figure 31. Note that the effectiveness of the AR coating depends dramatically on the index of refraction of the substrate material.

Figure 32 shows the effect of wavelength on the reflectance from an optical surface with a single AR coating layer. Although this coating is optimized for 550 nm, it's surprisingly insensitive to incident wavelength. The effectiveness of MgF_2 as an antireflection coating is increased dramatically with increasing refractive index of the component material (see Figure 31). This means that, for use on high-index materials, there is often little point in using more complex coatings.

For a further reduction in reflectance, multilayer AR coatings can be used. Multilayer coatings tend to be considerably more expensive and more susceptible to mechanical and optical damage.

Particularly when used with high-energy pulsed lasers, you must consider the damage threshold of optical coatings. The probability of damage occurring to the AR coating depends on a number of factors including: the coating and substrate material, defects or contaminants in the coating or substrate, polarization of the incident light, wavelength of the light, and peak value of the electric field caused by the incident light. Lens manufacturers

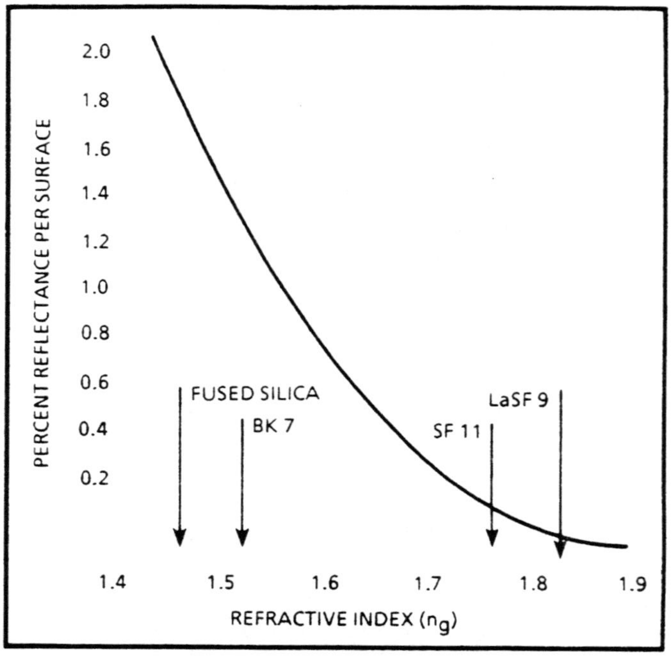

Fig. 31 Reflectance of coated surface of index n_g when coated with a quarter wavelength of magnesium fluoride (index = 1.38). (Melles Griot)

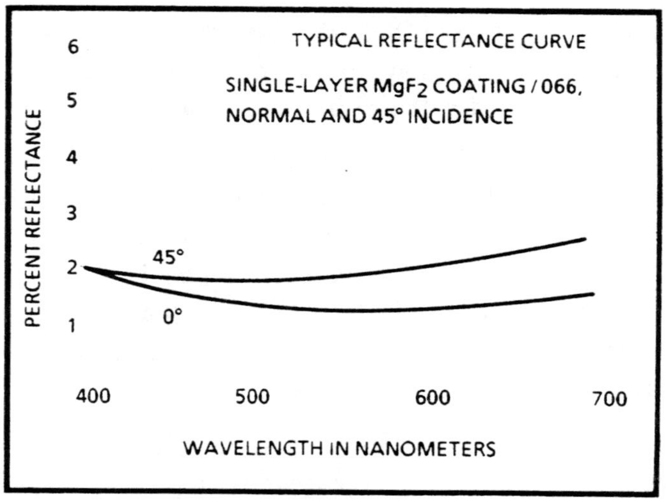

Fig. 32 Reflectance curve for MgF_2 on BK 7 optical glass. (Melles Griot)

usually will specify the nominal damage threshold for coated and uncoated optics that they supply. Modern AR coatings can be very resistant to laser-induced damage. For example, a hard single-layer oxide coating typically will withstand 5 joules/cm^2 in 1-nsec pulses from an Nd:YAG laser irradiating the surface at an incident angle of 45°. This represents an irradiance of 5 GW/cm^2.

EXERCISES

1. Explain the following terms, concepts, or instruments. Your explanations should convey the same meanings given in the text.
   a. Converging lens
   b. Diverging lens
   c. Power of a lens
   d. Focal length of a lens
   e. Lens systems
   f. Collimation
   g. Spherical aberration
   h. Chromatic aberration
   i. Coma
   j. Astigmatism
   k. Distortion
2. From the given cross-sectional diagrams identify the following lenses:

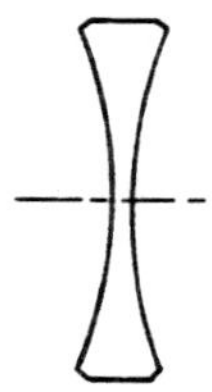

a.____________________

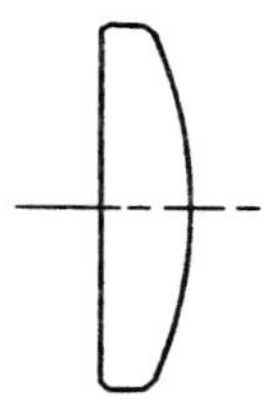

b.____________________

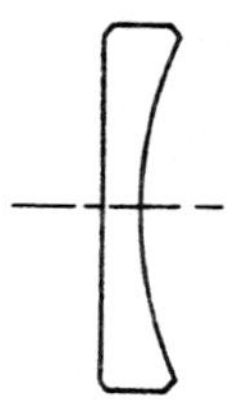

c.____________________

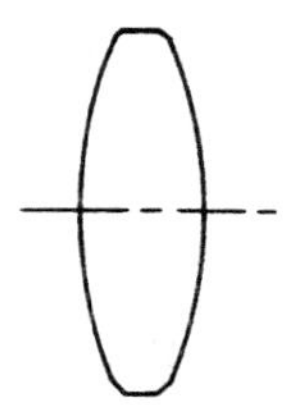

d.____________________

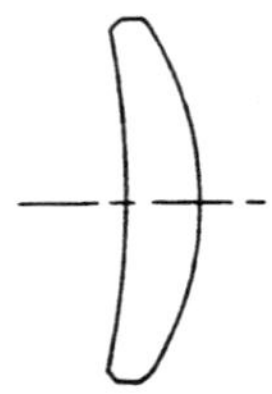

e.________________________

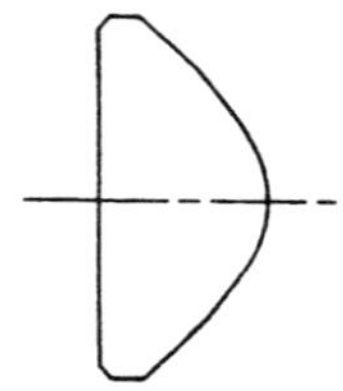

f.________________________

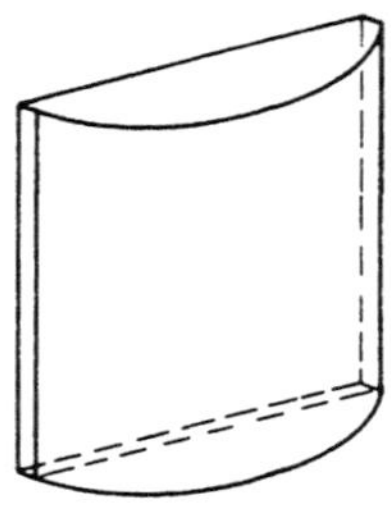

g.________________________

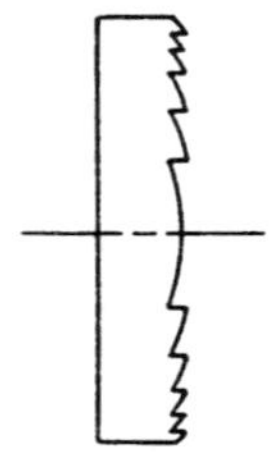

h.________________________

3. Explain, in your own words, the procedure for experimentally determining the focal length of a concave and a convex lens.

4. When working with high-power lasers, why is it necessary to use a beam expander made of a negative and a positive lens rather than one made from two positive lenses? Use accurately drawn ray diagrams to help explain your answer.

5. Using an accurately drawn diagram illustrate the difference between spherical aberration and chromatic aberration.

6. Depth of field is the range of distances over which a camera is in satisfactory focus. Use the concept of spherical aberration to explain why a camera has better depth of field when the lens aperture is small than it has when the lens aperture is large.

7. A 1.00-watt argon laser with an exit-beam diameter of 2.00 mm and a full-angle beam divergence of 1.20 mrad is shined onto a target 3.00 km away. Neglecting atmospheric loss, what is the average irradiance on the target? The output beam of the laser is directed into a Galilean beam expander comprised of a negative lens (f = –7.00 mm) and a positive lens f = +25.0 cm). Assume that the overall transmission of the beam expander/collimator is 92% and that it is focused properly. Calculate the average irradiance on the distant target.

8. The reflectance from a transparent surface is given by

$$T\left(\frac{n-1}{n+1}\right)^2$$

where T is the fractional amount of light reflected from the interface and n is the index of refraction of the transparent medium. (The other side of the interface is assumed to have an index = 1.00.) For a borosilicate crown glass lens, illuminated at normal incidence with a red HeNe laser beam, what is the reflective loss (in percent) per surface?

9. What is the thinnest coating of MgF_2 that can be applied to a borosilicate crown glass lens designed to work at normal incidence with a red HeNe laser beam? After the coating is applied, what is the approximate reflective loss in percent per surface? (See Fig. 31.)

10. Contact three vendors regarding the purchase of the following lens:

Type:	Plano-convex
Material:	Any high-quality optical glass transparent in the visible.
Focal length:	+20 cm.
Diameter:	At least 2 cm.
Coating:	Single or multilayer, "tuned" for use with the ruby laser line.

Obtain as much information as you can regarding the lenses, surface and bulk material properties, aberrations, AR coating efficiency, coating damage threshold, etc., and of course, cost.

If you were planning to use this lens to focus a 5-joule ruby laser (pulse width = 250 μsec), expanded beam diameter = 2.0 cm, onto a workpiece, which lens would you select? Are there other aspects of your application that you need to know to decide?

MATERIALS

Optical bench or table with optical mounts
2 Positive lenses (one short focal length < 5 cm, one long focal length 720 cm)
2 Negative lenses (one short focal length < 5 cm, one long focal length 720 cm)
Ground glass screen

HeNe laser
2 Optical stops (1 center stop, 1 edge stop)
Cross-line object
Illuminating lamp
Metric ruler or metric caliper

PROCEDURES

Determining the Focal Length of a Lens

You can determine the focal length of an unknown lens in several different ways. The simplest of these is described here. This experimental determination of focal length depends on the operational definition of focal length. The focal length of a lens is the distance from the center of the lens to the point where parallel collimated rays of light from a distant source are focused by the lens. This length normally is given in meters.

Set up the optical bench as shown in Figure 33. Place a convex (positive) lens of unknown focal length in the lens holder on the optical bench. Point the optical bench at some distant bright object, for instance, a lamp in a long corridor. The distance should be not less than 50 meters. Also place a ground

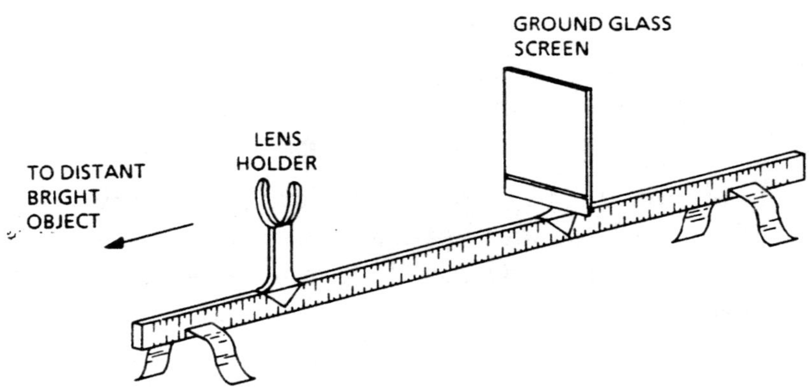

Fig. 33 Arrangement to determine focal length.

glass screen in its holder on the optical bench and form an image of the distant object produced by the lens on this screen. The difference between the readings of the lens holder and the ground glass screen holder will be the focal length of the lens. You might have to make a number of independent readings and average them for accuracy.

Determination of focal length of a concave (negative) lens takes a little more effort. You first must set up the meter bench as before and slide the ground glass screen until the bright object is in focus in front of a convex lens. Write down this position of the screen (S_1). Then place the concave lens of unknown focal length between the convex lens and this position. This is shown in Figure 34.

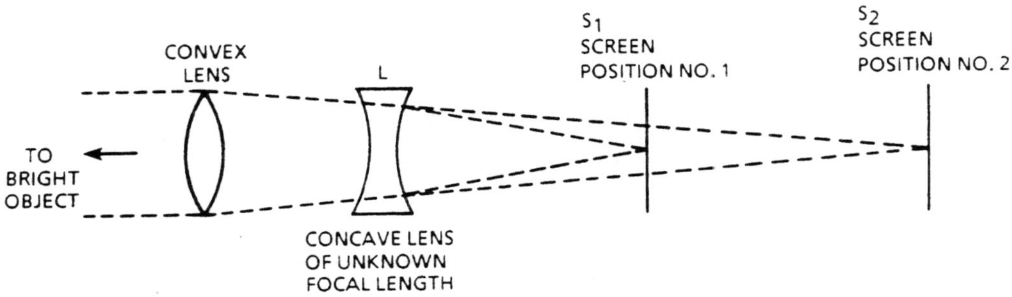

Fig. 34 Determining focal length of a concave lens.

When you have noted S_1 and inserted the concave lens, move the screen until the image is again in focus. This is shown as screen position S_2. Now use the lens system formula to calculate the focal length of the concave lens. This formula (rewritten for this purpose) is

$$\frac{1}{\textit{focal length}} = \frac{1}{\textit{Distance } LS_2} - \frac{1}{\textit{Distance } LS_1}$$

As before, take a number of independent readings for accuracy.

The Celestial Telescope Used as a Beam Expander

Mount on an optical bench the components necessary to produce a Keplerian beam expander/collimator. An arrangement similar to that shown in Figure 19a of this module is sufficient.

Shine a laser beam "backward" through the "telescope." Make any adjustments necessary to produce an expanded, collimated output beam. Record the focal lengths of the lenses and the diameters of the input and output beams. Repeat for the Galilean beam expander/collimator shown in Figure 19b.

Demonstration of Spherical Aberration and One Correction

The effect of spherical aberration is that light through the outer edge of the lens comes to a focus at a point on the axis nearer the lens than light through the center. This causes the image to appear blurred over the entire field.

Place the lamp, the cross-line object, the convex lens, and the ground glass screen on the optical bench or table as shown in Figure 35.

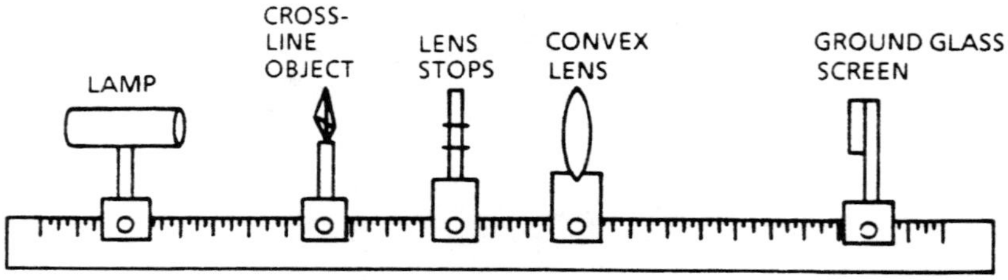

Fig. 35 Arrangement to demonstrate spherical aberration.

Place the "center-of-lens" stop—shown in Figure 36—in front of the lens and move the screen until the image of the cross line is sharply in focus.

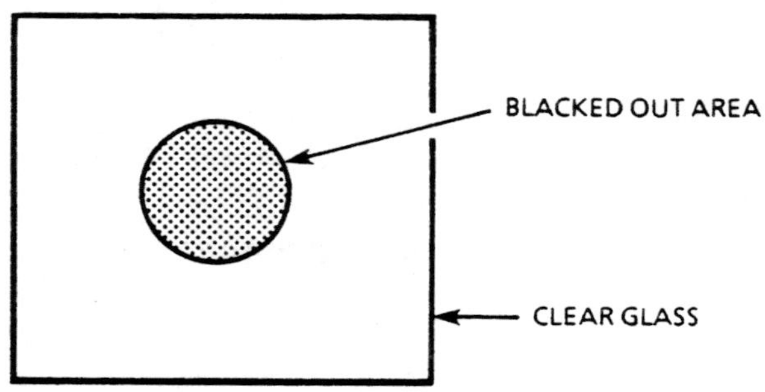

Fig. 36 Center-of-lens stop.

Next, move the center-of-lens stop and insert the "edge-of-lens" stop shown in Figure 37.

Move the ground glass screen back until the image is once again in focus. The amount of spherical aberration the lens has is indicated by the distance between these two focal points. Spherical aberration normally is corrected by using an edge-of-lens stop or a diaphragm so that only the center portion of the lens is used.

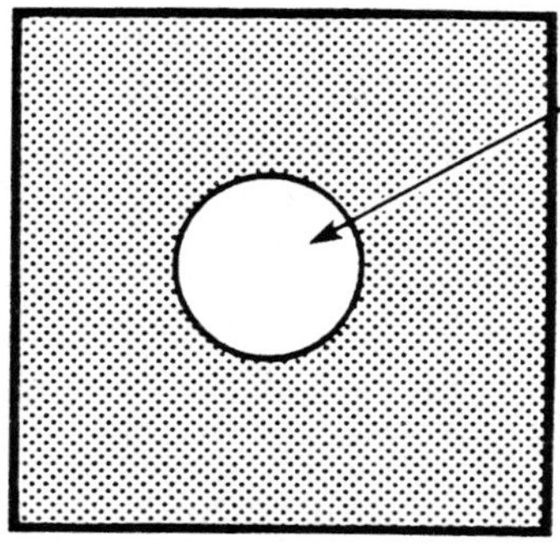

Fig. 37 Edge-of-lens stop.

Repeat the above procedure with a larger-diameter lens. Compare the spherical aberration of the two lenses.

REFERENCES

Hecht, E. Optics. 2nd edition. Addison Wesley, 1987.

Jacobs, Donald H. Fundamentals of Optical Engineering. McGraw-Hill Book Co., 1943.

Jenkins, Francis A., and Harvey E. White. Fundamentals of Optics. 4th edition. McGraw-Hill Book Company, 1976.

Kingslake, Rudolf. Applied Optics and Optical Engineering. Volume III. Academic Press, Inc., 1965.

The Newport Corporation Catalog No. 100. Newport Corporation, 18235 Mt. Baldy Circle, Fountain Valley, CA 92728-8020, (714) 963-9811.

Optics Guide 3. Melles Griot, 1770 Kettering Street, Irvine, CA 92714, (714) 261-5600.

Oriel Corporation Catalog Vol. III. Oriel Corporation, 250 Long Beach Blvd., Stratford, CT 06497-0872, (203) 377-8282.

Pedrotti, Frank L., and Leno S. Pedrotti. Introduction to Optics. Englewood Cliffs, NJ: Prentice-Hall, Inc., 1987.

Pollock, H. W. Applied Physics. 2nd edition. Prentice-Hall, 1971.

LASER/
ELECTRO-OPTICS

MODULE 6-9
GRATINGS

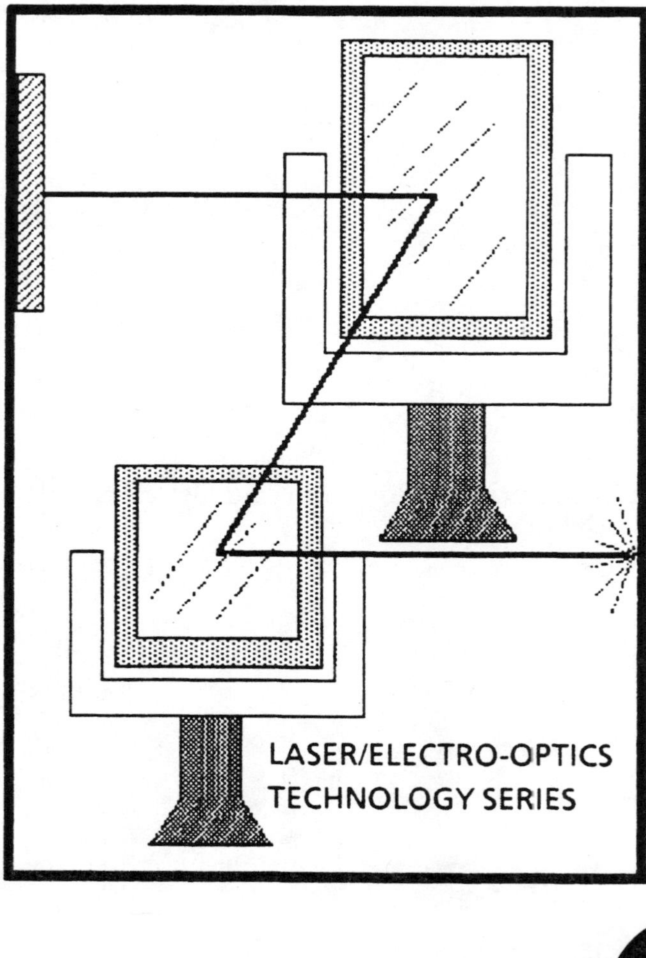

LASER/ELECTRO-OPTICS TECHNOLOGY SERIES

CORD

The Center for Occupational Research and Development
601 C Lake Air Drive
Waco, Texas 76710

Printed in the U.S.A.

ISBN 1-55502-024-0

GRATINGS

MODULE 6-9

A *diffraction grating* is a glass or polished metal surface on which a large number of very fine, parallel grooves or slits are cut. These parallel elements are used to diffract light, generally both separating and spreading different wavelengths or colors. Diffraction gratings are manufactured on flat or curved surfaces according to their use.

Diffraction gratings are used in spectrographic equipment (spectroscopes, spectrophotometers, monochromators, spectrographs, spectrometers, and so on) as the principal optical element to separate light into its component wavelengths (frequencies) by diffracting them at various angles. Gratings also are used in some lasers that are capable of emitting a coherent output at more than one wavelength. In this application the grating is used to select a desired narrow wavelength range for the laser output.

This module will acquaint you with the types of diffraction gratings available, how they work, and how and when to use them in optical experiments.

Before you study this module, you should have studied elements and operation of a laser, coherence, propagation, interference, and diffraction. You also will find a knowledge of light helpful. Algebra and trigonometry are required to successfully complete this module.

OBJECTIVES

When you complete this module, you should be able to do the following:

1. Write a statement of definition or explanation for each of the following terms or concepts.

Diffraction
Constructive interference
Destructive interference
Transmission grating
Reflection grating
Replica grating
Resolving power of a grating
Blaze angle
Efficiency
Free spectral range

2. Differentiate between ruled gratings and holographic gratings by writing a short explanation of the manufacturing processes of each.
3. Use the grating equation to calculate the wavelength of light incident upon a grating.
4. Use experimental data to calculate the groove spacing and resolving power of a grating.
5. Choose from manufacturers' data a grating designed to operate satisfactorily under given conditions.
6. Calculate the blaze angle of a reflection grating given the operating conditions.

DISCUSSION

A highly periodic arrangement of finely spaced, parallel apertures is called a diffraction grating. The very first grating was used by a German physicist and astronomer, Joseph Von Fraunhofer. It consisted of a grid formed by winding fine wires on two parallel screw threads. This early diffraction grating looked similar to the one shown in Figure 1. The system had many obvious disadvantages.

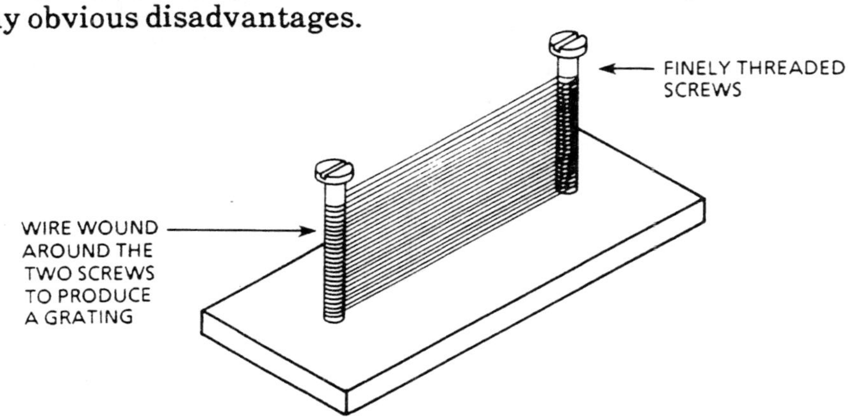

Fig. 1 Early diffraction grating.

Present-day gratings consist of equidistant line rulings on a flat glass plate (transmission grating) or on a flat mirror (reflection grating). Depending on the application, the number of lines per millimeter typically ranges from less than one hundred to several thousand. Because of this extremely fine spacing, the lines are carefully scribed with a diamond tool in a special machine called a "ruling engine."

Ruling a master grating is a very slow process. Ruling engines typically operate at a speed of 600 strokes per hour. So, to rule some large gratings an engine may have to operate 90 days or more without failure or even appreciable wear.

As you can imagine, master gratings produced in this way prove to be very costly. So these masters are used for making copies called "replicas." The replicas are made less expensively than the original by using a plastic or glass casting of the master grating.

Today holographically recorded diffraction gratings are gaining wide acceptance. Holographically ruled diffraction gratings eliminate some of the problems encountered when the groove spacing is not perfect. They can be obtained with rulings as fine as 6000 lines per millimeter, which gives them high resolution. However, the efficiency of such gratings is typically lower than that of conventionally ruled gratings.

You need an understanding of diffraction theory to visualize the operation of a diffraction grating. We won't go into a thorough discussion of diffraction theory, but a short review is appropriate at this time.

The bending of a wave front as the wave passes through an opening or around an obstruction is a common example of diffraction. As an opening becomes much larger than the wavelength, diffraction effects become less noticeable. All periodic waves exhibit diffraction effects, but under certain conditions, such as broad spectral bandwidth, it may be hard to see the effects. Figure 2 shows diffraction taking place when plane waves pass through an aperture.

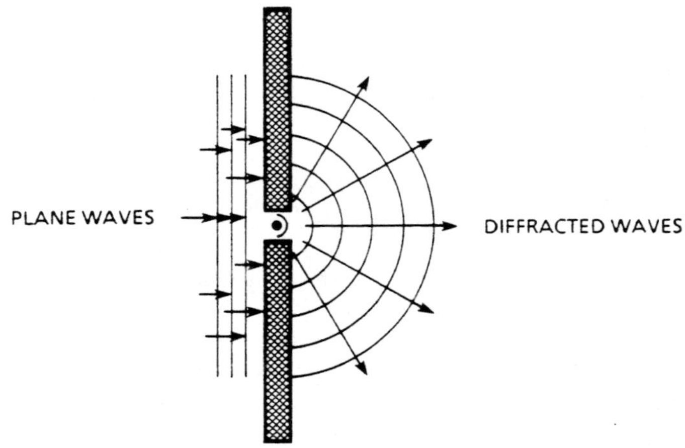

Fig. 2 Diffraction of plane waves.

When two such openings exist near each other, interference with other transmission waves will result. This effect is shown in Figure 3 where

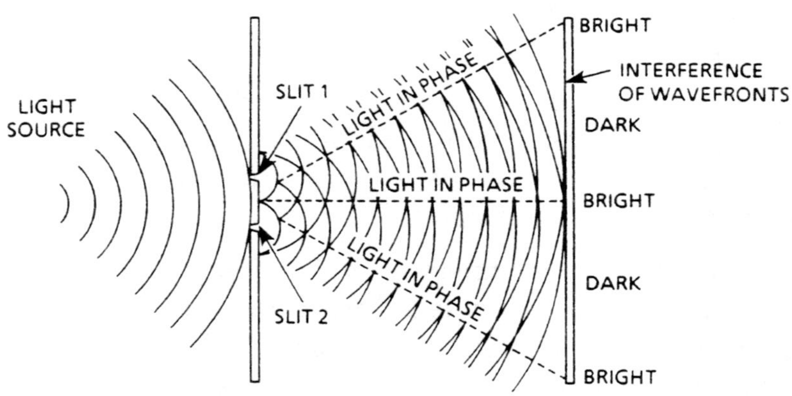

Fig. 3 Interference of light diffracted by two slits.

monochromatic light waves are incident on two adjacent slits. The two beams may travel different distances before meeting at the screen. So the crests of one may not necessarily coincide with the crests of the other.

Where crest and crest coincide on the screen, a region of brightness appears, and where a crest and trough coincide a region of darkness appears. The successive regions of brightness and darkness are called "interference fringe."

A graph of the intensity of the light on the screen after interference is shown in Figure 4. Notice the uniformity of the pattern.

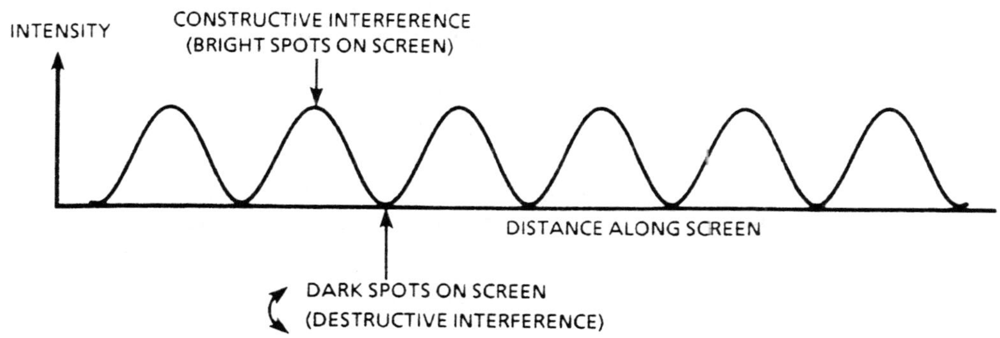

Fig. 4 Intensity of light produced by diffraction through two slits.

Figure 5 shows that light from two slits travels different distances to a given point P. The two rays arriving at P from S_1 and S_2 were in phase when they left the slits since both were derived from the same wavefront in the incident plane wave.

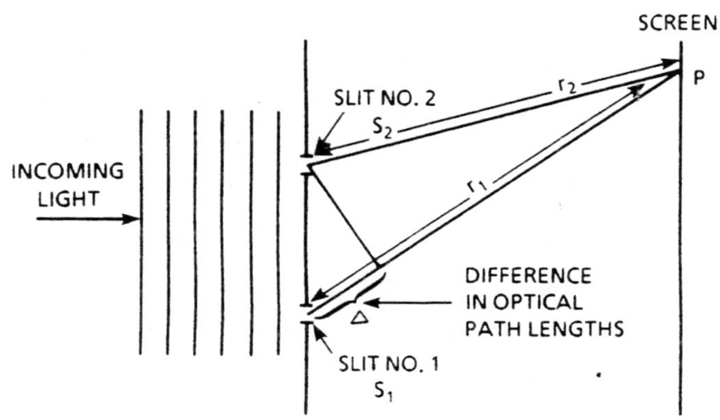

Fig. 5 Different distances traveled by light from two slits.

Because the rays travel different optical path lengths in reaching P, they arrive at P with a phase difference. The number of wavelengths in the path length difference ($r_1 - r_2 = \Delta$) determines the phase difference and therefore the nature of the interference at P.

When there are three narrow, equally spaced slits, the interference pattern, instead of becoming more complicated, actually becomes more distinct. The additional slit imposes an extra condition on the three beams of light if they are all to interfere constructively at the same place. This has the effect of narrowing the bright region and at the same time making it brighter, since more light is concentrated there. Figure 6a shows a three slit interference intensity pattern. Note the "secondary maxima" that appear between the "principal maxima." The number of secondary maxima increases as the number of slits increases, although, as shown in Figure 6b, the brightness of each secondary maximum decreases.

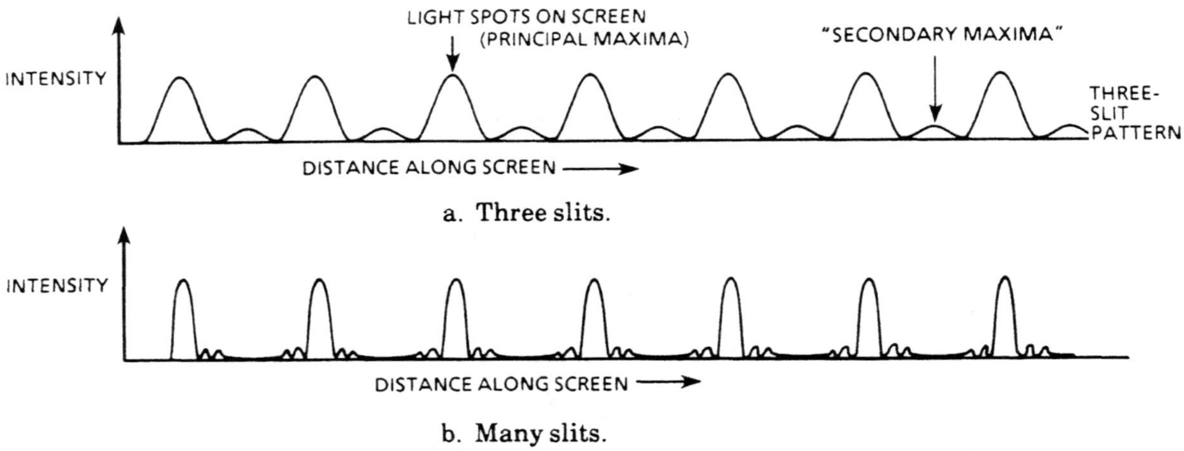

Fig. 6 Diffraction intensity patterns.

Increasing the number of narrow, equally spaced slits results in brighter and narrower fringes. Figure 7 shows the intensity pattern produced as additional slits are used. When there are a large number of slits (thousands of them per centimeter) a diffraction grating is formed.

The above discussion applies equally well for both reflection gratings and transmission gratings. The only difference is that, while the transparent slits are considered to be individual point sources of light in the transmission grating, small reflecting surfaces are considered to be the individual point sources in a reflection grating.

Transmission gratings seldom are used today. One reason for this is that light must pass through the glass or other material that forms the grating. So the light is subject to absorption and distortion caused by that material. Another reason is that reflection gratings yield more compact systems, because of the "folding" of the optical path through reflection.

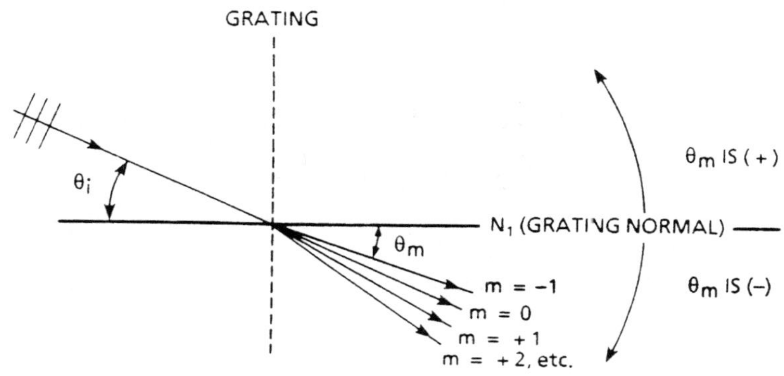

a. Transmission grating.

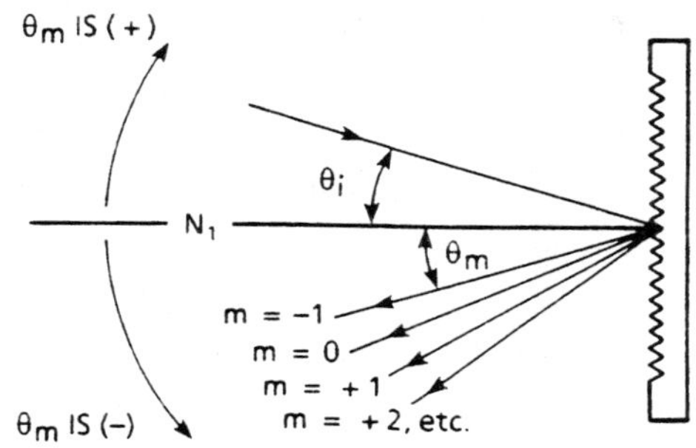

b. Reflection grating.

Fig. 7 Sign convention for grating equation (Equation 1) for both a transmission (a) and a reflection (b) grating.

The position of a bright fringe on the screen is a function of wavelength of light incident on the grating. As a result, different wavelengths are focused at different locations on the screen. If the grating is illuminated by "white" light, the light is separated by the grating into its component wavelengths much like it's separated by a prism.

The utility of a grating depends on the fact that there are unique sets of directions. For a given groove spacing and a given wavelength of light along these directions, the diffracted light from all the grooves is in phase.

The grating equation that governs the diffraction behavior of all gratings is:

$$m\lambda = a\,(\sin\theta_i + \sin\theta_m \qquad \text{Equation 1}$$

where: θ_i = Angle of incident rays measured with respect to the normal

θ_m = Angle of diffracted order m with respect to the normal. When θ_m is located on the opposite side of the grating normal from θ_i, then it is negative.

a = Distance between successive grooves

m = Diffracted order, usually a small integer, 0, ±1, ±2, . . .

λ = Diffracted wavelength

You can visualize this better from the geometry of Figure 8. Here a plane wavefront is incident on a grating surface at an angle θ_i measured with respect to the normal. Then it's diffracted at an angle θ_m measured with respect to the same normal. The groove spacing is designated by a. Notice that θ_i and θ_m are not necessarily equal due to the fact that light diffracts in all directions from the thin strips of polished surface between the lines.

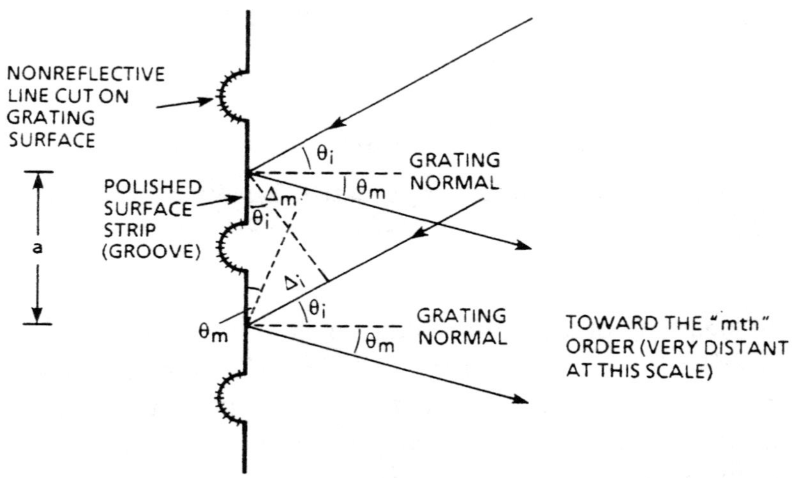

Fig. 8 Cross section of a small segment of a reflection grating surface.

You can see by this equation that, for a given order and grating spacing, there is a mathematical relationship between wavelength and the angles of the incident and diffracted rays. See in Figure 8 that the geometrical path difference between light diffracted from successive grooves is simply $\Delta = \Delta_i - \Delta_m = (a \sin \theta_i - a \sin \theta_m)$. The principle of interference dictates that only when the angles θ_i and θ_m are such that this difference equals the wavelength of the light, or an integral multiple thereof, will the light from all grooves interfere constructively. At all other angles, there will be some degree of destructive interference between the light rays of this wavelength coming from the grating.

So when light of wavelength λ hits a grating at a given incident angle, there's a unique set of diffraction angles at which the light will leave the grating reinforced. At all other angles its intensity will be reduced. This property lets us use gratings in instruments to measure wavelengths of light and to study the structure and intensity of spectral lines. Such instruments include spectroscopes, spectrophotometers, monochromators, spectrographs and spectrometers.

Further interpretation of Equation 1 shows that, for a given set of angles θ_i and θ_m and groove spacing a, the grating equation is satisfied by more than one wavelength ($m = 1, 2, 3$, etc.). To be reinforced, the light coming from successive grooves must differ in phase by a whole number of wavelengths. This happens first when the difference is one wavelength, in which case we speak of first-order ($m = 1$) diffraction. It happens again when the difference is two wavelengths ($m = 2$), called the second order, and so on.

The most troublesome aspect of the multiple-order behavior of gratings is that successive orders can overlap. For example, the violet wavelengths 400 nm of the second order ($m = 2$) will reinforce at the same position on the screen as the deep red wavelengths 800 nm of the first order.

We can prevent such overlapping of multiple orders by suitable filtering or by detector selectivity. More overlapping of wavelength regions occurs in the higher spectral orders than in the lower orders.

Blaze

It's possible with a reflection grating to concentrate most of the diffracted spectral energy into one spectral order and reduce the energy in all other orders. This distribution of energy among the orders depends on the angle between the reflecting element and grating surface. This angle is called the "blaze angle" ϕ. See Figure 9. Note that the light reflects preferentially toward the "blazed order." In other words $\gamma \approx \phi + \theta_i$. N_1 = grating normal, N_2 = blaze normal.

The diffraction spectral intensity curve shown in Figure 7 can be changed to that shown in Figure 10b when the grating is blazed for the first order.

Figure 10 shows how all orders are virtually extinguished and the available energy is concentrated mostly into a single order. This phenomenon occurs for only one wavelength, which is called the "blaze wavelength." Under these conditions, the specular reflection from the grating surface *and* the mth order diffraction occurs at the same, or nearly the same, angle. This

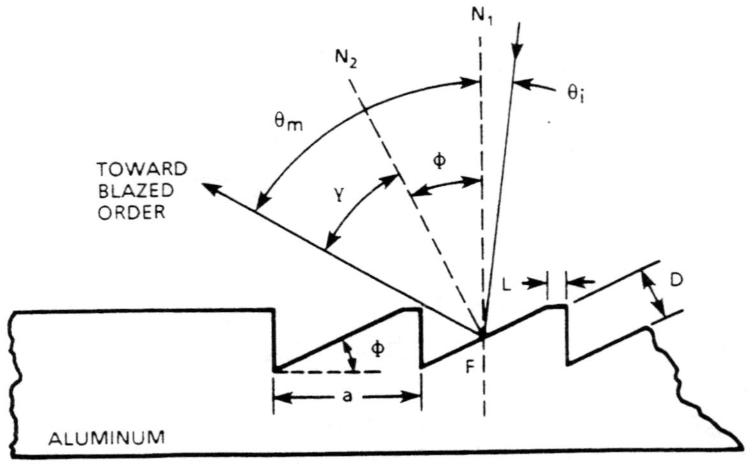

Fig. 9 Blazed groove profile.

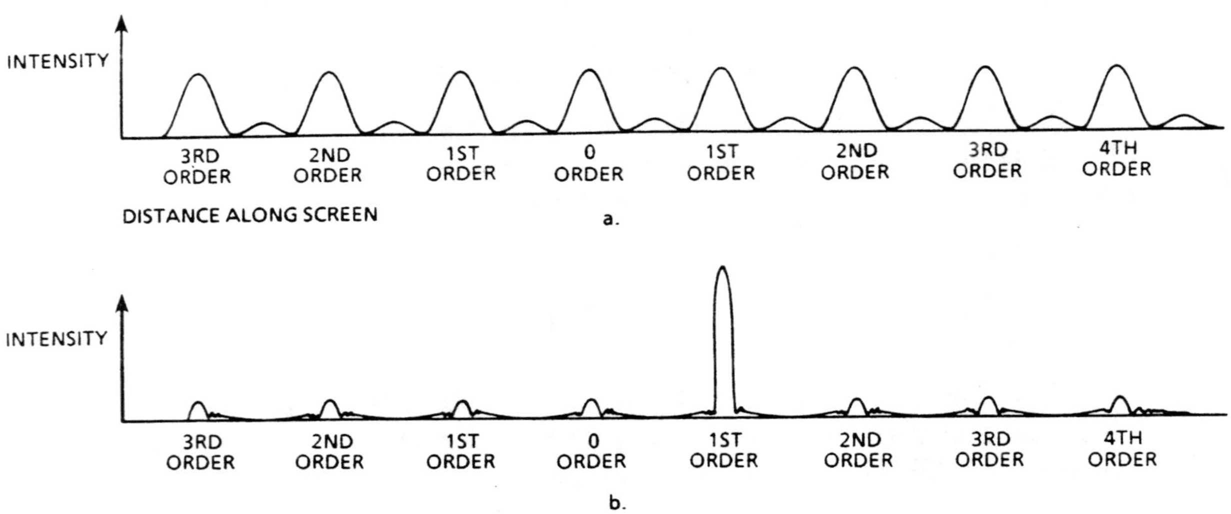

Fig. 10 Diffraction pattern from a multislit grating blazed from the first order.

causes most of the diffracted light to go into the blazed order. You can find the relationship between the blaze angle and blaze wavelength by applying the grating formula (Equation 1). When you apply the equation, you find that a grating blazed for a given wavelength in the first order also is blazed for half that wavelength in the second order and one-third of the wavelength in the third order, and so on.

When speaking of a grating blazed at a given wavelength, it is implied that the grating is blazed in a certain specified order for this wavelength and for Littrow use (light incident normal to the reflecting elements).

A grating blazed for use in low orders (1 or 2) is called an "echellette." A grating blazed for use in higher orders (>10) is called an "echelle" grating.

Resolving Power

The resolution or chromatic resolving power of a grating describes its ability to separate adjacent spectral lines. Resolution generally is defined as $R = \lambda / \Delta\lambda$, where $\Delta\lambda$ is the difference in wavelength between two equal intensity spectrum lines that are just separated.

The limit of resolution of a grating is theoretically $R = mN$, where m is the grating order and N is the total number of grooves illuminated on the grating. To increase the resolution we could think of increasing N by obtaining a finer-ruled grating. However, you must know that, if you do this, certain orders disappear and the maximum value for m decreases. You get a more useful expression for resolution by substituting for m from the grating equation and arriving at Equation 2.

$$R = mN = \frac{Na\left(\sin\theta_i \pm \sin\theta_m\right)}{\lambda} \qquad \text{Equation 2}$$

Since $(\sin\theta_i + \sin\theta_m)$ can have a maximum value of 2, the maximum resolving power at any wavelength turns out to be equal to $2\,Na / \lambda$, where N is the number of slits and a is the slit spacing. So the product Na is the width of the grating. At 500 nm a 15-cm grating will have a maximum resolution of:

$$R_{max} = \frac{2w}{\lambda}$$

$$R_{max} = \frac{2\,(15\ cm)}{500\ nm}$$

$$R_{max} = 600{,}000$$

where: w = Width of illuminated portion of grating

Efficiency

Two forms of efficiency are considered in connection with diffraction gratings. The absolute efficiency is the ratio of power diffracted at a particular wavelength (λ) in the order of interest to the power incident on the grating at that wavelength (λ). The relative efficiency takes into account the reflectivity of the grating's metallic surface and is the ratio of the energy diffracted by the grating at a particular wavelength (λ) in the order of interest to the energy reflected by a mirror under the same working conditions. A relationship exists between the two efficiencies such that:

Absolute efficiency = Relative efficiency x reflectance of a mirror.

The mirror mentioned in the formula must have the same coating as the grating. And it must work in the same angular conditions as the grating. The efficiency of a grating is generally greatest at the blaze wavelength in ruled gratings or at a particular wavelength of interest for holographic gratings. Figure 11 shows an efficiency curve for both ruled gratings and holographic gratings.

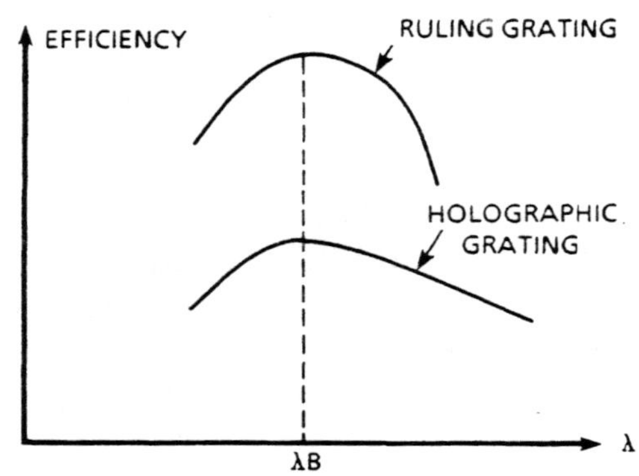

Fig. 11 Efficiency curve of a holographic grating is generally lower but flatter than that of a ruled grating.

Gratings also can be either plane or concave. Plane gratings are usually less expensive and easier to control in manufacture. Concave gratings, however, offer the added advantage of focusing. This eliminates one element such as a lens from the optical system.

Groove spacing is a wide open variable in gratings. For useful diffraction, the spacing must be at least $^2/_3$ of the longest wavelengths of interest. But in some cases it can be as much as 100 wavelengths or more.

Rulings as fine as 3600 lines per millimeter can be made. This is a spacing of about half the wavelength of visible light.

One of the most widely used applications of the diffraction grating is in spectroscopy. The spectroscope is used to study the optical spectra of incandescent sources to determine their composition, temperature, or motion.

Figure 12 shows a simple grating spectroscope that's used for analyzing the spectrum of a light source assumed to emit a number of discrete wavelengths or "spectral lines." The light from source S is focused by lens L_1 onto slit S_1 placed in the focal plane of lens L_2. The collimated light emerging from

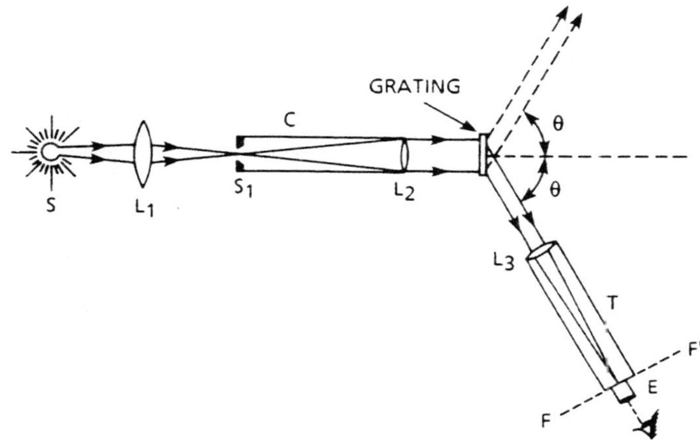

Fig. 12 A simple transmission grating spectroscope used to analyze wavelengths of light emitted by source S.

collimator C falls on the grating G. Rays of light associated with a particular interference maximum occurring at angle θ form a parallel beam that enters the camera or field lens L_3. The image of the slit thus is focused onto plane F-F' and is examined with the magnifying eyepiece E.

A typical pattern that you might see with a spectroscope is shown in Figure 13. The lines actually appear as different colors, and their spacing and individual color tell much about the properties of the source.

Fig. 13 Typical spectrograph pattern.

Instruments much like that of Figure 12 are used to learn many facts about the age and composition of planets and stars. The light falling on a telescope lens from the stars may be sent through a grating. Then the spectrum formed by the starlight is studied to tell important facts about bodies billions of miles away.

Two special ways of using a grating as a spectrograph are important enough to deserve mention. The first is the autocollimating or "Littrow" scheme. As shown in Figure 14a, the slit source is at a focus of a collimating lens positioned close to the grating. The grating is rotated on an axis parallel to the grooves to select the diffraction wavelength. The chosen light returns through the lens and is refocused near the original slit. This type of spectrograph is generally more compact than the others.

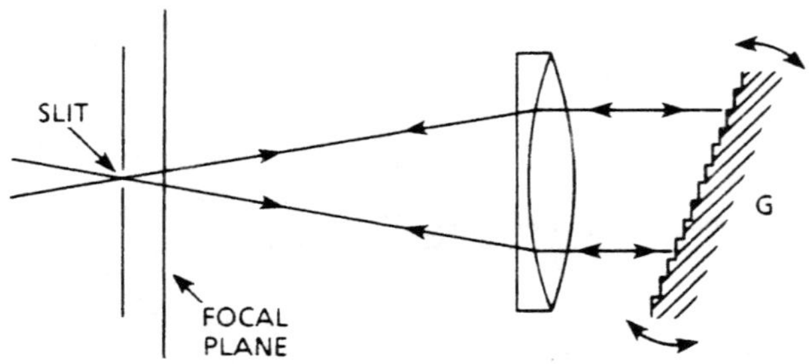

a. Littrow autocollimating spectrograph.

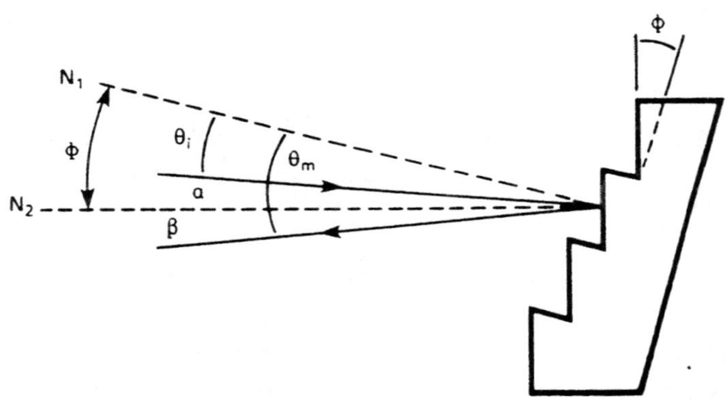

b. Geometry of the Littrow configuration for a blazed reflection grating.

Fig. 14 "Littrow" scheme.

Compare the geometry in Figure 14b of the Littrow configuration for a blazed reflection grating to Figure 9. Note that the incident and diffracted rays are nearly parallel to the blaze normal N_2 such that $\alpha \approx \beta \approx 0$ and $\theta_i \approx \phi, \theta_m \approx -\phi$. Note also that $\phi - \theta_i = -\theta_m - \phi$ (θ_m is negative by sign convention). Therefore $\phi = {}^1\!/\!2(\theta_i - \theta_m)$. In this configuration the grating equation reduces to $m\lambda = 2a \sin \phi$.

The spectrometer shown in Figure 15 uses two mirrors to control the light as the grating is rotated for a spectral scan.

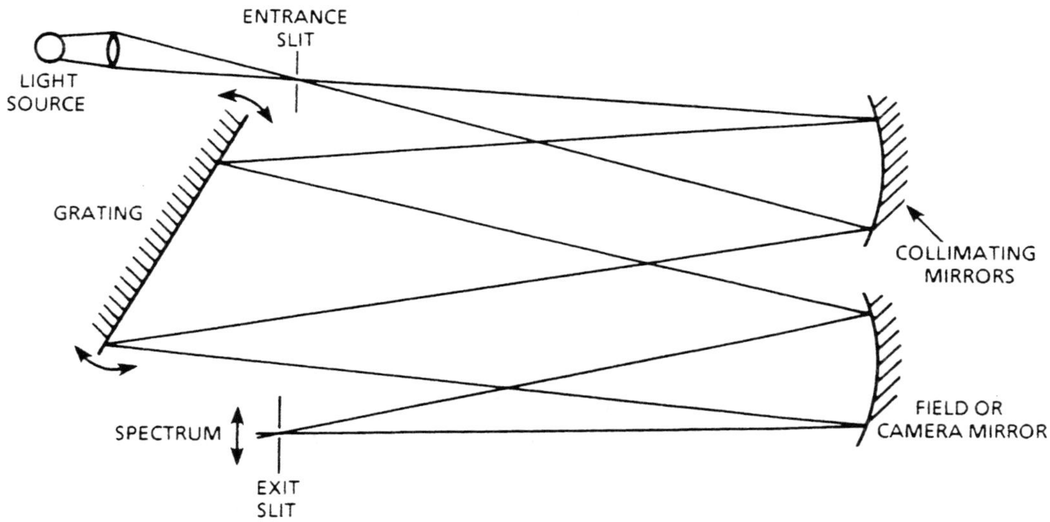

Fig. 15 Double-slit spectrometer.

Field stops and stray light baffles must be used to prevent reflection from one slit to another and to control scattered light within the system. These types of instruments are sometimes used as monochromaters to select or filter out one single wavelength of light from many wavelengths present.

Another important use of the diffraction grating is in wavelength tuning and selectivity in laser cavities. There are two ways of using a grating as part of a laser cavity. A plane diffraction grating can be used as the "total" reflector (Figure 16) or as the "output" reflector (Figure 17) in a laser cavity, or as an intermediate reflector to "fold" the cavity (not shown).

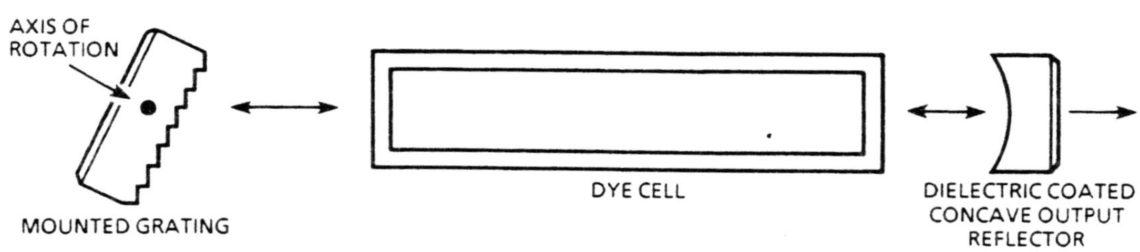

Fig. 16 Littrow configuration grating as total reflector.

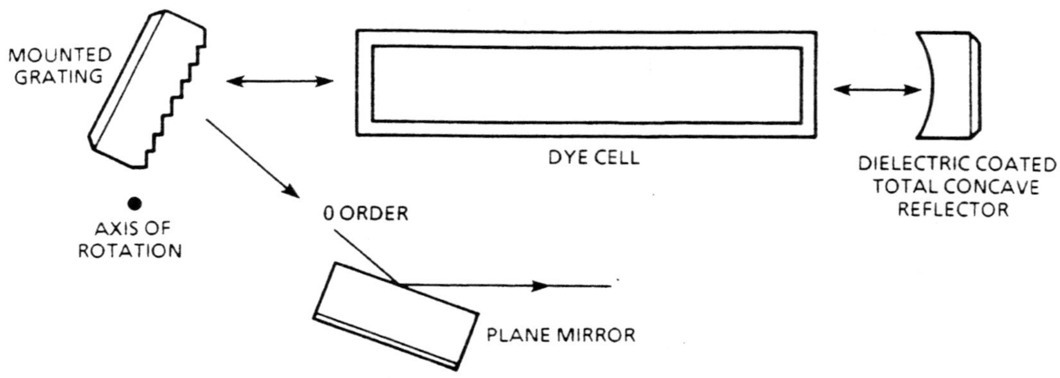

Fig. 17 Grating as output reflector.

You can tune the output wavelength of the dye laser by turning the diffraction grating end reflector about an axis parallel to the grooves. Using a grating as one of the reflectors in a dye laser causes the spectral line width of the output to be reduced to a narrow region around the diffracted wavelength.

Care and Cleaning of Gratings

Good reflectivity is achieved in modern gratings by applying an evaporated aluminum or other metallic coating to the replica of the master grating. The aluminum coating is rather soft, as well as very thin, and is placed on a plastic replica. Both are easily damaged.

Foreign matter such as dust and fingerprints seriously alters the grating's performance. Fingerprints and oral spray are ever-present hazards. Gratings should be given maximum protection from contamination, and the utmost care in necessary handling. Safe cleaning techniques are very limited and should be attempted by experts.

Table 1 is a composite summary of the properties of gratings that you must consider when you select a grating.

Table 1. Properties of Diffraction Gratings

Physical Property	Classically Ruled Plane Grating	Holographically Recorded Plane Grating
Number of grooves per millimeter (n)	• Available from 20 to 600 g/mm. • Optimum groove density generally is dictated by mechanical limitation.	• Available from 600 to 6,000 g/mm. • Limiting factor for number of lines is availability of different wavelengths. • A holographic plane grating should be considered in all cases where you want high density of grooves without prohibitive noise levels.
Size of grating	• Size generally is determined by financial and design considerations. • Minimum desired resolution can be factor in deciding size.	• Size generally is determined by financial and design considerations. • Minimum desired resolution can be a factor in deciding size. • Holographic gratings are available in sizes much larger than the maximum size of ruled gratings. • For the user who is in a position to select the number of grooves per millimeter, a smaller grating of higher groove density may be advantageous.
Working spectral region ($\lambda_1 - \lambda_2$)	• Available for UV, visible, IR. (This is the maximum spectral bandwidth that can be obtained in a given order without having overlap from adjacent orders.)	• Available primarily for UV, visible and near infrared. • At shorter wavelengths holographic gratings may offer advantages for many applications.
Free spectral range	• Depends on number of grooves per millimeter. With working angle conditions fixed, the free spectral range will increase as groove spacing becomes finer.	• Same as classically ruled gratings. • Finer spacings (as a consequence, larger free spectral ranges) are available for holographic gratings.

Table 1. Properties of Diffraction Gratings (Continued)

Physical Property	Classically Ruled Plane Grating	Holographically Recorded Plane Grating
Blaze wavelength (λB)	• A ruled grating can be blazed at almost any angle (θ) for maximum efficiency at any particular λB $2 \sin \theta = m_{\lambda B} / a$ • Select blaze wavelength near the center of working spectral region. • A grating blazed at λB in 1st order also is blazed at λB / 2 in 2nd order.	• Groove shape and depth are controlled to maximize efficiency of the grating at any given wavelength.
Efficiency	• Efficiencies at blaze wavelength generally range from 60 to 99% depending on the working spectral region. • In the far UV region, efficiency is considerably lower because of lower reflectivity of all coating materials. • Efficiency given in the "Specification Registered Certificate" is for λB and Littrow use. • As a rule of thumb,the efficiency curve drops to about 70% of blaze efficiency for $\lambda = \lambda B \pm \lambda B / 2$ (in order m).	• Maximum efficiency depends upon the ratio λ / a. • Very high efficiency can be obtained in most cases, although it is generally lower than that of the ruled gratings. • To get the best possible efficiency, the exact working spectral region should be specified.
Ghost and stray light	• Can result from unevenly spaced grooves.	• Physical process involved in production of holographic gratings results in total absence of ghosts. • Stray light level is below that of the best ruled gratings. • A holographic grating should be considered when the level of stray light and ghosts levels is vital, e.g. Raman spectroscopy and high-resolution work.

Table 1. Properties of Diffraction Gratings (Continued)

Physical Property	Classically Ruled Plane Grating	Holographically Recorded Plane Grating
Resolution	• Depends on width of the grating, working angles, and wavelength • The finer the groove spacing, the higher the resolution. • Actual resolution attained also depends on the stray light level.	• Depends on width of the grating, working angles, and wavelength. • The finer the groove spacing, the higher the resolution. • Actual resolution attained also depends on stray light level. • Holographic gratings should be preferred for high-resolution work because their stray light level is lower than that of ruled gratings.
Coatings	• Coatings such as Al, Au, Pt, MgF_2, LiF and SiO_2 are available for maximum efficiency within a given spectral range and for mechanical and environment protection.	• Same as classically ruled diffraction grating.
High-energy laser beams	• Standard ruled gratings are not manufactured to endure high-energy laser beams.	• Holographic gratings are not recommended for high-energy laser beams.
Care	• A grating is as vulnerable as an unprotected first-surface mirror. As a consequence, it should not be touched in any way. Fingerprints and oral spray must be avoided at all times.	• Same as classically ruled gratings.

EXERCISES

1. Define or explain the following terms or concepts:
   - Diffraction
   - Constructive interference
   - Destructive interference
   - Transmission grating
   - Reflection grating
   - Replica grating
   - Resolving power of a grating
   - Blaze angle
   - Efficiency
   - Free spectral range
2. In your own words briefly explain the manufacturing processes used for ruled gratings and for holographic gratings.
3. Calculate the wavelength of light incident normally upon a grating with 14,300 lines per inch when a second-order image is produced at a 33° angle.
4. Calculate the groove spacing and the maximum resolving power of a reflection grating that has a width of 5 centimeters. The grating will diffract light of 546 mm to a first-order image at an angle of 15° when the light is incident along the normal to the grating surface.
5. A tunable dye laser is to operate in a frequency range of 400 nm to 800 nm. A reflection grating is to be used as a tunable end reflector. The grating must be one inch in diameter and be approximately one-half inch thick. It should be blazed to operate in the first order at a wavelength of 600 nm. There should be approximately 15,000 lines per inch on the grating. Use catalogs from three manufacturers to compare gratings with the given specifications. Choose a grating from one of the manufacturers to recommend for purchase. Give the reasons why you recommended this grating.
6. A blazed reflection grating will be used as the HR in a dye laser (λ = 560 nm). The grating will have 1000 lines/mm and is to be used in 3rd order. What is the blaze angle ϕ? (**Hint:** See Figure 14b.)
7. Consider the same dye laser and Littrow grating as in Exercise 6. Approximately what angle should the grating be rotated through to tune the laser from 560 nm to 590 nm? Which way should the grating be rotated? Draw a sketch.

MATERIALS

HeNe laser
Transmission grating
Reflection grating
Screen
Optical bench and mounts for gratings
Mercury arc or helium arc lamp

PROCEDURES

Measurement of Grating Spacing and Resolving Power

1. Arrange a HeNe laser so that the beam is incident normally on a transmission grating as shown in Figure 18. With this arrangement, the grating equation (Equation 1) reduces to $m\lambda = a \sin \theta_m$.

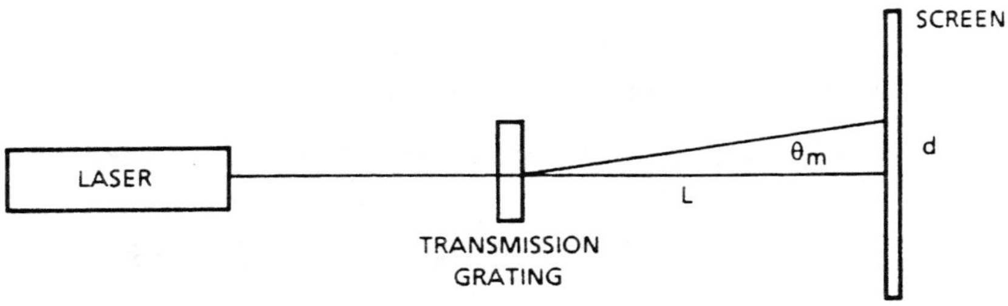

Fig. 18 Experimental arrangement.

2. Using trigonometry, determine θ_m by measuring L and d as shown on Figure 18. Do this for at least the first three orders. Record your results in a data table you prepare for that purpose.
3. Calculate the groove spacing (a) using the reduced grating equation. Average the results obtained for each order used.
4. Repeat the above procedures using a reflection grating instead of the transmission grating. Make appropriate changes in the experimental arrangement.
5. Assuming that the entire grating surface is illuminated, calculate the resolving power for each grating using Equation 2. Assume the incident light to be from a HeNe laser.

Determination of Wavelength of a Light Source

1. Arrange either of the gratings used in the first experiment so that light from a light source other than the HeNe laser can be incident upon the grating and then transmitted or reflected to a screen. Light sources such as a mercury arc, sodium arc or sodium flame will suffice.
2. Using the average value for the groove spacing as determined in the first experiment, make the measurements necessary to calculate the wavelength of one spectral line in the incident light.
3. Compare your value for the wavelength of this light with accepted values given in charts or handbooks.
4. Compute the percent error of your measurements.

REFERENCES

"Handbook of Diffraction Gratings—Ruled and Holographic." Metuchen, NJ: Prepared by the staff of JoBin-Yvon, Inc.

Jenkins, F., and H. White. Fundamentals of Optics, 4th edition. New York: McGraw-Hill, 1976.

Kingslake, Rudolph, ed., Applied Optics and Optical Engineering, Vol. V., New York: Academic Press, 1969.

Pedrotti, Frank L., and Leno S. Pedrotti. Introduction to Optics, Chapter 20. Englewood Cliffs, NJ: Prentice-Hall, 1987.

LASER/
ELECTRO-OPTICS

MODULE 6-10
POLARIZERS

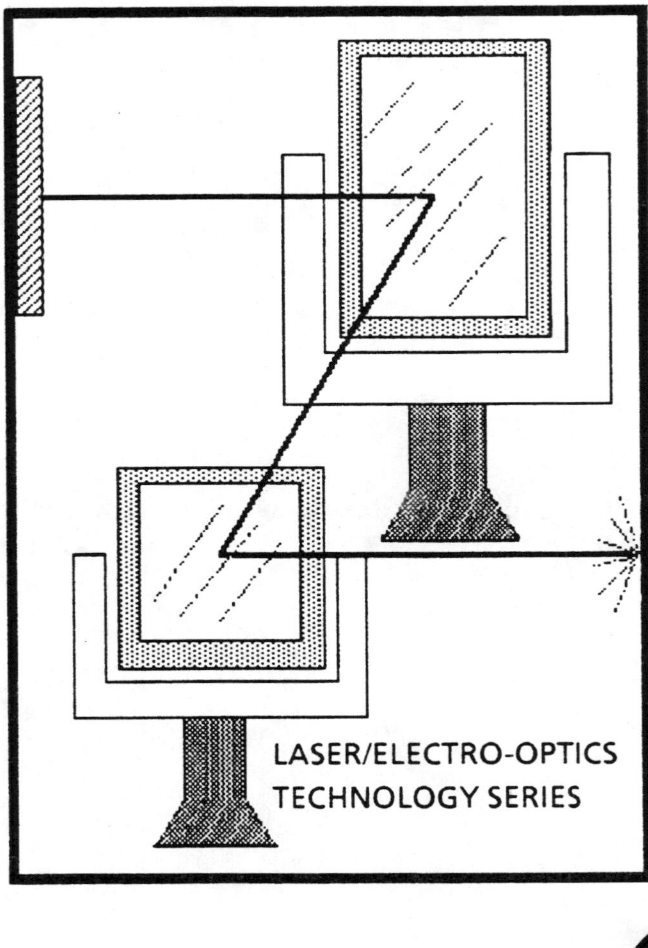

LASER/ELECTRO-OPTICS
TECHNOLOGY SERIES

ORD

The Center for Occupational Research and Development
601 C Lake Air Drive
Waco, Texas 76710

Printed in the U.S.A.

ISBN 1-55502-024-0

POLARIZERS

MODULE 6-10

Unpolarized light consists of wave vibrations in many planes perpendicular to the direction of travel. A polarizer is a device that filters out or blocks all planes of vibration, except those desired, which in the ideal case may be a single vibration plane. This vibration plane is transverse to the direction of travel of the beam.

A polarizer is only a selector. It doesn't create transverse vibrations but rather preferentially selects certain transverse vibrations and rejects others. This selection is accomplished using one or more of the principles of absorption, reflection, and refraction. This module will acquaint you with the types of polarizers, how they work, and how to use them.

To better understand this module, you should understand the theory of polarized light.

OBJECTIVES

When you complete this module, you should be able to do the following:

1. Write statements of definition or explanation for each of the following terms, concepts or components:
    - Transverse wave
    - Plane polarization
    - Vertical polarization
    - Circular and elliptical polarization
    - Dichroic polarizer
    - Reflection polarizer
    - Double-refraction polarizer

Quarter-wave plate
Half-wave plate

2. Compare and contrast, in your own words, the following polarizers discussed in the text.
   Polaroid polarizer
   Thin-film polarizers
   Glan-Foucault polarizer
   Glan-Thompson polarizer
   Nicol polarizer
   Rochon prism polarizer
   Wollaston prism polarizer
3. Using suggested procedures, polarize a light beam and plot a curve of intensity versus angle as an analyzer is rotated through a full circle.
4. Experimentally determine the plane of polarization of the following:
   Glan-type polarizer
   Wollaston polarizer
   Reflection-type polarizer
5. Calculate Brewster's angle for flat plates surrounded by various media.
6. Calculate the transmission of a series of linear (plane) polarizing filters using the law of Malus.

DISCUSSION

Light is a form of wave motion, but it travels so fast (186,326 miles per second) that we can't see the ways in which the moving waves vibrate. This wave motion must be known to describe the operation of polarizers. So, first we'll discuss some important features of light waves.

Even though we can't see light waves, experiments have shown that the electric and magnetic fields in light waves vibrate at right angles to the direction of travel of the beam. Light therefore is said to consist of transverse waves. For example, if the direction of travel of a given light beam is east, the electric vibrations may be up and down, or north and south, or along any other line perpendicular to the east-west axis.

Figure 1 shows a "snapshot" of a plane-polarized wave traveling from left to right, made up of transverse electric (E-vector) and magnetic (B-vector) components.

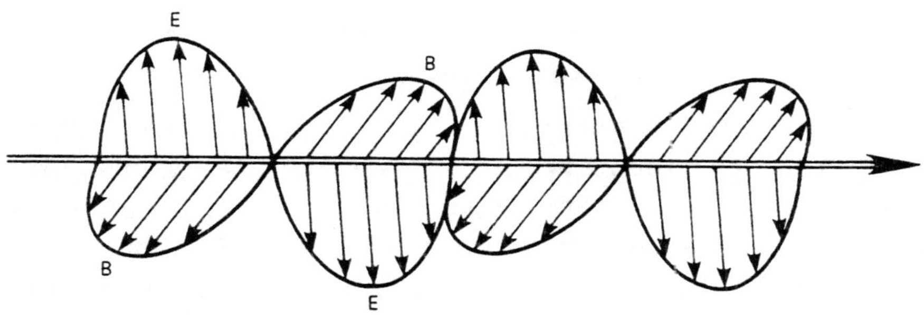

Fig. 1 Snapshot of plane polarized wave of light.

The plane containing the E-vector and the direction of propagation is called "the plane of polarization," which in Figure 1 is shown as the vertical plane. In polarization work, the direction of the E-vector is specified, so only the electric component will be shown and considered during the remainder of this module. In the plane-polarized transverse wave it's necessary to specify two directions, that of the wave disturbance (the E-vector) and that of propagation. In plane-polarized transverse waves, we can thus expect a lack of symmetry about the axis of propagation. This lack of symmetry in the limit is polarization, but let's establish that definition more clearly.

Electromagnetic waves in the radio and microwave range readily exhibit this lack of symmetry. Such a wave, generated by the swinging of charge up and down in the dipole that forms the transmitting antenna, has an electric field vector parallel to the dipole axis. This is shown in Figure 2.

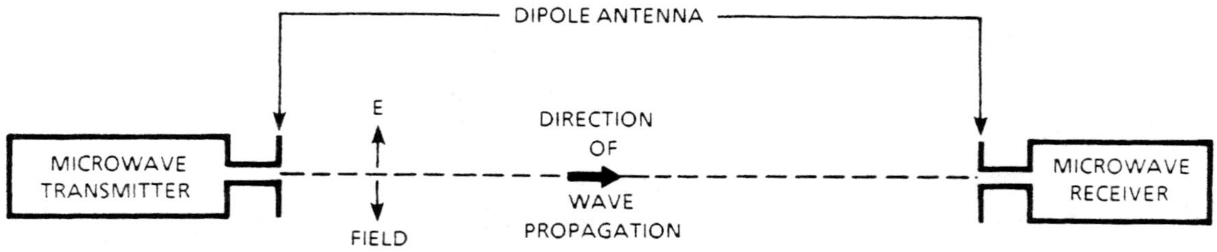

Fig. 2 Generation of microwave E field.

When this plane-polarized wave falls on a second dipole connected to a microwave detector, the alternating electric component of the wave will cause electrons to surge back and forth in the receiving antenna, producing an indication on the detector. If we turn the receiving antenna through 90° about the direction of propagation, the detector reading drops to nearly zero. In this orientation the electric field vector can't cause charge to move along the dipole axis because it points at right angles to this axis.

Most sources of visible light differ from the microwave source shown above in that the elementary radiators—that is, the atoms and molecules—all act independently. The light propagated in a given direction consists of independent wavetrains whose planes of vibration are randomly oriented about the direction of propagation. This situation is shown in Figure 3 looking along the direction or propagation (into the paper).

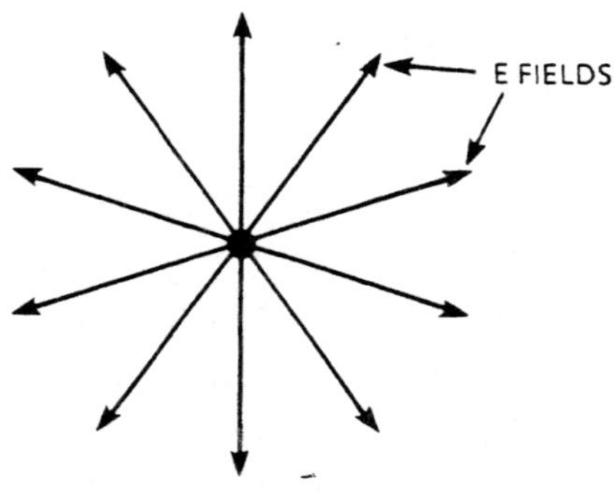

Fig. 3 Unpolarized light.

Such light, though still transverse, is unpolarized. A prime purpose of polarizers is to sort out the different planes of vibration of such radiation.

Polarized light is light whose transverse vibration has a simple pattern. Figure 4 shows a light wave traveling along the Z-axis of a right-handed coordinate system. The electric vibrations are horizontal, and accordingly the light is said to be polarized linearly and horizontally.

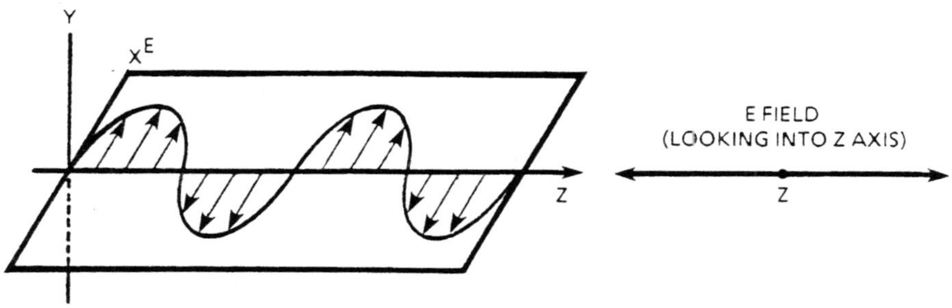

Fig. 4 Light polarized linearly, horizontally.

The right portion of Figure 4 shows the sectional end-on pattern, which corresponds to a time-exposure photograph taken by a camera looking at a cross section of the beam. The camera is situated on the axis of the beam and aimed straight at the light source. When the transverse vibrations are vertical, the cross-sectional pattern is indicated by a vertical line, and the light is said to be polarized linearly and vertically. This case is shown in Figure 5.

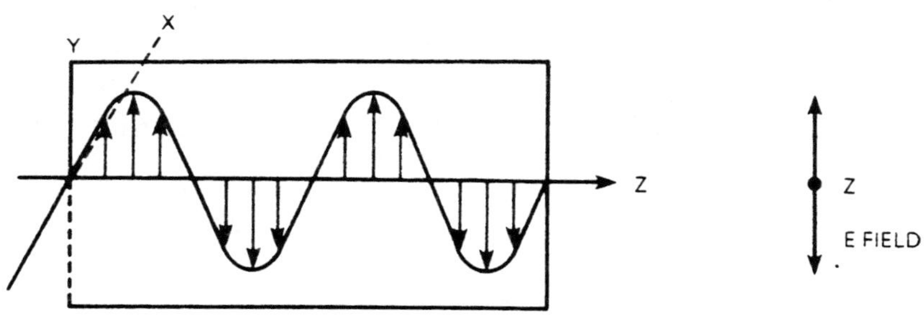

Fig. 5 Light polarized linearly, vertically.

Don't infer from the figure that that light "wiggles" side to side. Rather the pattern should be interpreted as a plot of the electric field strength within the light wave as a function of distance (or time).

With the proper phase relationship between vertically and horizontally polarized beams, the snapshot pattern is a helix and the sectional pattern is a circle. This case is shown in Figure 6.

By the most common convention, if the light is viewed *toward the source* and the electric field rotates clockwise, the polarization is called "right-hand

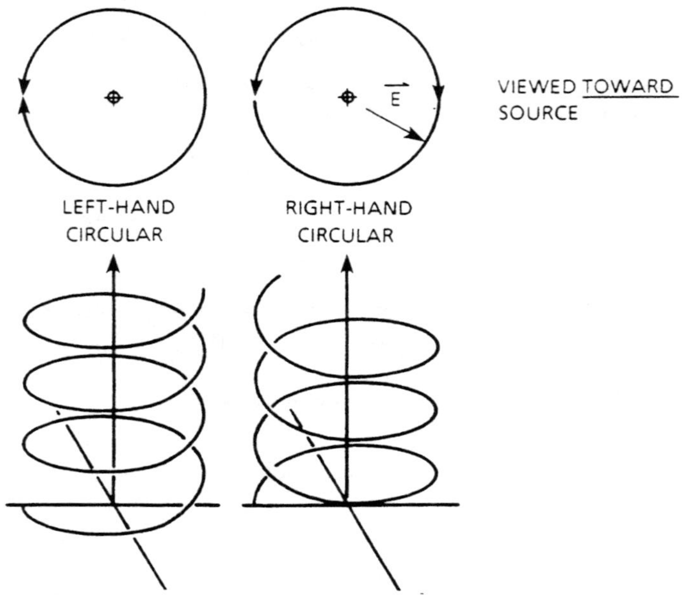

Fig. 6 Creation of circularly polarized light.

circular." Similarly, a counterclockwise rotation would be called "left-hand circular polarization."

Figure 7 shows how two plane-polarized waves of equal amplitude and at right angles to each other can create circularly polarized light. The two waves of Figure 7 differ in phase by 90°. Where one wave has maximum values, the other is zero. Values of the resultant amplitude of the approaching wave for

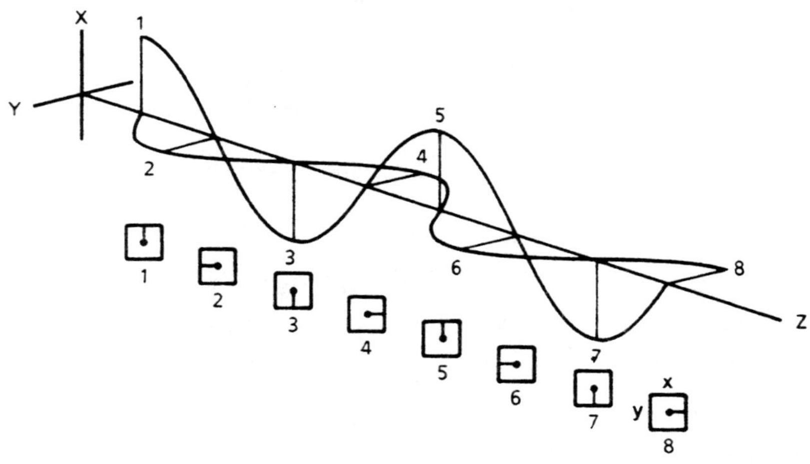

Fig. 7 Creation of circularly polarized light. For clarity, only the electric field vectors are shown.

positions 1 through 8 as seen by an observer looking toward the oncoming wave, are indicated in the small squares under the wave pattern. Note that an observer looking toward the approaching wave will see the resultant vector rotate counterclockwise with time and form left-hand circularly polarized light.

In a similar manner, if the amplitudes of the horizontal and vertical electric field components are not equal, so called elliptically polarized light can be produced. As an alternative, the phase difference between the components can be changed to produce elliptical polarization. You can find complete discussion of such effects in most introductory physics texts under the topic of Lissajous figures.

Polarizers are devices that change light from the unpolarized state of Figure 3 to one or more of the states shown by Figures 4, 5, and 6. The remainder of this module will be concerned with how polarizers operate and how they are used.

Types of Polarizers

In the laboratory the usual method of producing polarized light is the common, three-step method of first producing ordinary unpolarized light, then reducing this to two orthogonally polarized components, and finally eliminating one of these components. The component that remains is polarized light. This process is inefficient, but inexpensive and simple.

A polarizer is really a divider-and-selector. It doesn't create transverse vibrations. It merely divides all existing vibrations into two classes. There are many types of polarizers. That's because so many types of optical manipulations can be used. The processes of absorption, reflection, refraction, and scattering can be used to resolve a beam into polarized components.

Dichroic Polarizers

Dichroic materials have the property of selectively absorbing one of the two orthogonal components of ordinary light. Dichroism is exhibited by a number of minerals and by some organic compounds.

In the early nineteenth century, dichroism was discovered. And one of the crystals exhibiting it, tourmaline, was a commonly used polarizer for some time. Although this particular crystal has many shortcomings, the same principle of dichroism has been used to produce perhaps the most frequently used polarizer, namely Polaroid type-H sheets.

The simplest analogy to a dichroic polarizer is given by the wire grid as shown in Figure 8. We can explain the dichroic polarizer theory simply in terms of this device. The wire-grid polarizer consists of an array of slender wires arranged parallel to one another. The metal wires provide high conductivity for electric fields parallel to the wires. Such fields produce electric currents in the wires, like the microwave dipole receiver antenna. The energy of the fields is converted to energy of currents in the wires. This current then is converted to heat because of the small but significant electrical resistance of the wires.

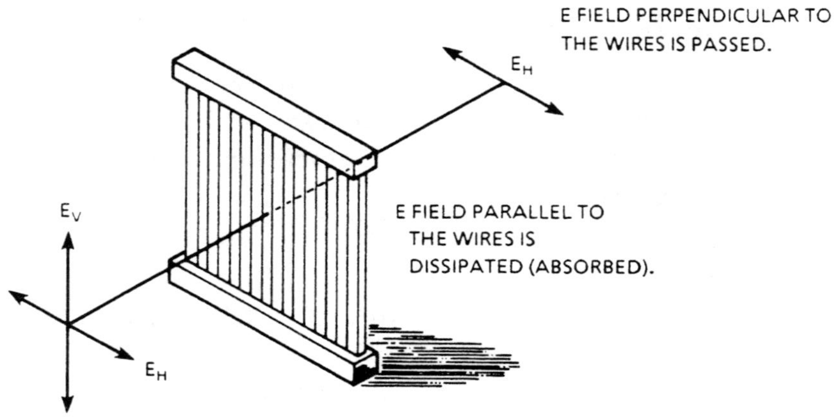

Fig. 8 The wire-grid polarizer.

Because of the nonconducting spaces between the wires, no current can flow perpendicular to them. So electric fields perpendicular to the wires produce virtually no currents and lose little energy. The wire grid, when placed in an unpolarized beam, drains the energy out of one component and lets the other component pass with almost no attenuation at all. The transmitted component is the one whose electric field vibrates in a direction perpendicular to the wires.

An example of light passing through a dichroic material is shown in Figure 9. The randomly polarized light entering T_1 is reduced to simple vertical polarization, which passes unaffected through T_2 when its orientation is the same as T_1. If T_2 is rotated 90° it has the effect of two wire-grid polarizers at right angles to each other, and virtually no light passes.

Although the wire-grid polarizer is easy to understand, it's too difficult to make to be practical. The wires have to be on the order of a wavelength of light apart, and this is difficult and very expensive.

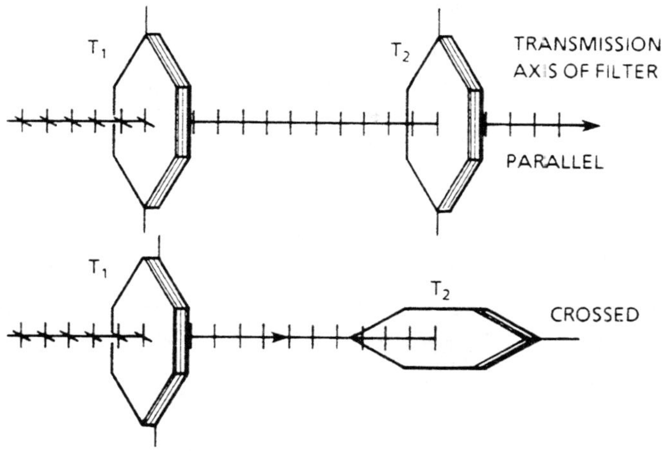

Fig. 9 Operation of a dichroic polarizer.

A better way had to be found. This happened in 1938 when E. H. Land invented H-Polaroid sheet. Think of the H-Polaroid sheet as a chemical version of the wire grid. Instead of long thin wires it uses long thin polyvinyl alcohol molecules that contain many iodine atoms. These long, straight molecules are aligned almost perfectly parallel to one another and about 3 Å apart. Because of the conductivity provided by the iodine atoms, the electric vibration component parallel to the molecules is absorbed. The component perpendicular to the molecules passes on through with little absorption.

The way Land produced this type of polarizer is shown in Figure 10. A sheet of polyvinyl alcohol was heated and then quickly stretched to many times its length. During the stretching operation most of the long polymer molecules were turned so as to have nearly the same direction, namely, the direction of the stretching force.

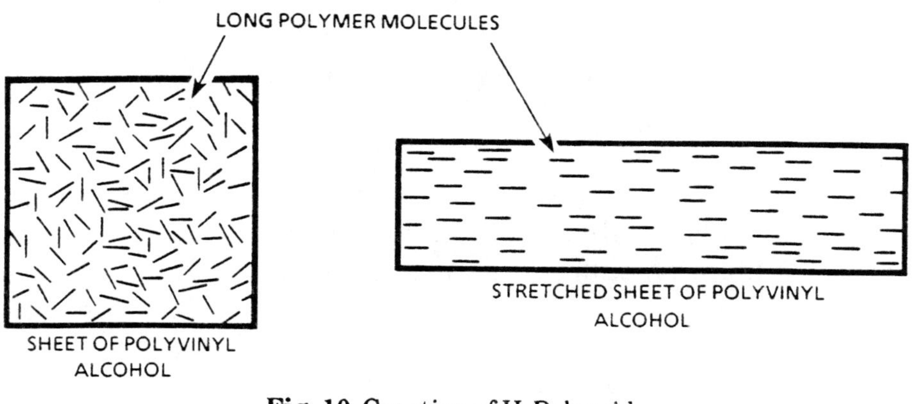

Fig. 10 Creation of H-Polaroid.

The next step was to dip this stretched sheet into a liquid solution that was rich in iodine. The iodine atoms diffused into the polyvinyl-alcohol layer and attached themselves to the thin, nearly parallel molecules. So the iodine atoms imitated their host and formed long thin chains.

These iodine chains in Polaroid act like wires. They absorb electric field vibrations parallel to their aligned axis and transmit freely the vibrations perpendicular to that axis. So the transmission axis is perpendicular to the stretched direction.

Figure 11 shows the effect of Polaroid material on light when two sheets are lined up parallel and when two are used perpendicular to each other.

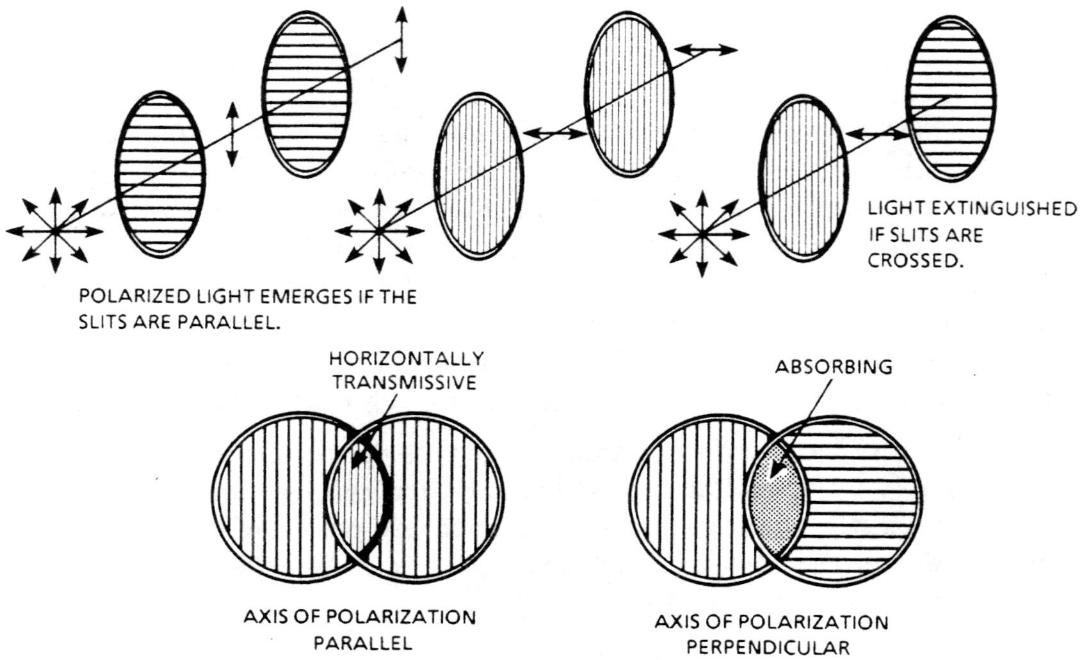

Fig. 11 Effect of polarized material on light.

Many other kinds of dichroic sheet polarizers can be made. Some use dyes that consist of long thin molecules, and some use long thin crystals. However, all of these types are similar in theory to the Polaroid sheet mentioned above.

TM and TE Polarizations

Before we continue our discussion of the methods of polarizing light, it's helpful to define two particularly important polarizations, TM (transverse magnetic) and TE (transverse electric). These two orthogonal (mutually perpendicular) polarizations are shown in Figure 12. Note that the dots in Figure 12b represent an electric field vibrating in and out of the page.

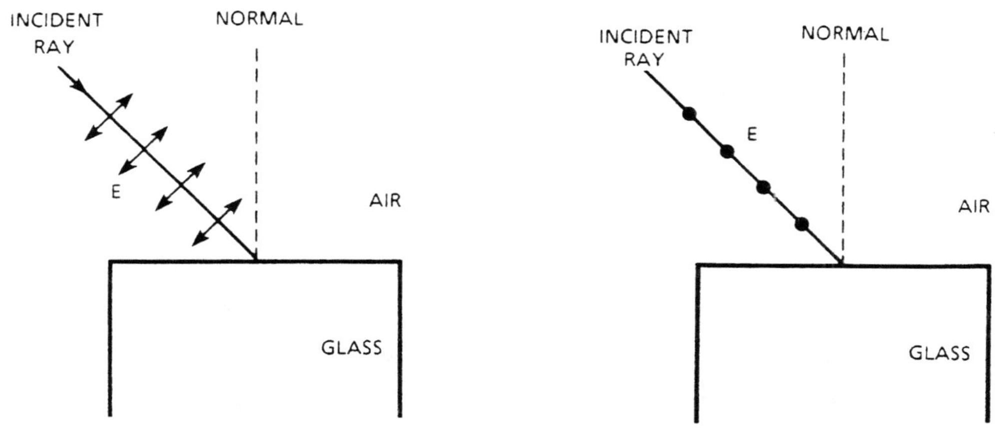

a. TM polarization

a. TE polarization

Fig. 12 Two orthogonal polarizations of light.

TM and TE are defined with respect to the so-called "plane of incidence" (POI) which is the plane containing the normal to the surface and the incident ray. So in Figure 12, the POI is the plane of the page. Note that in Figure 12a the electric field is *in* the POI. This means that the magnetic field must be transverse (across or perpendicular) to the POI. On the other hand, Figure 12b shows the TE polarization with the electric field transverse (perpendicular) to the POI. In both figures for the TM and TE polarizations, *only the electric field* is shown to save confusion.

Reflection-type Polarizers

Perhaps the simplest method of polarizing light is the one discovered by Malus in 1808. If a beam of unpolarized white light is incident at one certain angle on the polished surface of a plate of ordinary glass, it is found upon reflection to be linearly polarized.

When the plate is mounted perpendicular to the beam, no polarization results; all components of the light are transmitted equally and the transmitted beam is found to be unpolarized. About 8% is reflected, (4% at each surface) and this light is also unpolarized.

When the incident ray is tilted so that the symmetry of the reflection process is destroyed, the transmitted and reflected beams are found to be partially polarized. This situation is shown in Figure 13a.

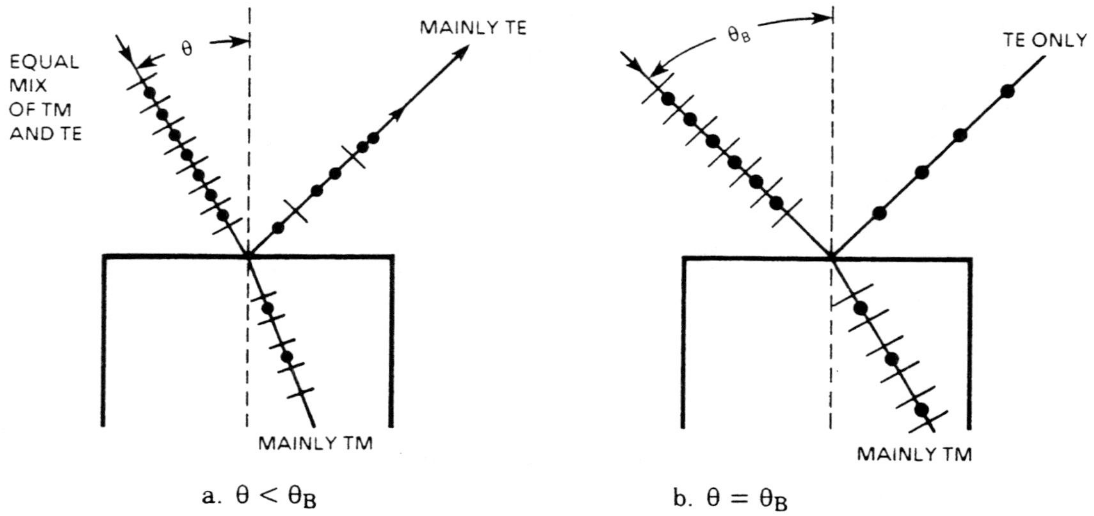

Fig. 13 Reflection of TM and TE polarizations from a plane surface.

When the angle of incidence is further increased to the so-called "Brewster angle," the TM polarization is 100% transmitted. It leaves the reflected beam as pure TE polarization as shown in Figure 13b. You can find Brewster's angle from the expression

$$\theta_B = \tan^{-1}\left(\frac{n_2}{n_1}\right) \qquad \text{Equation 1}$$

where: n_2 = Index of refraction of the glass plate.

n_1 = Index of refraction of the medium surrounding the plate.

For example, for a glass plate ($n_2 = 1.52$) in air ($n_1 = 1.00$), Equation 1 gives

$$\theta_B = \tan^{-1}\left(\frac{1.52}{1.00}\right) \approx 56.7^\circ$$

If you have studied *Introduction to Lasers*, you will recall that this is the angle at which a laser Brewster window is oriented on the end of a plasma tube. This is common, for example, in most argon or krypton ion lasers.

Sometimes reflection plate polarizers are composed of several plates and are called "stacked plate" or "Brewster stack" polarizers. Using many plates further increases the purity of the transmitted polarization. Brewster stacks are generally of two types—linear stacks or tent stacks. The latter has the advantage of an undisplaced beam. (See Figure 14.)

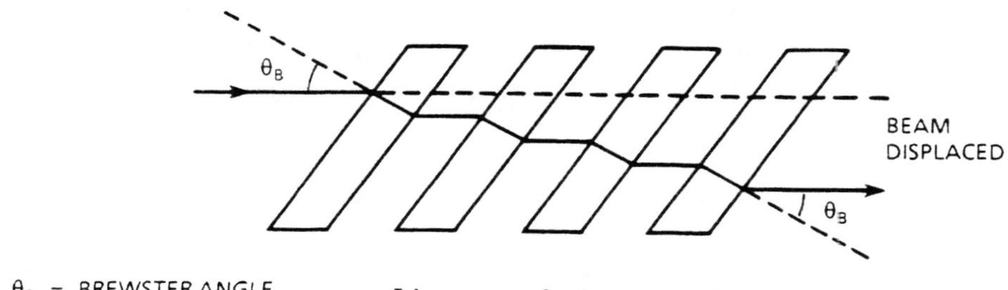

a. Linear stacked array polarizer

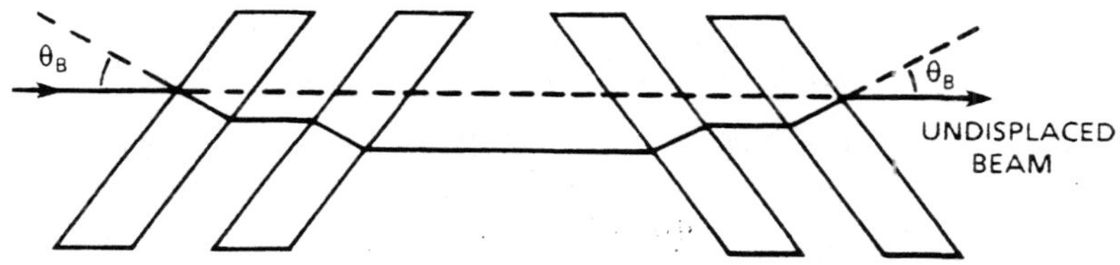

b. Tent polarizer

Fig. 14 Types of stacked plate polarizers.

Polarizers of the reflection type have several attractive features. They can be designed for use in any part of the spectrum. Also, they are easy to make and can have very high thresholds for laser-induced damage. On the other hand, reflection-type polarizers seldom are used to polarize light in the visible range because they are outperformed by the sheet polarizers discussed previously. Only in certain parts of the infrared and ultraviolet range are reflection-type polarizers superior, or in high-power laser applications where other types of polarizers would suffer laser-induced damage.

Other types of polarizers based upon reflection (and transmission) also are becoming important for high peak power laser applications. These polarizers generally are constructed in the form of thin dielectric coatings. So they are called "thin-film polarizers." There are two principal types.

The first thin-film polarizer, called a MacNielle biprism polarizer, uses a thin-film multilayer dielectric between two prisms. The second is more generally used because of its reduced cost and capability for larger size. It's the thin-film multilayer stack placed on a suitable substitute. This combination then is placed at an appropriate angle in the beam.

Simply stated, these polarizers, similar in many ways to thin-film filters, exhibit vast differences in reflection and transmission characteristics for the two types of polarization, TM and TE, at a specific selected wavelength. These ratios between the transmission of TM and TE polarization can reach several hundred or even a thousand.

Increased discrimination can be achieved by using several plates in combination. However, these thin-film polarizers are fairly sensitive to angular orientation. For a further description of these novel polarizers, see the referenced article by Austin.

Double-refraction Polarizers

Polarization was discovered with the aid of crystals that exhibit double refraction. It was discovered by the Dutch scientist Huygens while he studied some crystals of the doubly refracting mineral calcite ($CaCO_3$). Double refraction is a complicated subject that requires an understanding of crystallography. Because the crystal line structure is so complicated, we won't explain the reasons for the double refraction. Rather we'll concentrate on the effect of calcite on the light beam.

If a polished crystal of calcite is laid over some printed letters, the image appears double. This double bending of a beam through calcite is called *double refraction*. Figure 15 shows a beam of unpolarized light falling on a calcite crystal. The single beam splits into two at the crystal surface.

If you do experiments at various angles of incidence, you will find that one of the beams of Figure 15 obeys Snell's law of refraction at the crystal surface, just like any ray passing from one medium to another. This beam is represented by the "ordinary" or O-ray in Figure 15.

The other refracted beam does not obey the laws of refraction. This beam is called the "extraordinary" or E-ray. You can understand the difference between the waves represented by the O- and E-rays with respect to the laws of refraction in these terms:

1. The O-wave travels in the crystal with the same speed V_o, in all directions. In other words, the crystal has, for this wave, a single index of refraction, n_o, just like an isotropic solid.
2. The E-wave travels in the crystal with a speed that varies with direction. In calcite it varies from V_o to a larger value V_e. The E- and O-rays of

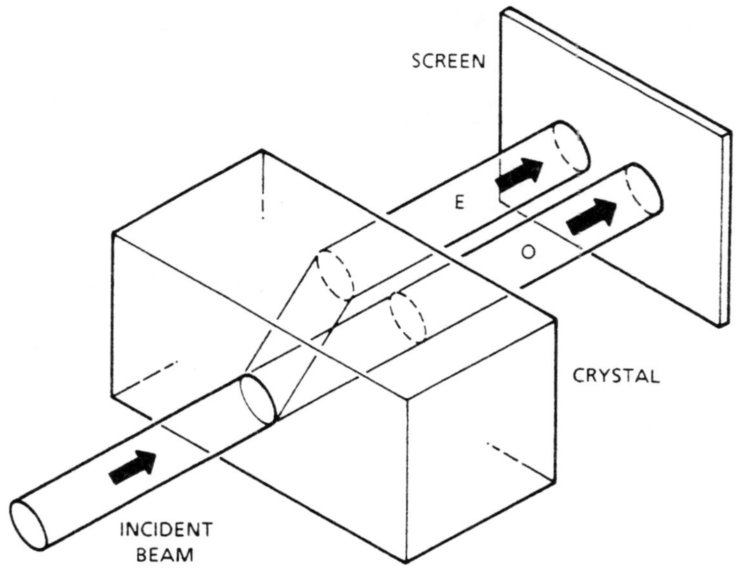

Fig. 15 Double refraction of unpolarized light.

Figure 15 are polarized at right angles to each other and are ~ 100% polarized.

Simply stated, a double-refracting crystal resolves light into two components by causing one type of vibration to travel one path and the other type to travel another path.

It should be clear that constructing a polarizer out of calcite is easy. In fact any smooth-faced calcite crystal is already a polarizer. The natural crystal structure of double-refracting crystals causes them to polarize light. When inside the crystal, light travels in slightly different directions, depending on its polarization. The polarizations emerge at slightly different places. So they constitute two side-by-side beams of linearly polarized light with mutually perpendicular vibration directions. But unless the incident beam is very slender or the crystal is very large, the two emerging beams will overlap a good deal. Several geometries have been formed with calcite that overcome this problem.

The Glan-Foucault Polarizer

One design of a calcite polarizer is that invented by Glan and Foucault. It's shown in Figure 16. Two right-triangular pieces of calcite are used. The two pieces are mounted side by side and held so that there's a thin air gap between the respective hypotenuses as shown in Figure 16.

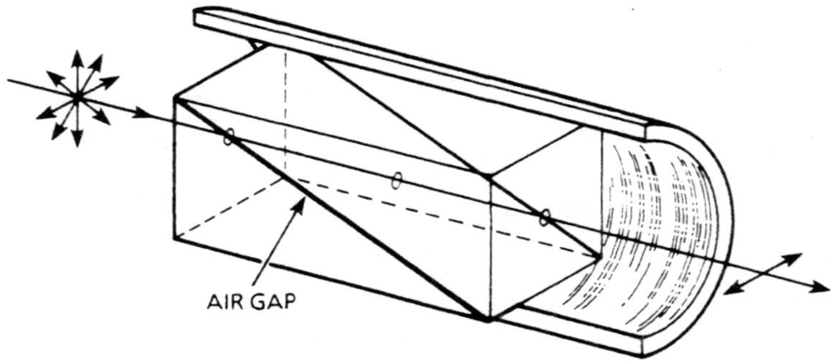

Fig. 16 Glan-type polarizer.

When a beam of unpolarized light strikes one prism along the normal to the end face, it enters the prism and forms two beams there, namely the O-ray and the E-ray. The inventors chose an angle for the hypotenuse so that one beam is totally internally reflected and the other is transmitted. This is shown in Figure 17.

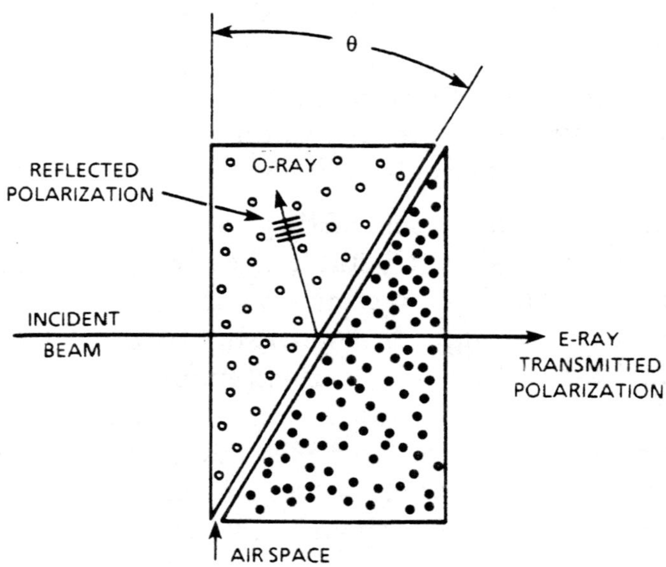

Fig. 17 How a Glan polarizer works.

The reflected beam strikes some black paint on the side of the first prism and is absorbed. The transmitted beam proceeds onward, into the air gap, and through the second prism. The degree of polarization of this beam is practically 100%.

Since calcite is transparent from the ultraviolet (230 nm) to the infrared (5000 nm), and since the Glan-type polarizer contains nothing but calcite, this polarizer can be used throughout a large spectral range. One drawback however, is that the incident beam must strike the entrance face no more than 7° off normal.

The Glan-Thompson Polarizer

If the two prisms of the Glan-Foucault polarizer are cemented together and the air gap is eliminated, the device is called a Glan-Thompson polarizer. It has an advantage over the Glan-Foucault polarizer in that it can accept incident light far in excess of 7°. One drawback of this type is that the cement used is normally opaque to certain wavelengths and subject to damage at high light intensities.

The Nicol Polarizer

Another useful polarizing device made from calcite is the Nicol polarizer. The Nicol polarizer, like the Glan polarizers, removes one of the refracted rays by total reflection. As shown in Figure 18, the Nicol polarizer also is made of two triangular wedges of calcite. In this case however, the wedges are cemented together with Canada balsam which has an index of refraction about midway between the index of the O- and E-rays. The angle over which incident light can enter a Nicol polarizer and successfully pass through is about 14°.

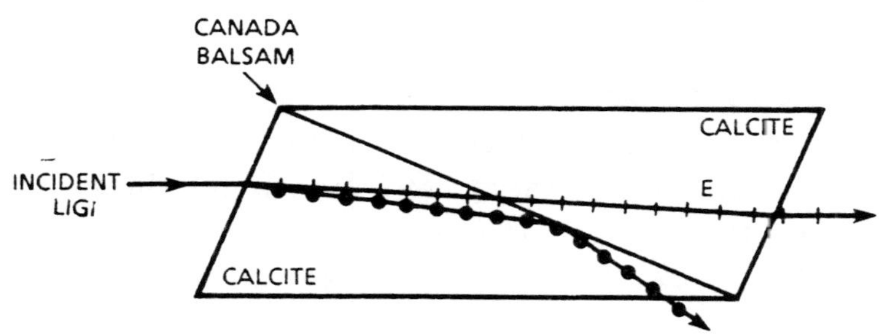

Fig. 18 Nicol polarizer.

The Nicol polarizer performs well. But it has a limited spectral range because of the absorption of ultraviolet light by the Canada balsam.

Rochon and Wollaston Polarizers

The double-refraction type of polarizers mentioned above are designed to transmit one of the beams and absorb the other. Sometimes you need a polarizing device that transmits both beams and separates them widely enough so that each beam can be used independently. This device then serves as both a polarizer and a polarizing beam splitter. This is true to the Rochon prism shown in Figure 19.

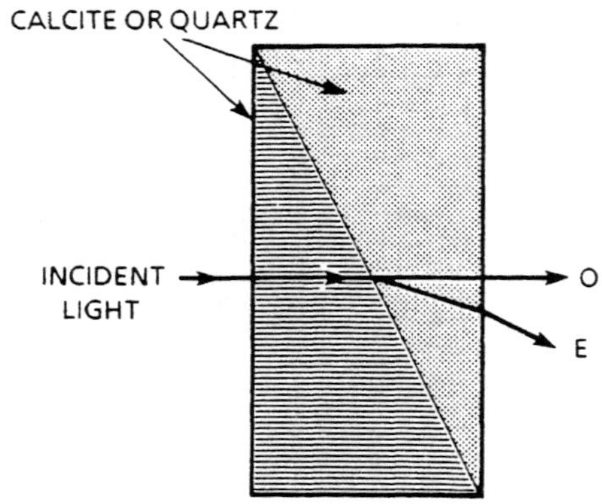

Fig. 19 Rochon prism.

In the first half of the Rochon prism, both the ordinary and the extraordinary rays travel with the same velocity. In the second half, the ordinary ray continues at the same velocity. But the extraordinary ray travels more rapidly and therefore is deviated by an amount that depends on the angle of the interface.

The Wollaston polarizer shown in Figure 20 behaves similarly. But it deviates the beams in opposite directions. The ordinary ray in the first half of the prism becomes the extraordinary ray in the second half, and vice versa. So one ray is deviated by the same amount that a Rochon prism would deviate the extraordinary ray, and the other is deviated by very nearly the same amount.

The Rochon prism is used in optical instruments to produce a plane-polarized beam. The O-beam is used and the E-beam is screened off at a sufficiently large distance from the prism.

The Wollaston prism is used where a comparison of intensities is desired. So both beams are used. The intensities of the two output beams will be equal if the input beam is unpolarized, but they will differ if the input beam is polarized to any measurable extent.

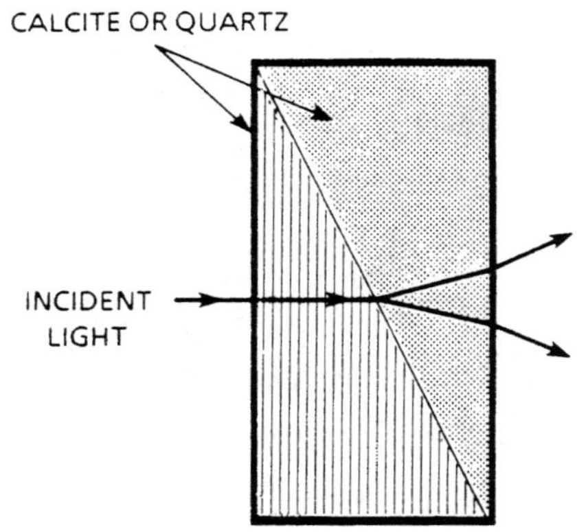

Fig. 20 Wollaston prism.

Retarders–Polarization Converters

Retarders are not polarizers in the strict sense of the word. They have no effect on unpolarized light. But they're very useful in changing one form of polarized light into another form.

The principle of operation of a retarder is easily stated. It divides an incident, polarized beam into two components, changes the phase of one relative to the other, then recombines them again. Because of the resulting phase shift, the exit light has a different polarization form. Since the key step is the introduction of a relative phase shift, some people call the device a "phase shifter" or a "wave plate."

A typical retarder consists of a thin crystalline plate as shown in Figure 21. Let's assume that a linearly polarized beam is incident on the plate

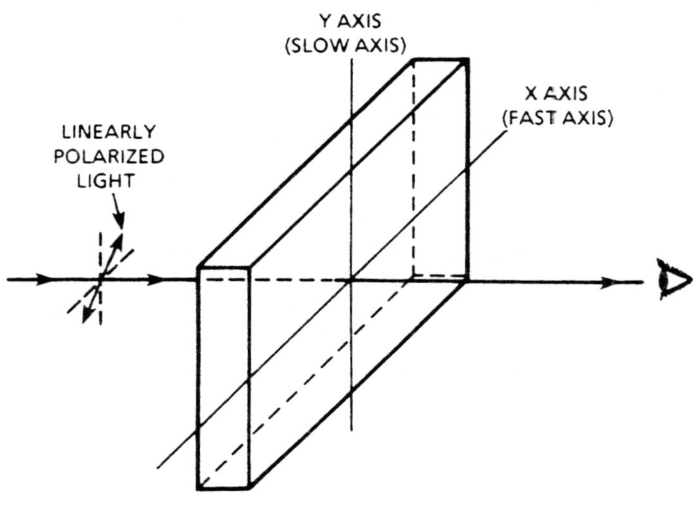

Fig. 21 180° retarder.

from the left as shown. You can consider the beam to have a vertical vibration and a horizontal vibration direction. The important fact is that the two vibrations will travel at different speeds so that, when they emerge, one has been retarded or slowed compared to the other.

We're interested only in relative retardation. In calcite the beam with the vertical vibration direction is slowed. This vertical direction is called the"slow axis." The direction of vibration of the faster-traveling beam is conversely called the "fast axis."

The thickness of the retarder and so the time the light spends inside the crystal determines the amount of phase shift. If the phase shift is 90°, the retarder is a quarter-wave plate. It can convert linearly polarized light into circularly polarized light, and vice versa. Figure 7 showed circularly polarized light. Figure 22 shows the effect of a quarter-wave plate on this light. Note that one of the waves has been slowed by 90°.

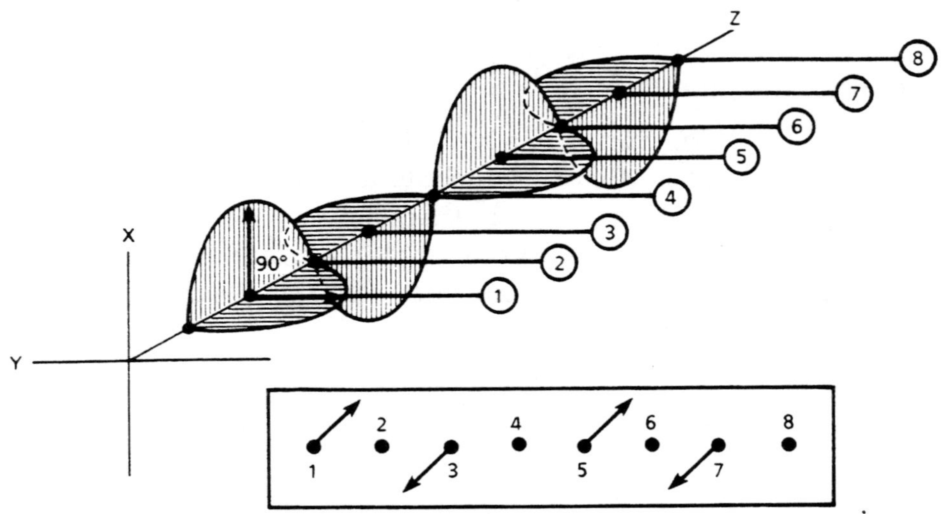

Fig. 22 Linearly polarized wave.

A little thought will convince you that the quarter-wave plate changed the circular polarization of Figure 7 to the linear polarization of Figure 22 by simply delaying one of the waves by 90°. The 180° retarder, or half-wave plate, can be used to convert right-circularly polarized light into left-circularly polarized light, or vice versa.

In practice, retarders are not made of calcite. Calcite's birefringence is large and the plate necessarily must be very thin and therefore fragile. Usually quartz or mica is used.

Uses of Polarizers

Polarizers are used in our everyday lives as well as in the optics laboratory. We'll give some examples of popular uses.

Polarizing sunglasses take advantage of the fact that in nature "glare" consists mainly of light having a horizontal vibration direction. The reason for this is that sunlight comes downward and hits mostly horizontal surfaces (such as the oceans). Reflection from such a surface results in polarization with a predominant horizontal vibration. (See Figure 13a.) Figure 23 shows that in polarizing sunglasses the transmission axis is vertical and thus reduces the glare reflected from horizontal surfaces.

Fig. 23 Polarizing sunglasses.

Solutions of various sugars and other organic substances have the property of rotating the plane of polarization of light that passes through them. The extent of this rotation is measured with a polarimeter. As shown in Figure 24, this instrument is comprised of a source of light (usually a sodium lamp), a polarizer, a tube to hold the sample and an analyzer. An analyzer is simply another polarizer that can be rotated.

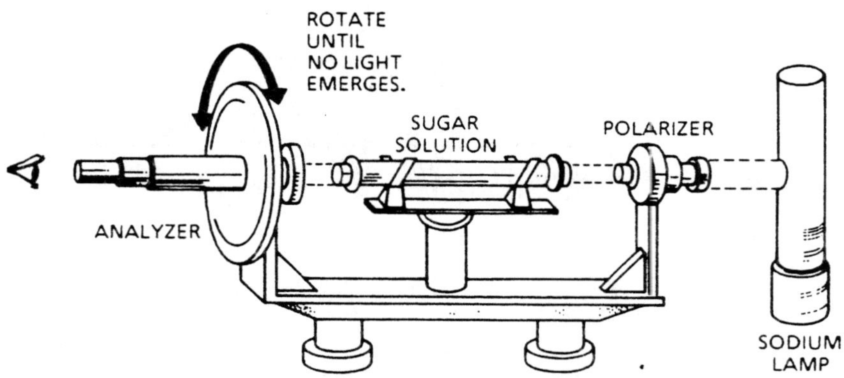

Fig. 24 Polarimeter.

To use the device, the tube is first filled with water, the lamp turned on, and the analyzer rotated until no light is seen to emerge from it. The polarizer and analyzer then are said to be "crossed." The water in the tube then is replaced by the solution under test. The analyzer again is rotated until it's again crossed. The change in the plane of polarization is thus determined. Such information provides clues to the structure of these compounds. The amount and direction of polarization indicate the strength, concentration and type of the solution.

A further example of the use of polarization is the Kerr cell that was used in high-speed photography and in modulating beams of light. This cell depends on polarization effects induced in a liquid as shown in Figure 25.

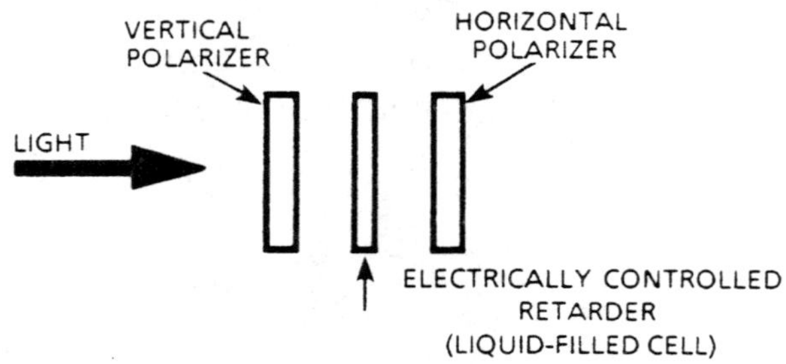

Fig. 25 Kerr cell.

Light is vertically polarized by the first polarizer and, unless it is changed by 90° in the retarder, it can't pass through the second polarizer. This arrangement forms a shutter that can open and close in 10^{-12} seconds. A similar effect occurs in some solid materials and is called the "Pockels effect." Pockels cells are used, for example, as laser Q-switches and have all but replaced the more cumbersome Kerr cells.

The Law of Malus

The irradiance of plane polarized light passing through a polarizer is given by the so-called "law of Malus,"

$$I = I_o \cos^2 \phi \qquad \text{Equation 2}$$

where:

I = Irradiance transmitted through the filter

I_o = *Maximum* transmitted irradiance

ϕ = Angle between the polarizer's axis and the plane of polarization of the incident light. (See Figure 26.)

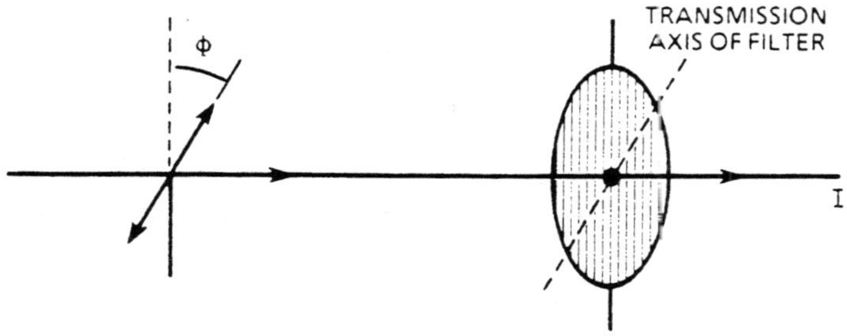

Fig. 26 Law of Malus.

For example, if the maximum transmission of a polarizing filter is 90% and it's oriented at 30° to the plane of the incident lights' polarization, then

$$I = 0.90 \cos^2 30° = 0.675$$

So the transmission of the filter is 67.5%.

Note that, if the area of the beam is unchanged as it passes through the filter, Equation 2 can be written as,

$$P = P_0 \cos^2 \phi \qquad \text{Equation 3}$$

where P and P_0 are in power units rather than irradiance.

Furthermore, you can compute the transmission of a series of linear polarizers by successive applications of the law of Malus. More complicated combinations of polarizing filters (quarter-wave plates, half-wave plates, and so on) can be handled easily by the Jones vector theory, a powerful matrix technique introduced in many optics texts.

EXERCISES

1. Define or explain each of the following terms, concepts or components.
   - Transverse wave
   - Plane polarization
   - Vertical polarization
   - Horizontal polarization

Circular and elliptical polarization
Dichroic polarizer
Reflection polarizer
Double-refraction polarizer
Quarter-wave plate
Half-wave plate

2. Compare and contrast, in your own words, the following polarizers discussed in the text.
    Polaroid polarizer
    Thin-film polarizers
    Glan-Foucault polarizer
    Glan-Thompson polarizer
    Nicol polarizer
    Rochon prism polarizer
    Wollaston prism polarizer

3. Explain in your own words how to determine the plane of polarization for each of the following:
    Glan-type polarizer
    Wollaston polarizer
    Reflection-type polarizer

4. Compute Brewster's angle for a glass plate (n = 1.50) surrounded by water (n = 1.33).

5. A glass plate is found to have a Brewster's angle of 55.77° when surrounded by air (n = 1.00). Find the index of refraction of the glass plate.

6. A plane polarizer has a maximum transmission of 89.5%. When 100% plane polarized light is directed through the filter, the transmission is found to be 57.2%. What's the angle between the plane of polarization and the filter's transmission axis?

7. What will the reading on the optical power meter be for the setup shown below?

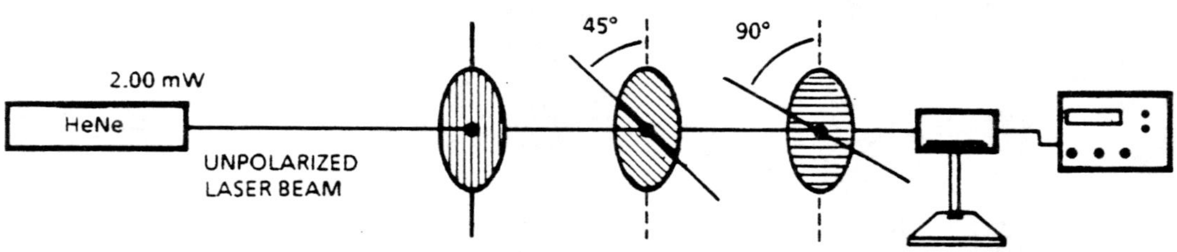

Assume that each polarizer has a maximum transmission of 92.0%. Also assume that the"perfect" linear polarizer would transmit one half of the incident unpolarized light irradiance striking it.

MATERIALS

Sodium light source
2 Polaroid sheet-type polarizers
2 Polarizer mounts (rotatable)
Glass plate with tilt stand (index of refraction ≈ 1.50)
Glan-type polarizer
Mount for Glan polarizer
Ground glass screen
Screen holder
Wollaston prism
Prism holder
2 Optical benches
Optical power meter

PROCEDURES

Polaroid-type Polarizers

1. Set up the sodium light source, optical bench, and screen as shown in Figure 27 and measure the intensity.

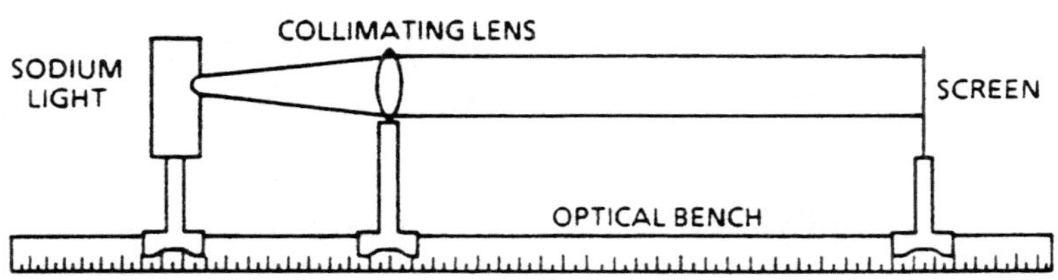

Fig. 27 Experimental arrangement.

2. Insert one of the Polaroid sheet-type polarizers in the beam. Note the measure in intensity of the light on the screen with a sensitive light meter. (Only one polarization is present now.) The preferred plane of polarization should be marked on the polarizer.

3. Insert the second Polaroid into the beam and line the filter transmission axes up so that they are parallel. Note the intensity change on the detector. This change is not due to polarization effect, but rather to the fact that polarizers also absorb, reflect, scatter, etc.

4. Rotate the second Polaroid 90° and measure the intensity with the optical power meter. Plot an intensity curve for a full 360° rotation of the second Polaroid. Be careful to take into account any "background" reading due to ambient room light.

Glan Polarizer

1. Leave the second Polaroid in place. Replace the first one with a Glan polarizer as shown in Figure 28.

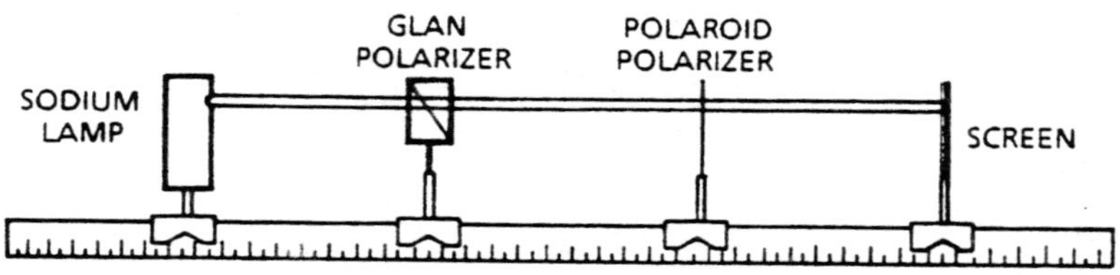

Fig. 28 Polarization experimental arrangement.

2. By rotating the Polaroid polarizer determine the plane of polarization produced by the Glan polarizer.

Wollaston Prism

1. Replace the Glan polarizer with a Wollaston prism.
2. Position the prism so that one of the emergent beams passes through the Polaroid.
3. Note the plane of polarization in this beam.
4. Repeat Step 2 for the other emergent beam.

Polarization by Reflection

1. Change the setup to resemble Figure 29.
2. Move the second optical bench so that light reflecting off the glass plate at 56.3° will go through the Polaroid and hit the screen.

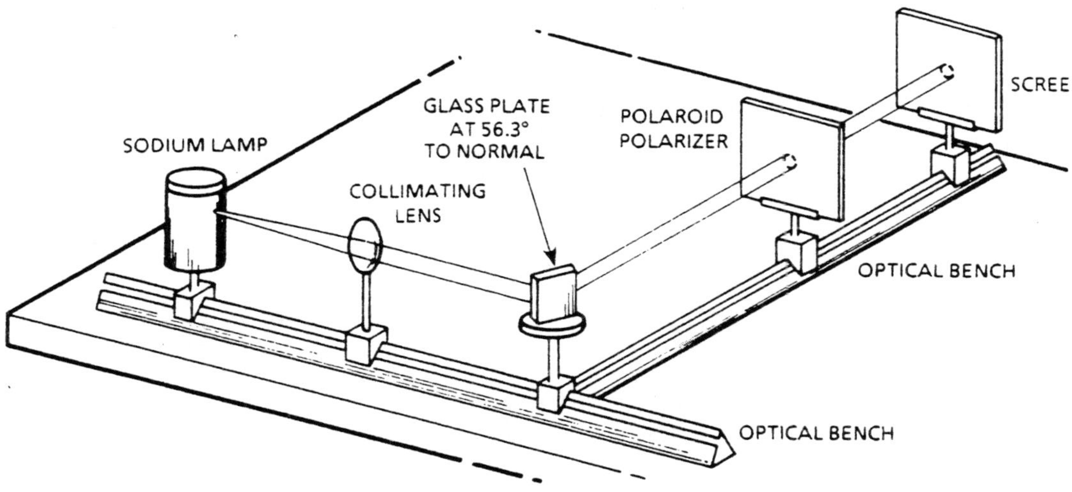

Fig. 29 Experimental arrangement.

3. By rotating the Polaroid, verify that the reflected light is polarized and determine the plane of polarization.

REFERENCES

Austin, Russell. "Thin Film Polarizing Devices," Electro-Optical Systems Design. Feb. 1974.

Jacobs, Donald. Fundamentals of Optical Engineering. McGraw Hill Book Co., 1943.

Jenkins, Francis A., and Harvey E. White. Fundamentals of Optics, 4th Edition. New York: McGraw-Hill Book Company, 1976.

Pedrotti, Frank L., and Leno S. Pedrotti. Introduction to Optics. Englewood Cliffs, NJ: Prentice-Hall, Inc., 1987.

LASER/
ELECTRO-OPTICS

MODULE 6-11
NONLINEAR MATERIALS

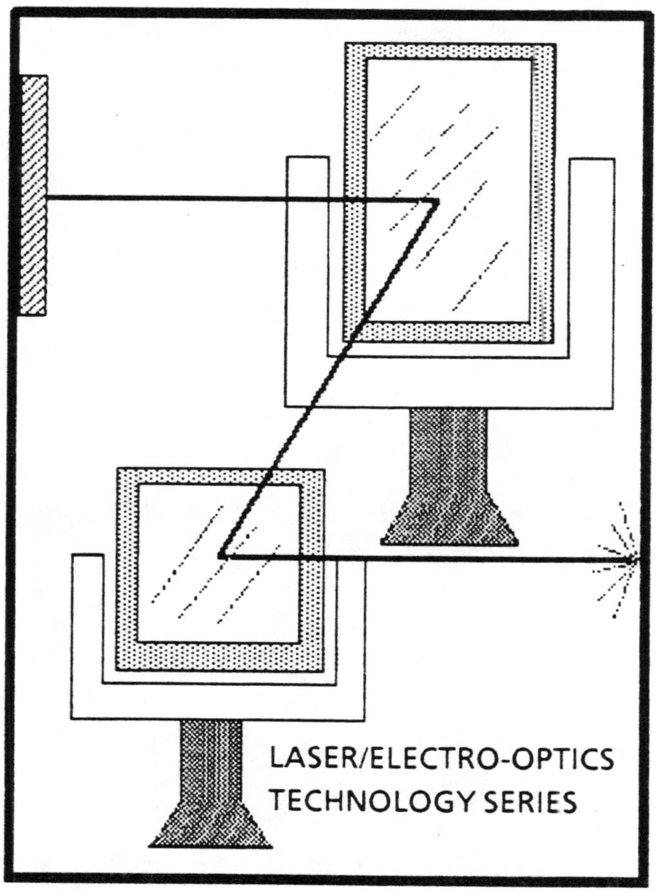

ORD

The Center for Occupational Research and Development
601 C Lake Air Drive
Waco, Texas 76710

Printed in the U.S.A.

ISBN 1-55502-024-0

NONLINEAR MATERIALS

MODULE 6-11

Nonlinear optical materials are materials in which the intensity of light input, including its frequency, is not related to the intensity of light output by a simple proportionality constant. Because of this nonlinear behavior, an intense light beam propagating through a nonlinear optical material will produce new effects that can't be seen with weak light beams. For example, an intense light beam propagating through a nonlinear material can generate harmonics, or overtones, of the original light frequency. This means that the red beam from a ruby laser can be changed to an ultraviolet beam.

The interaction of intense light beams with nonlinear materials has opened a large field of potential applications in optical communication systems, where many well-known radio-frequency techniques such as mixing, heterodyning, and modulation now can be transferred to the domain of optical frequencies. For these reasons, nonlinear materials play an ever-increasing role in laser applications.

This module will familiarize you with the fundamental physical ideas that underlie some of the more commonly encountered nonlinear phenomena and how these phenomena depend on the properties of the nonlinear materials. The experiment at the end of this module will give you practical experience in using a nonlinear material to generate the visible second harmonic of the invisible 1.06-μm Nd:YAG laser wavelength.

For a better understanding of the ideas in this module, you should have knowledge of algebra and some trigonometry. Before beginning the experiment at the end of this module, be sure to review laser safety.

OBJECTIVES

When you complete this module, you should be able to do the following:

1. Describe the nonlinear behavior of a system by describing its linear and nonlinear behavior using appropriate equations.
2. Explain second harmonic generation of light with the aid of the induced charge polarization.
3. Discuss in general terms the needs for phase matching in second harmonic light generation.
4. Describe two experimental techniques used for phase matching.
5. List the various materials and experimental parameters that determine the efficiency of second harmonic generation through the use of the correct mathematical equations.
6. Differentiate between passive and active nonlinear materials.
7. List and rate the importance of some of the more common nonlinear materials.
8. Define the field-dependent nonlinear index and its implications.
9. Produce the second harmonic of the 1.06-μm radiation from an Nd:YAG laser using a crystal of KDP or ADP.
10. Knowing the type of SHG crystal, its size and operating parameters, estimate the theoretical conversion efficiency.

DISCUSSION

Our objective is to develop a familiarity with the behavior and application of nonlinear materials. We'll do this by first describing the behavior of a linear system or material. Next, we'll look at its nonlinear behavior and discuss the internal properties of the material that make it behave in a nonlinear fashion. Finally, we'll describe in a very elementary fashion the fundamental physical idea behind one of the more common nonlinear effects and how this effect depends on the properties of the nonlinear material. We conclude this section with a brief discussion of the more common nonlinear materials.

Let's start then with a brief discussion of linear materials and linear effects. It is a fact that our everyday experiences tell us that we live in an essentially linear world. When you first encounter the mathematical expressions of elementary physical laws, you may be tempted to conclude that nature is linear. For example, you learn that gas pressure in a container will double if the temperature of the gas is doubled; or that when a given force on a spring produces a stretching of the spring of one centimeter then doubling the force will produce a stretching of exactly two centimeters. The example of the spring is shown schematically in Figure 1.

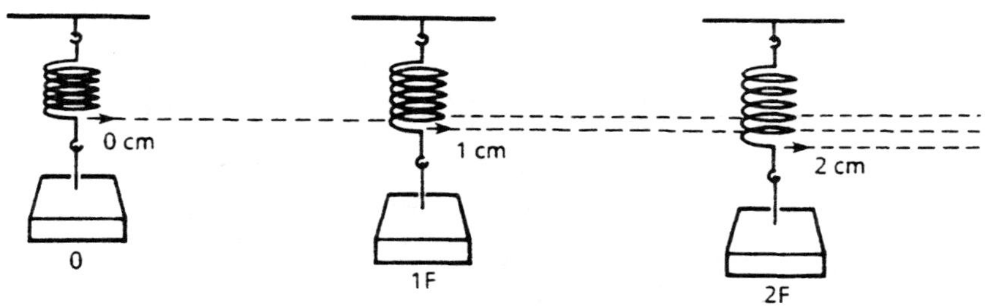

Fig. 1 Linear displacement of a spring.

From these and other examples you may conclude that the response to small disturbances in all physical systems follows a linear law. Doubling the cause doubles the effect, tripling the cause triples the effect, and so on.

Consider once more the linear spring. The mathematical expression relating the force F on the spring to the stretching of the spring in the x-direction is

$$F = kx \qquad \text{Equation 1}$$

where the proportionality constant k is the spring constant.

In practice, as long as the stretching of the spring is small compared to the total length of the spring, the force can be calculated with Equation 1. However, when the distance through which the spring is stretched exceeds a certain limit, the magnitude of the force F no longer is related to the distance x in a simple linear manner. Then it's more accurately expressed by the nonlinear relation.

$$F = kx + k'x^2 + k''x^3 + k'''x^4 + \ldots \qquad \text{Equation 2}$$

where k', k", k"' are higher order constants and are much smaller than the constant k.

The magnitude of the force as a function of the distance the spring is stretched is shown in Figure 2, where the linear as well as the nonlinear behavior of the spring is illustrated.

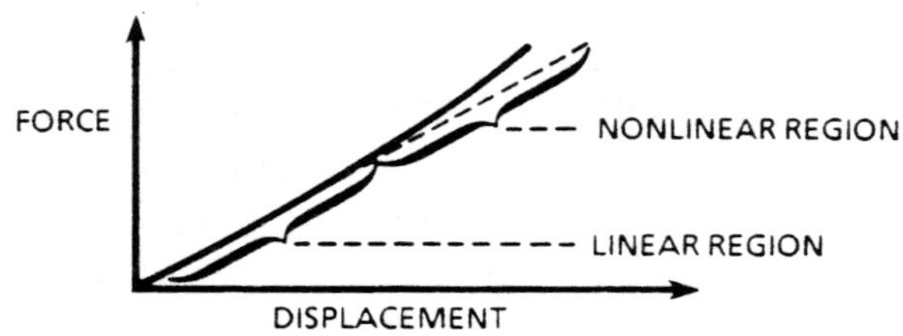

Fig. 2 Linear and nonlinear behavior at a stretched spring.

Now we've stated what we mean by "linear" and "nonlinear" behavior of a system (the spring). Let's next discuss the nonlinear behavior of certain materials when intense light beams propagate through them. These light beams are of course electromagnetic waves that are made up of an electric field and a magnetic field. Both fields are locked in step at right angles to each other and oscillate together at the frequency of the light.

Consider what happens when such an oscillating field interacts with an atom, which you can think of as a positively charged nucleus surrounded by a shell of electrons. The electrons in the outermost shell are more loosely bound, and their radial displacements are governed by the same force equation that describes the stretching of a spring. These outermost electrons can redistribute themselves in step with the electric field. In this way the atom becomes charge-polarized. This means that positive and negative parts separate slightly to form a small electric dipole with a so-called dipole moment pointing in the direction of the impressed field. The dipole moment is the magnitude of the displaced charge times the distance between the positive nucleus and the center of the displaced electron shell.

The number of electric dipoles per unit volume multiplied by the dipole moment of one atom gives the induced macroscopic charge polarization of the material. Don't confuse this charge polarization induced by the electric field of the light beam with the polarization of a light beam with a polarizing filter.

In any given material, the magnitude of the induced charge polarization depends on the magnitude of the applied electric field E. Specifically, we can express the charge polarization P in a series of powers of E and write

$$P = aE + dE^2 + d'E^3 + \dots \qquad \text{Equation 3}$$

where: a = Polarizability of the material

d, d', etc. = Higher-order nonlinear optical coefficients that are much less than a.

The similarity between Equation 3 and Equation 2 should be clear.

The electrons in the outermost shell not only are subject to the pulling of the applied external field, but also are subjected to the pull of the internal field due to the positive nucleus. The magnitude of this internal electric field is enormously large, somewhere in the neighborhood of 10^9 volts/cm. For comparison, the electric field of sunlight at the surface of the earth is only about 10 volts/cm. It turns out that, when the applied electric field is many orders of magnitude less than the atom's internal electric field—which is always the case for ordinary light sources—the linear approximation $P = aE$ of Equation 3 is very accurate.

With the development of the laser in 1960 we were able to build light sources capable of producing optical electric fields of 10^6 to 10^7 volts/cm. When some optical materials are irradiated by such intense fields, the contribution to the charge polarization by the higher-order terms in Equation 3 no longer can be neglected.

Let's see what happens when we use only the first two terms in Equation 3. For simplicity, let's also assume that the electric field points in the x-direction. Then the total induced charge polarization in the x-direction is

$$P_x = aE_x + dE_x^2 \qquad \text{Equation 4}$$

For plane polarized light of frequency f and amplitude E_{xo}, the electric field at some given point in an optical material is described by

$$E_x = E_{xo} \cos(2\pi ft) = E_{xo} \cos(\omega t) \qquad \text{Equation 5}$$

where $\omega \equiv 2\pi f \equiv$ the "angular frequency" in rad/sec

and the induced charge polarization is

$$P_x = aE_{xo} \cos(\omega t) + dE_{xo}^2 \cos^2(\omega t) \qquad \text{Equation 6}$$

Using the trigonometric identity

$$\cos^2\theta = \frac{1 + \cos 2\theta}{2} \qquad \text{Equation 7}$$

Equation 6 can be written as:

$$P_x = aE_{xo} \cos(\omega t) + \frac{dE_{xo}^2 \cos 2(\omega t)}{2} + \frac{dE_{xo}^2}{2}$$

or

$$P_x = P_x(\omega) + P_x(2\omega) + \textit{constant term (DC)} \qquad \text{Equation 8}$$

An inspection of the above equation reveals that the first term $P_x(\omega)$ is a charge polarization of angular frequency ω. The second term $P_x(2\omega)$ is a charge polarization of angular frequency 2ω, twice the fundamental frequency ω. Finally, the third term has no frequency dependence whatsoever. It provides a "DC" charge polarization. The three charge polarization terms are shown in Figure 3. Equation 8 then tells us that a rapidly oscillating electric field in the neighborhood of 10^{14} to 10^{15} Hz (optical frequencies) will induce two types of oscillating charge polarization; one of frequency ω and another of frequency 2ω. The material consists, then, of a collection of oscillating dipoles that absorb and reradiate light waves. The total energy in the beam of light is not significantly altered.

One effect of the induced charge polarization and subsequent reradiation is a decrease in the speed of the light in the medium. The decrease in speed is reflected in the increased value of the index of refraction n, which is given by the ratio of the speed of light in vacuum c divided by the speed of light in the medium v, that is,

$$n = \frac{c}{v}$$ Equation 9

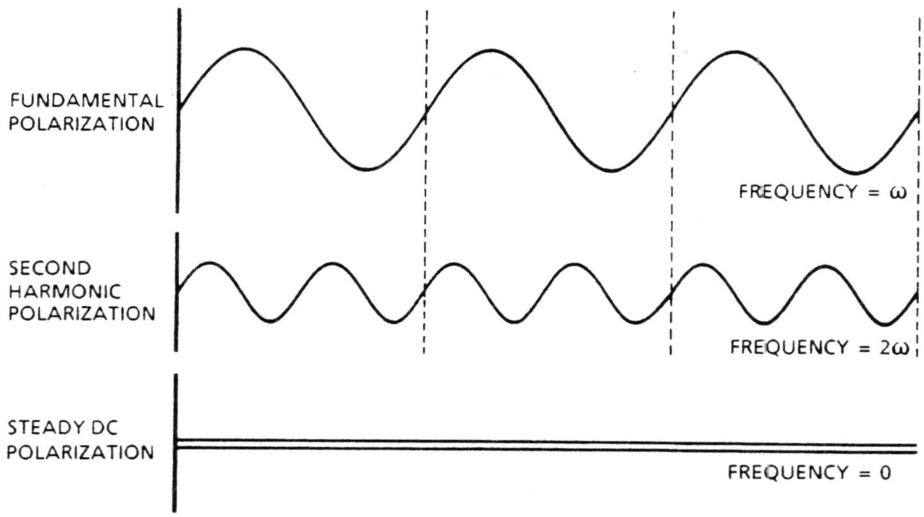

Fig. 3 Frequency associated with charge-polarization terms.

Another effect, predicted by Equation 8, is the reradiation of energy at the frequency 2ω, twice the frequency of the incident radiation. This doubling of frequency is known as "second harmonic generation" (SHG) or "frequency doubling."

Frequency doubling was observed for the first time in 1961 by Professor Peter Franken and some graduate students at the University of Michigan. They irradiated a quartz crystal with the beam from a ruby laser that operated at 694.3 nm. A very small amount of the light striking the crystal was converted to light with a wavelength of 347.2 nm. This wavelength lies in the ultraviolet region of the spectrum and is of course exactly half the wavelength and twice the frequency of the incident laser light.

The details of this famous experiment are shown in Figure 4. As you might expect, this experiment initiated a search for materials in which this effect occurs strongly. As a result, the experiment has been duplicated many times with a host of different nonlinear materials.

What are the properties of a crystal that are responsible for efficient SHG? To answer this question let's first note that, whatever these properties may be, they manifest themselves in the expansion coefficient d of the second-order term in Equation 3.

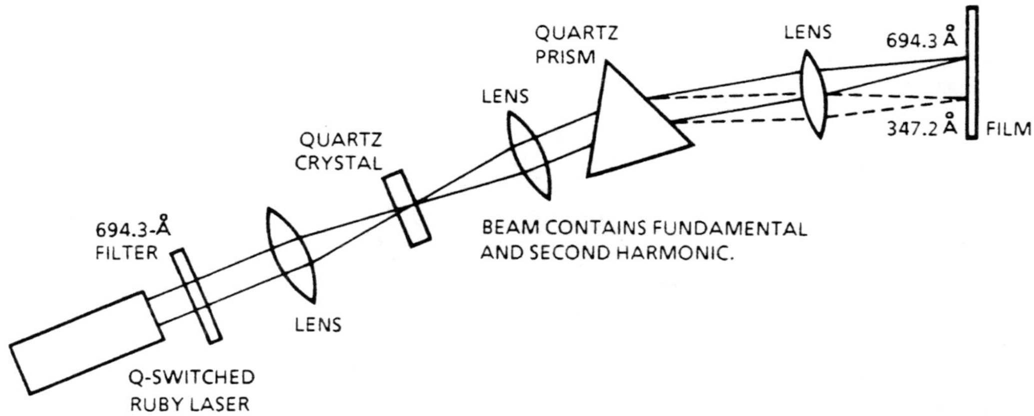

Fig. 4 Experimental arrangement for second harmonic generation.

To find out how d is governed by the internal properties of the crystal, we need to examine the oscillatory motion of the dipoles. This is best done by adding the three curves in Figure 3. The resultant charge polarization is shown in Figure 5, along with the applied electric field.

The curve for the total polarization shows mainly the contributions of the fundamental polarization and steady DC polarization terms. The contribution of the second harmonic term would cause a slight change in the slopes of the rise and fall of the total polarization, but not in the location of the maximum, zero, and minimum points. So its effect has not been shown in the second curve in Figure 5.

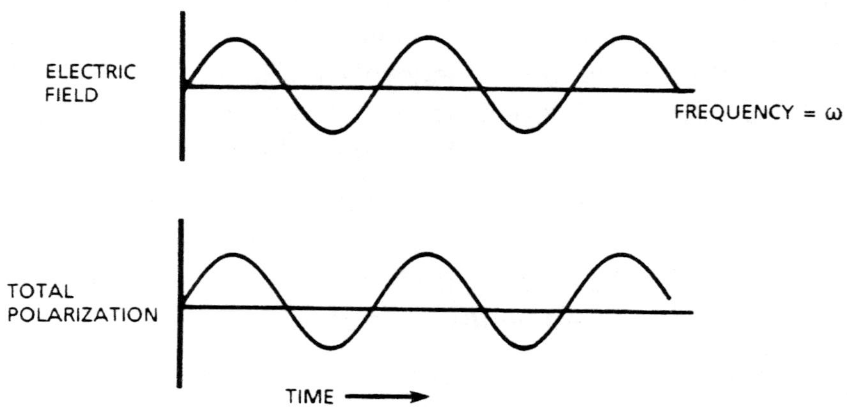

Fig. 5 Impressed electric field and resulting total polarization.

It's clear from Figure 5 that the individual dipoles, whose oscillatory motions lead to the charge polarization, do not faithfully reproduce the sinusoidal electric field. Note especially that the total polarization is asymmetric in the upward direction. This is because the outer electrons of the atoms in a crystal are more easily pushed in one direction than in the opposite direction.

Figure 5 shows that an intense electric field in the up-direction is more effective in polarizing the medium than the same field in the down-direction. Such a situation can occur only in a crystal with no internal symmetry. Only about 10% of the crystals found in nature fall in this category.

The distorted charge polarization wave produced by the intense radiation field travels at the same speed as the incident light wave. This light, which is reradiated by the oscillating dipoles, contains, as we saw above, two discrete frequencies ω and 2ω. Since the refractive index for the light of frequency 2ω is usually larger than it is for light of frequency ω, the speed of the component with frequency 2ω will be just a bit slower than the component with frequency ω, in accordance with Equation 9.

The difference in propagation speed between the fundamental frequency and the second harmonic frequency means that the second harmonic light radiated from one part of the crystal, say P(1), will not be in phase with the second harmonic light radiated from another part of the crystal P(2). As a result, the two waves from P(1) and P(2) will interfere destructively at some point P(3). The distance that it takes for a second harmonic charge polarization wave and its radiated light to get 180° out of phase is called the "coherence length." This distance is usually on the order of 10 μm. If the crystal length happens to be equal to an odd multiple of the coherence length, virtually no second harmonic radiation will emerge from the crystal.

This problem of destructive interference can be overcome by making use of the double-refraction birefringence exhibited by certain crystals. It turns out that in such crystals the velocity of light depends on the direction of polarization of the electric field, the propagation direction, and the wavelength.

For example, the speed of light in the nonlinear material potassium dihydrogen phosphate (KDP) for the horizontally and perpendicularly polarized electric field components is shown in Figure 6. As you can see, the ordinary fundamental that we have taken to be the light from a ruby laser at 694.3 nm, polarized perpendicular to the plane of the page, travels at exactly the same speed as the extraordinary second harmonic light at 347.2 nm, when both light waves propagate in a direction that's 50° to the optic axis of the crystal. The optic axis of a doubly refracting crystal is that direction along

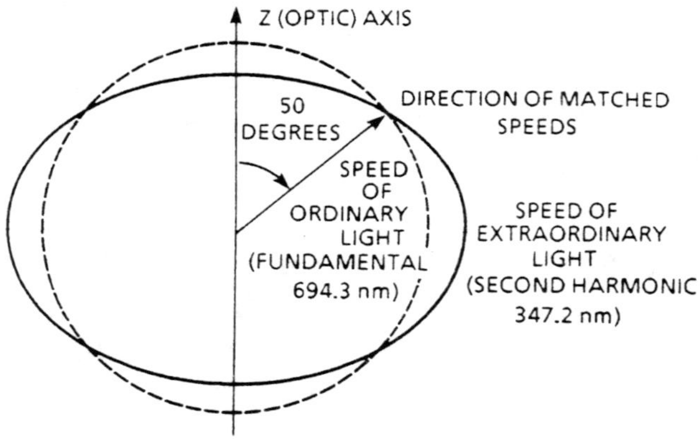

Fig. 6 Velocities of ordinary and extraordinary light in KDP.

which the ordinary and extraordinary waves of the *same* frequency propagate at the same speed.

When the direction of propagation of the fundamental is chosen so that the index of refraction of the fundamental and second harmonic are the same, both waves will stay in phase. The crystal then is said to be "phase-matched" or "index-matched." The second harmonic wave produced at any given point in the beam now will interfere constructively with the second harmonic wave produced further "downstream" in the beam. As we will see later, when the light travels along the index-matching direction, the efficiency for SHG can be improved very significantly.

Phase-matching also can be achieved in some crystals by varying their temperature. The reason is that the extraordinary index is in general more temperature-dependent than the ordinary index. So by varying the temperature you can adjust the birefringence of the crystal until phase matching is obtained. The exact temperature for phase matching depends on the exact chemical composition of the crystal. This effect was discovered by accident when it was noted that different crystals of lithium niobate required different phase-matching temperatures under otherwise similar experimental conditions.

For the niobates in which the phase-matching is achieved by controlling the temperature, the apparatus shown in Figure 7 is used to determine the phase-matching temperature for any given crystal. A laser such as an Nd:YAG operating at 1.06 µm is used to illuminate the crystal. The intensity of the second harmonic is monitored and displayed with a pen recorder as the temperature of the crystal is tuned through the phase-matching temperature.

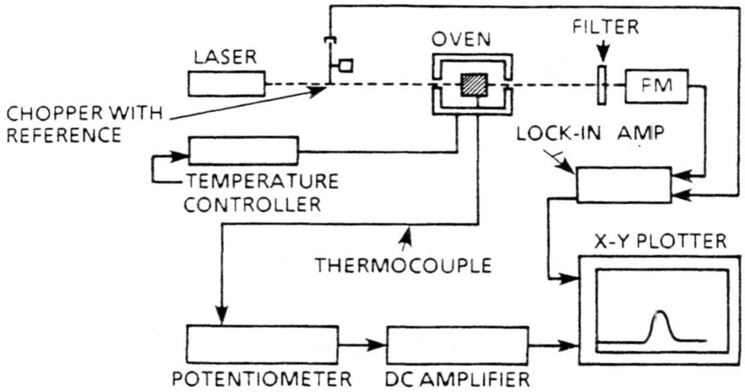

Fig. 7 Apparatus for determining phase-matching temperature in birefrigent crystals.

It turns out that a perfect crystal should give a plot of second harmonic power versus temperature that has a $[(\sin x) / x]^2$ dependence. However, because of slightly varying chemical compositions at different sites of the crystal, different portions of the crystal will phase-match at slightly different temperatures. The result is a broadening of the central peak and a reduction in height. Furthermore, the side lobes are obscured. A typical curve of second harmonic power versus temperature for an ideal crystal and a nonideal crystal is shown in Figure 8.

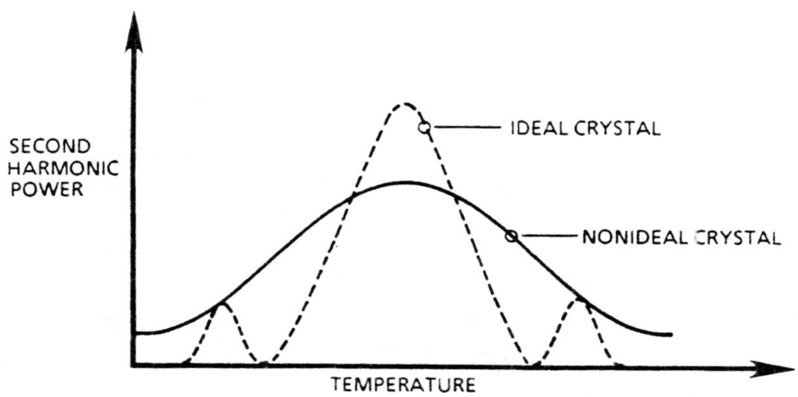

Fig. 8 Second harmonic power output versus temperature.

The conversion efficiency, η_{SHG} defined as the ratio of the emerging second harmonic power P(2ω) to the incident power P(ω) is one of the most useful measures of the performance of a nonlinear crystal. It is given by

$$SHG\ conversion\ efficiency \equiv \eta = \frac{P_{2\omega}}{P_{\omega}} = 2\left(\frac{\mu_o}{E_o}\right)^{\frac{3}{2}} \frac{\omega^2 d^2 \ell^2}{\eta^3} \frac{\sin^2(\Delta k\ell/2)}{(\Delta k\ell/2)^2} \left(\frac{P_{\omega}}{A}\right)$$

Equation 10

where: $\Delta k \equiv k_{2\omega} - 2k_{\omega}$

$k = {}^{2\pi}/_{\lambda}$

E_o and μ_o = Permissivity and the magnetic permeability of free space, respectively.

η = Index of refraction.

ω = Angular frequency of the incident light.

d = Second-order coefficient.

ℓ = Length of the crystal.

A = Beam area.

The factor Δk represents the amount of phase mismatch between the second harmonic wavefronts generated at different points in the crystal. If Δk is zero (proper phase-matching) then the "interference" term,

$$\frac{\sin^2(\Delta k\,\ell/2)}{(\Delta k\ell/2)^2} = 1$$

reaches a maximum value that in turn maximizes the efficiency of the SHG process. Note that the "interference" term is squared, making the power converted into the second harmonic very sensitive to phase mismatches.

The efficiency of second harmonic generation is directly proportional to the intensity $I(\omega) = P_{\omega}/A$ of the incident fundamental radiation. This is why second harmonic generation often is done in an intracavity configuration since the higher power of the fundamental produces much higher intensities than are achievable in external convertors.

Example A: Compute the Theoretical SHG Conversion Efficiency with a Doubling Crystal External to the Laser Cavity

Given: Q-switched, Nd:YAG laser, output power = 1.2 MW, with a 2.00-mm-diameter beam inside KDP SHG crystal. Length of optical path inside crystal is 2.35 cm. Assume proper phase-matching and $n \simeq 1.49$.

Find: Estimated SHG efficiency.

Solution: *Equation 10,*

$$\eta = 2\left(\frac{\mu_o}{E_o}\right)^{\frac{3}{2}} \frac{\omega^2 d^2 \ell^2}{\eta^3} \left(\frac{P_\omega}{A}\right)$$

Note: if phase-matched

$$\frac{\sin^2(\Delta k\ell/2)}{(Ak\ell/2)^2} = 1$$

work in mks (SI) units:

$$\mu_o = 1.26 \times 10^{-6}\ H/M$$

$$E_o = 8.85 \times 10^{-12}\ F/M$$

$$\omega = 2\pi f = 2\pi\left(\frac{c}{\lambda}\right) = 2\pi \frac{(3.00 \times 10^8\ m/sec)}{(1.06 \times 10^{-6}\ m)}$$

$$= 1.78 \times 10^{15}\ rad/sec$$

$$A = {}^1\!/\!_4\,\pi d^2 = {}^1\!/\!_4\,\pi\,(2.00 \times 10^{-3})^2 = 3.14 \times 10^{-6}\ m^3$$

$$d = 0.45\left(\frac{1}{9} \times 10^{-22}\right) = 5.0 \times 10^{-24} \frac{coul^3}{J \bullet volt^3}$$

(See Table 1)

$$\eta = 2\left(\frac{1.26 \times 10^{-6}}{8.85 \times 10^{-12}}\right)^{\frac{3}{2}} \left(\frac{(1.78 \times 10^{15})^2\,(5.0 \times 10^{-24})^2\,(2.35 \times 10^{-2})^2}{(1.49)^3}\right)\left(\frac{1.20 \times 10^6\ \omega}{3.14 \times 10^{-6}\ m^2}\right)$$

$$\eta = 2\left(1.42 \times 10^5\right)^{\frac{3}{2}}\left(1.32 \times 10^{-20}\right)\left(3.82 \times 10^{11}\right)$$

$$\eta = 5.40 \times 10^{-1} = 54\%.$$

Occasionally, in the case of intracavity generation, an alternative definition of efficiency is used to specify performance. This definition relates the output of the second harmonic (for a cavity designed for maximum second harmonic output) to the output of the fundamental (with the cavity designed for maximum fundamental output). Since mirrors can be fabricated with essentially 100-percent transmittance at one frequency and 100-percent reflectance at some other frequency, it's entirely feasible to design a cavity that retains the fundamental while coupling out all of the second harmonic. In this type of system, according to the above definition of efficiency, it's possible to achieve 100-percent conversion. Bear in mind, however, that the conversion refers to the available intensity (obtainable as output) and not to the actual intensity at the crystal itself.

Before we conclude our discussion of second harmonic generation and describe other nonlinear effects, we need to discuss one important complexity that we've ignored. We stated earlier that all nonlinear effects arise directly from higher-order polarization induced in the nonlinear material by the incident radiation. For simplicity we assumed that the induced polarization would have the same direction as the electric field of the incident radiation. This was the assumption made when we presented Equations 3 and 4.

In reality, the second-order polarization P (2ω) is a vector quantity, each component of which may depend on all three components of the incident electric field. For example, you can see that the x-component of the second-order polarization, $P_x(2\omega)$, given by Equation 11 is dependent on all components of the

$$P_x^{(2\omega)} = d_{xxx}^{(2\omega)} E_x^{(\omega)} E_x^{(\omega)} + d_{xyy}^{(2\omega)} E_y^{(\omega)} E_y^{(\omega)} + d_{xzz}^{(2\omega)} E_z^{(\omega)} E_z^{(\omega)}$$

$$+ 2d_{xzy}^{(2\omega)} E_z^{(\omega)} E_y^{(\omega)} + 2d_{xzx}^{(2\omega)} E_z^{(\omega)} E_x^{(\omega)} + 2d_{xxy}^{(2\omega)} E_x^{(\omega)} E_y^{(\omega)}$$

Equation 11

electric field (E_x, E_y, E_z) in a rather complicated way. In addition; before, there was only one second-order coefficient (d) in the relation between P and E (Equation 3 or Equation 4), now there are six coefficients (d_{xxx}---d_{xxz}) for P_x alone. As you might guess, there are also six coefficients in the equation for $P_y(2\omega)$, and six for $P_z(2\omega)$. Some values for nonlinear optical coefficients of common crystals are given on Table 1. You can find values for other coefficients in selected references given at the end of this module. Estimates for the damage threshold of some common SHG crystals are given in Table 2.

Table 1. The Nonlinear Optical Coefficients of a Number of Crystals*

Crystal	$d_{ijk}^{(2\omega)}$ in units of $1/9 \times 10^{-22}$ (mks)
$LiIO_3$	$d_{31} = 0.46 \pm 0.1$
$NH_4H_2PO_4$	$d_{36} = d_{312} = 0.45$
(ADP)	$d_{14} = d_{123} = 0.45 \pm 0.02$
KH_2PO_4	$d_{36} = d_{312} = 0.45 \pm 0.03$
(KDP)	$d_{14} = d_{123} = 0.45 \pm 0.03$
KD_2PO_4	$d_{36} = d_{312} = 0.42 \pm 0.02$
(KD*P)	$d_{14} = d_{123} = 0.42 \pm 0.02$
KH_2ASO_4	$d_{36} = d_{312} = 0.48 \pm 0.03$
	$d_{14} = d_{123} = 0.51 \pm 0.03$
Quartz	$d_{11} = d_{111} = 0.37 \pm 0.02$
$AlPO_4$	$d_{11} = d_{111} = 0.38 \pm 0.03$
ZnO	$d_{33} = d_{333} = 6.5 \pm 0.2$
	$d_{31} = d_{311} = 1.95 \pm 0.2$
	$d_{15} = d_{113} = 2.1 \pm 0.2$
CdS	$d_{33} = d_{333} = 28.6 \pm 2$
	$d_{31} = d_{311} = 14.5 \pm 1$
	$d_{15} = d_{113} = 16 \pm 3$
GaP	$d_{14} = d_{123} = 80 \pm 14$
GaAs	$d_{14} = d_{123} = 107 \pm 30$
$BaTiO_3$	$d_{33} = d_{333} = 6.4 \pm 0.5$
	$d_{31} = d_{311} = 18 \pm 2$
	$d_{15} = d_{113} = 17 \pm 2$
$LiNbO_3$	$d_{31} = d_{311} = 4.76 \pm 0.5$
	$d_{22} = d_{222} = 2.3 \pm 1.0$
Te	$d_{11} = d_{111} = 730 \pm 230$
Se	$d_{11} = d_{111} = 130 \pm 30$
$Ba_2NaNb_5O_{15}$	$d_{33} = d_{333} = 10.4 \pm 0.7$
	$d_{32} = d_{322} = 7.4 \pm 0.7$
Ag_3AsS_3	$d_{22} = d_{222} = 225$
(Proustite)	$d_{36} = d_{312} = 135$

* These coefficients are defined with $1 = x, 2 = y, 3 = z$. From Yariv.

Table 2. Approximate Damage Thresholds for Common SHG Materials at λ_o = 1.06 μm for Q-switched Operation.

Material	Approximate Damage Threshold in (GW/cm^2)
ADP	0.50
$Ba_2NaNb_5O_{15}$	0.001
CDA	0.50
CD*A	0.36
KDP	0.20
KD*P	0.50
$LiNbO_3$	0.02
$LiIO_3$	0.05

Precise values will depend on many factors including laser pulse width. Data courtesy of Quantum Technology, Inc., Sanford, Florida.

Common substances that exhibit SHG are potassium dihydrogen phosphate (KDP), barium titanate ($B_aT_iO_3$) and lithium niobate ($L_iN_bO_3$). In barium titanate, for example, only five of the second-order coefficients are non-zero, leading to the following simplified forms for $P_x(2\omega)$, $P_y(2\omega)$ and $P_z(2\omega)$:

$$P_x(2\omega) = 2\, d_{xxz} E_x E_z$$

$$P_y(2\omega) = 2\, d_{yyz} E_y E_z$$

$$P_z(2\omega) = d_{zxx} E_x^{\,2} + d_{zyy} E_y^{\,2} + d_{zzz} E_z^{\,2}\,.$$

If, in addition, the incident electric field is polarized in the x-direction, ($E_y = 0, E_z = 0$), the equations above reduce to only one:

$$P_z(2\omega) = d_{zxx} E_x^{\,2} \qquad \text{Equation 12}$$

Equation 12 tells us that an electric field polarized in the x-direction, incident on barium titanate, will produce a second harmonic charge-polarization in the z-direction. The orientation of this crystal in terms of its optic axis and the electric field polarization of the incident beam is shown in Figure 9.

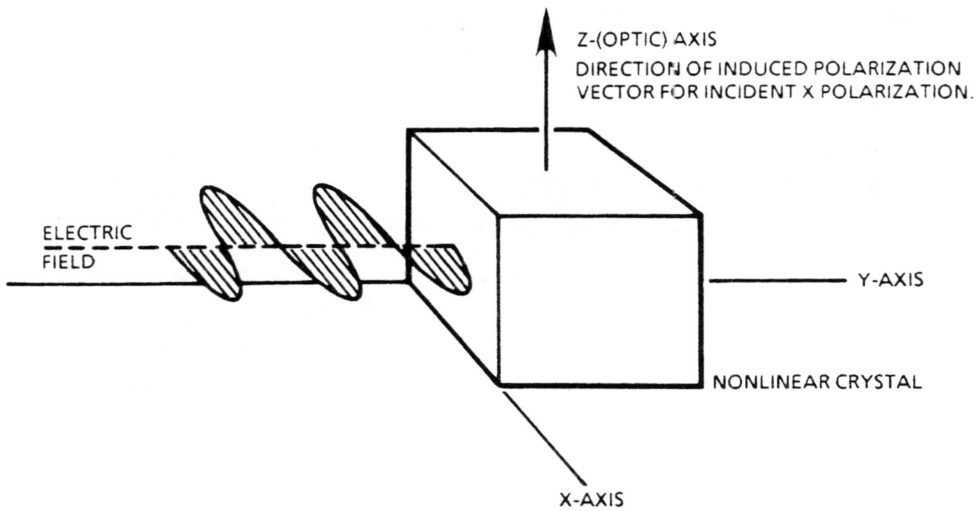

Fig. 9 Typical electric-field-optic-axis orientations in a second-harmonic crystal.

Now let's turn our attention from SHG, which was the first nonlinear effect observed with a laser beam, to some other important nonlinear effects. Depending on the particular nonlinear material, these effects can be cataloged into two groups; passive and active.

Materials producing *passive* optical effects are those that act essentially as catalysts without imposing their characteristic internal resonance frequencies onto the particular nonlinear effect. Passive nonlinear effects include harmonic generations, frequency mixing, optical rectification, and self-focusing of light.

The latter results from a field-dependent change in refractive index as shown in Equation 13.

$$n = n_o + n_2 E^2 \qquad \text{Equation 13}$$

where:

n	=	Effective refractive index.
n_o	=	Ordinary refractive index.
n_2	=	Nonlinear refractive index.
E	=	Electric field magnitude.

The quantity n_2 comes from other considerations of polarizabilities of materials as we discussed for second harmonic generation. This nonlinear behavior can lead to intensity-dependent lensing (focusing or defocusing) of propagating beams. If focusing occurs, the decreased beam diameter can cause power densities (W/cm^2) exceeding the damage threshold of the material and result in breakdown failure or permanent damage to the material as well as spreading of the beam.

Materials producing *active* nonlinear effects do impose their characteristic resonance frequencies onto an incident beam of light. The optical nonlinear effects produced by these active materials include two-photon absorption, stimulated Raman, Rayleigh, and Brillouin scattering.

We won't discuss the numerous active nonlinear materials, nor will we describe the related effects in this module. If you're interested, you will find it worth your time to read the paper by J. A. Giordmaine in the April 1964 issue of the *Scientific American*, where some of these effects are described in an appropriately elementary fashion.

We will briefly describe some of the passive nonlinear effects.

Sum and Difference Frequency Generation

Two light beams of different frequencies ω_1 and ω_2 are mixed in a nonlinear crystal. The light leaving the crystal then will have two additional frequencies $\omega_1 + \omega_2$ and $\omega_1 - \omega_2$. Generation of these additional frequencies requires intense unidirectional sources of light. Although lasers commonly are used in this application, their monochromaticity and coherence play no essential role. In fact, sum and difference frequencies have been obtained with very intense incoherent light sources.

You can understand this particular phenomenon best in terms of the quantum theory of radiation, according to which light propagates in discrete packages of energy called photons. For example, two photons can be annihilated to produce a new photon that carries the energy of the two that disappeared. Since the frequency is proportional to its energy, the new photon will exhibit the sum frequency of the two annihilated photons.

In the interaction of a photon of frequency ω_1 with a photon of frequency of ω_2 to produce a photon of frequency $\omega_3 = \omega_1 - \omega_2$, the ω_1 photon is annihilated. In its place appears another ω_2 photon. The ω_1 photon gives up sufficient energy to create one ω_3 photon. The total energy is conserved. This particular interaction will produce as many new ω_2 photons as ω_3 photons.

Optical Rectification

In Equation 8 we saw that a DC signal should be produced across a nonlinear crystal when an intense polarized laser beam passes through the crystal. This optical rectification has indeed been observed.

You can detect the DC signal by connecting conductors to opposite sides of a piezoelectric crystal that is transmitting the light output from a laser. The DC signal appears as a voltage pulse that's proportional to the intensity of the laser beam. This effect is, of course, analogous to that of a semiconductor diode used to rectify an AC electrical signal.

Beam-trapping

As we said previously, the index of refraction of a dielectric material is actually a function of the light intensity. This has been observed in a variety of liquids and solids.

One manifestation of this is the well known Kerr effect, where a strong applied electric field aligns the molecules of a liquid and thereby produces birefringence. A second effect is electrostriction which is due to a strain experienced by dielectrics when they are subjected to an electric field. The strain depends on the square of the electric field. Because of this field-induced strain, electrostriction involves a change in the density of the material. But the index of refraction depends on the density, which in turn implies that the index of refraction is intensity-dependent.

In liquids, both the Kerr effect and electrostriction are of equal importance in changing the refractive index. In solids, however, molecules can't be aligned easily so that electrostriction is the dominant cause of an intensity-dependent index of refraction, except at very short pulse lengths.

If a laser beam is focused so that the focal volume lies within a piece of glass, the beam, if sufficiently intense, can increase the index of refraction so much that the slender, intense beam is trapped because of total internal reflection. The trapped beam has been observed to travel through transparent materials along a thin filament and leave a trail of ionization damage in the form of microscopic bubbles and cracks.

Because they want to understand and apply the many different nonlinear effects, scientists started a search for materials in which these effects occur strongly. They have learned how to find these nonlinear materials and also how to calculate and predict their properties.

Before we discuss some of the more common nonlinear materials, let's briefly list some properties common to all useful nonlinear optical crystals.

Foremost on the list is the fact that nonlinear materials belong to a group of non-centrosymmetric crystals. This means that the internal motion of electrons in an oscillating electric field is not symmetric. The material should furthermore exhibit a large index of refraction since there is, as a rule, a strong dependence of nonlinear optical properties on the refractive index. It also should have a large birefringence for easy index-matching.

From a practical point of view the material should have a low susceptibility for optical quality, can be optically polished, is not hygroscopic, and has a wide range of spectral transmission. This is a big order! Yet, certain materials such as lithium niobate and barium sodium niobate tend to exhibit several of these properties. So the challenging problems encountered when growing these materials now are receiving a lot of attention. The nonlinear properties of literally hundreds of nonlinear materials are cataloged already. Let's turn our attention then to a few of the more common ones.

ADP and KDP

Ammonium dihydrogen phosphate (ADP) and potassium dihydrogen phosphate (KDP) were among the first nonlinear crystals used for the generation of phase-matched second harmonic light. Because of their similarity we'll discuss them together.

Both materials first were used in piezoelectric applications such as ultrasonic transducers. Since these materials are easy to grow in aqueous solutions that produce high-optical-quality single crystals, you can purchase specimens as large as 10 cm on one side. The optical transmission band extends from the ultraviolet to the near infrared, typically 200 nm to 1500 nm.

ADP will deteriorate when heated to temperatures of over 100°C and has a tendency toward cracking upon cooling. But KDP is relatively stable and can be heated and cooled. Both materials are sufficiently resistant to laser damage for most applications.

Isomorphs, which are substances of different chemical composition that crystallize in the same form as ADP or KDP, also have been used to produce nonlinear optical effects. The best-known of these is deuterated KDP, which is designated KD*P and called: "K-D-star-P." The shortcomings of KDP and ADP are a relatively small nonlinear coefficient and poor optical transmission in the infrared region of the spectrum.

Lithium niobate ($LiNbO_3$)

This is one of the more important of the nonlinear materials that have appeared in recent years. It's transparent from 400 nm to 5 µm, and its nonlinear coefficient is about ten times larger than that of KDP. This means that lithium niobate is two orders of magnitude more efficient than KDP, and SHG efficiencies close to 100 percent are possible. It's one of the few nonlinear materials available in large sizes in commercial quantities.

Since it has a very large birefringence in the visible and near-infrared region, it allows phase-matching of the fundamental and harmonic waves. Phase-matching usually is accomplished by heating the crystal to the phase-matching temperature for the particular laser being used.

Crystals are available in all sizes up to a few centimeters in diameter and more than 10 centimeters in length. The crystals are clear as water and insoluble in water. This material is used in second harmonic generation; parametric oscillation, which is an application of sum and difference frequency generation; and a host of other applications.

One problem with lithium niobate is that it has a very low damage threshold. In some samples, this damage disappears by itself soon after the laser beam is turned off. But in other samples the characteristic "tracks" can remain for days. Fortunately, the damage can be reversed by heating the crystal to about 200°C. This is one reason for using temperature-controlled phase-matching.

Barium sodium niobate ($BaNaNb_5O_{15}$)

This nonlinear material—which also is known as "banana"—is similar to lithium niobate. But it doesn't suffer as much from optical damage when its temperature is maintained above room temperature. Barium sodium niobate has a nonlinear coefficient that's about three times larger than lithium niobate.

$BaNaNb_5O_{15}$ is optically transparent from 370 nm to about 5 µm. Some samples don't have a good optical quality due to a marked brown discoloration. Phase-matching for the second harmonic generation of 1.06-µm radiation from an Nd:YAG laser occurs around 100°C. The exact temperature depends on the stoichiometric composition.

This material has been used for very efficient generation of the second harmonic of 1.06-µm radiation and in the building of parametric oscillators.

Proustite (Ag_3AsS_3)

Proustite is a naturally occurring crystal found in mineral deposits. It also can be produced synthetically. The one characteristic that distinguishes this material from all other nonlinear materials is its transmission band, which extends from about 600 nm to beyond 13 µm. This and its large birefringence make it a prime candidate for the study of phase-matched interactions between the infrared region and the visible range. Furthermore, the appropriate coefficients that are a measure of its efficiency for producing nonlinear effects are relatively large. In fact, they are about 300 times larger than the analogous coefficient of KDP.

The material has been used in experiments that involve the mixing of the 10.6-µm radiation from a CO_2 laser with a visible laser. Proustite crystals of good optical quality and dimensions of several centimeters are grown in an aqueous solution. Some disadvantages of this material are its inability at times to accept dielectric coatings and its resistance to optical polishing.

Aside from the few materials listed above, a host of other materials are available to the experimenter, and new nonlinear materials constantly are added to the already large list. An excellent source that lists most of these materials is the *CRC Handbook of Lasers.*

This concludes the discussion of nonlinear materials and nonlinear effects. The field of nonlinear optics is in its infancy, with a promise for a bright future. As electro-optic technology expands and the applications of coherent light waves become more numerous, nonlinear materials and effects will be of great importance in exploiting the intense laser beam.

EXERCISES

1. Describe linear and nonlinear behavior of a crystal through the use of appropriate equations.
2. Describe second harmonic generation using the concept of induced charge polarization and the need for phase-matching. Indicate two experimental methods for achieving phase-matching conditions.
3. List the various material and experimental parameters that determine the efficiency of second harmonic generation through the use of the correct mathematical equation.
4. Differentiate between passive and active nonlinear materials.
5. List and rate the importance of some of the more common nonlinear materials.

6. Describe mathematically the field-dependent nonlinear index. Explain how this quantity is made manifest phenomenologically in damaging optical components subjected to high-peak-power laser radiation.

7. Compute the approximate length of a KD*P SHG crystal to be used with a 2.0-mw Q-switched Nd:YAG laser that has a beam diameter within the crystal of 5.00 mm. Assume that we want 20% conversion efficiency when properly phase-matched.

8. According to Equation 10, if the incident fundamental power is doubled, what increase should we expect in the power of the second harmonic? What increase in conversion efficiency should we expect? Discuss.

9. A 200-mJ Q-switched Nd:YAG laser with a pulse width of 20 nsec is doubled with a 2.00-cm-long KDP crystal. The beam diameter is 4.00 mm. What will be the approximate irradiance produced by the second harmonic assuming proper phase-matching?

10. Use Equation 3 including up to the third term in the expression for the induced charge polarization,

$$P_x = \alpha E_x + dE_x^2 + d'E_x^3$$

Assume an oscillating electric field like in Equation 5. Using appropriate trigonometric identities derive an expression for $P_x(x)$ in terms of the angular frequency of the incident light ω. Group the terms according to the frequency dependence: DC term, fundamental term, second harmonic term, third harmonic term. Compare this new expression to Equation 8 and discuss the physical significance of each term.

MATERIALS

Q-switched Nd:YAG laser
SHG crystal with mount
Crystal mounting, alignment device with angular readout
Screen
Energy meter
Photodetector and oscilloscope (to measure pulse width)
Lens of about 10-cm focal length
530.0-nm filter
1.06-µm filter

PROCEDURES

Before you start the experiment, you should review safety procedures that apply to the Nd:YAG laser. You will assemble the apparatus and observe the second harmonic of the 1.06-μm radiation from the Nd:YAG laser on a screen.

The main object of this experiment is to familiarize you with one nonlinear material and one nonlinear effect. Specifically, in the experiment you will observe the second harmonic radiation at 530 nm that's produced with a crystal of KDP and ADP when the beam from a 1.06-μm Nd:YAG laser is focused into the crystal. Use care not to damage the crystal.

The experimental setup is quite simple. The output from the Q-switched laser is focused into the crystal, which is mounted on an angular orientation mount. If you know the phase-matching direction for the crystal, mount it accordingly. If you don't know the phase-matching direction, you can find it by trial and error.

To do this, start with an arbitrary orientation of the crystal and observe the screen as the laser is fired. If a green flash of light splashes on the screen, the crystal is mounted in a position that isn't too far removed from the desired phase-matching orientation. If you don't see a green light, change the orientation of the crystal slightly and fire the laser again. Repeat this procedure until a bright green flash is visible on the screen every time the laser is Q-switched. Use care when viewing the green light. **Never try to observe the second harmonic by direct viewing of the light as it exits the crystal.**

Adjust either the angle and/or the temperature of the doubling crystal to maximize the SHG efficiency. Using the energy meter and photodetector, measure both the power of the fundamental and the power of the second harmonic. Compute the conversion efficiency.

Compare your measured efficiency to the estimated theoretical value given by Equation 10.

OPTIONAL:

If you have time, investigate the dependence of light intensity in the second harmonic as a function of crystal orientation. To do this, carefully rotate the crystal about a vertical axis, degree by degree, on either side of the orientation for which you got optimum phase-matching. Measure the intensity at each angle with a sensitive energy meter. Plot the intensity versus angle. Compare this plot with Figure 8 and explain why the curves are similar.

Next, when the crystal has been oriented for maximum output, vary the input intensity from the YAG laser. Measure the resultant change in output. You can vary the input intensity with appropriate filters. Plot results as I_{YAG} input, I_{SHG} output at appropriate filter wavelengths.

REFERENCES

Adhav, R. S., and A. D. Vlassopoules. "Guide to Efficient Doubling," Laser Focus, May 1974.

Baldwin,G.C. An Introduction to Nonlinear Optics. New York: Plenum Press, 1969.

Bass, M.; P. A. Franken; A. E. Hill; C. W. Peters; and G. Weinreich. "Optical Mixing," Physical Review Letters, V. 8, 1961.

CRC Handbook on Lasers, Chemical Rubber Co., Cleveland, OH.

Giordmain, J. A. "Nonlinear Optics," Physics Today, January 1969.

__________. "Nonlinear Optical Properties of Liquids," Physical Review Letters, Volume 138, 1965.

__________. "The Interaction of Light with Light," Scientific American, April 1964.

Hulme, K. F.; O. Jones; P. H. Davies; and M. V. Hobden. "Synthetic Proustite: A New Crystal For Optical Mixing," Applied Physics Letters, V. 10, 1967.

Kaminow, I. P. "Parametric Principles in Optics," IEEE Spectrum, 1965.

Weinberg, D. L. "Tunable Optical Parametric Amplifiers and Generators," Laser Focus, April 1969.

Yariv, A. Introduction to Optical Electronics. San Francisco: Holt, Rinehart, and Winston, Inc., 1971.